# RELIGION: *THE SOCIAL CONTEXT*

T H I R D    E D I T I O N

## MEREDITH B. McGUIRE
*TRINITY UNIVERSITY*

Wadsworth Publishing Company
Belmont, California
A Division of Wadsworth, Inc.

*Editor:* Serina Beauparlant
*Editorial Assistant:* Marla Nowick
*Production:* Ruth Cottrell
*Print Buyer:* Randy Hurst
*Copy Editor:* Betty Duncan
*Cover:* Michael Rogondino
*Signing Representative:*
*Compositor:* Kachina Typesetting, Inc.
*Printer:* Malloy Lithographing, Inc.

*This book is printed on acid-free paper that meets Environmental Protection Agency standards for recycled paper.*

2 3 4 5 6 7 8 9 10 — 96 95 94 93 92

**Library of Congress Cataloging in Publication Data**
McGuire, Meredith B.
   Religion, the social context / Meredith B. McGuire. — 3rd ed.
      p.   cm.
   Includes bibliographical references and indexes.
   ISBN 0-534-16968-6
   1. Religion and sociology.   I. Title.
BL60.M36   1991
306.6—dc20                                                    91-32909
                                                                    CIP

ISBN 0-534-16968-6

# CONTENTS

# PREFACE

Writing this third edition, I have been impressed with the growing usefulness of the sociology of religion as a "window" for understanding many far-reaching sociopolitical developments in the modern world. Religion has figured prominently in recent events in the Middle East, Latin America, and Eastern Europe. In the United States, religion has exhibited considerable political assertiveness and various religious groups have articulated important social-moral issues facing the nation.

In this edition, I continue to emphasize that religion exists in a social context, is shaped by that social context, and, in turn, often influences it. This book presents only selected key issues in the sociology of religion and is not intended to be a comprehensive text; thus, it is ideally suited to be a core of essays around which instructors can build a course and assign related articles, monographs, or readings to complement their own emphasis. My primary audience consists of undergraduates taking a first course in sociology of religion; this material should prove useful to seminary and graduate students and to interested general readers.

Several chapters include an extended example in addition to the main explicative portion. The purpose of such examples is to illustrate a sociological interpretation of relevant religious phenomena, such as new religious movements and the role of religion in social change in Latin America. Also, some extended examples (such as those on religious conflict in Northern Ireland and the relationship between religion and healing) illustrate the complexity and extensive historical roots of contemporary phenomena.

Some of these essays reflect interpretations of phenomena on which I have done extensive empirical research: spiritual healing, religious movements, and religion in Ireland. All the extended examples represent a more personal stance than the discussions in the main text and, as such, invite the reader to encounter the sociological issues intellectually and perhaps disagree. These interpretations are intended to illustrate useful sociological concepts in a stimulating and provocative way and to show how sociological theories are relevant to interesting aspects of social life. Several of these interpretive essays are far from neutral; my sympathies clearly lie with the exploited and oppressed (specifically women, minorities, natives of colonized lands, and victims of political or economic suppression), and my personal concern is to explore the role of religion in their oppression and liberation.

In reorganizing the final chapter, although I include some of the theoretical discussion surrounding the "secularization thesis," I have included relatively little of the survey data on the vicissitudes of "official" religions in the United States. I have retained only the aspects of the earlier discussion (in the first edition) that I believe are still important in any interpretation of religion in modern society. Some instructors may therefore want to supplement this text with data from Gallup or NORC opinion researches or from some of the recent surveys about mainline religious beliefs and practices. I personally doubt that opinion research reliably pertains to the real theoretical issues of "secularization" *per se*. Students are, however, often interested in the diversity of religious opinion, and several of these studies do include important data on church-oriented religiosity in this society.

This edition includes extensive, current international and cross-cultural material, ending with a focus on religion in a global perspective. Not only do I believe such cross-cultural understanding is an important part of a broad education, but also I dare to hope that greater understanding and an appreciation of the global setting of our individual lives will promote peace. At the same time, I have deliberately chosen a disproportionate number of brief examples from religion in the United States and other Western, industrialized societies, because most students more readily comprehend these examples without needing extensive historical and cultural background explanation.

Ending each chapter is an annotated list of recommended readings— articles, books, and selections from recent books of readings. Two appendices suggest further useful materials. Appendix A outlines major resources for a literature search in sociology of religion. Students preparing term papers or theses will find these resources essential. Several useful new bibliographies and encyclopedias are now available. The bibliography of this text itself reflects an extensive and reasonably current literature review. Appendix B includes recommended film resources for teaching sociology of religion. I have found good films to be almost essential in teaching sociology of religion because they provide concise glimpses into religious situations beyond the often limited experiences of students. Field learning experiences can also help, but films are typically more manageable and more broadly informative.

With gratitude, I acknowledge those who have given thoughtful suggestions for this third edition, as well as all whose assistance made possible two earlier editions. Thanks to Madeleine Adriance, Cathy Albanese, Eileen Barker, John Donahue, Mary Howes, John Martin, Christopher Nolan, and Ole Riis. Also very welcome were the comments of reviewers Tony Battaglia, California State University, Long Beach; Stephen Beach, Avila College; M. G. Dunn, University of Southern Mississippi; Phillip E. Hammond, University of California at Santa Barbara; and James Richardson, University of Nevada at Reno. The continuing lively response from my students at Trinity University, as well as at Montclair State College and many other universities where I have lectured, has been particularly useful. Finally, I want to again thank Kieran, Rachel, and Daniel McGuire for their encouragement and for helping to keep my writing projects in perspective.

# RELIGION: *THE SOCIAL CONTEXT*

# CHAPTER ONE

# The Sociological Perspective on Religion

---

Religion is one of the most powerful, deeply felt, and influential forces in human society. It has shaped people's relationships with each other, influencing family, community, economic, and political life. Religious beliefs and values motivate human action, and religious groups organize their collective religious expressions. Religion is a significant aspect of social life, and the social dimension is an important part of religion.

Sociologists are interested in studying religion primarily for two reasons. First, religion is very important to many people. Religious practices are important parts of many individuals' lives. Religious values influence their actions, and religious meanings help them interpret their experiences. Sociologists seek to understand the meaning of religion to believers themselves. Second, religion is an important object for sociological study because of its influence on society and society's impact on religion. Analysis of this dynamic relationship requires examining the interdependence of religion and other aspects of society. Often this means questioning taken-for-granted ways of understanding social action. From the earliest foundations of the discipline, sociologists have sought to understand the larger society through examining religion and its influence.

## The Religious Contours of a Community

Before probing religious phenomena in depth, we might note the broad contours of religion as they appear in one community.[1] This highlighting of religion suggests some of the questions a sociologist might ask in surveying, analyzing, and interpreting the place of religion in contemporary society. For example, what are some of the religious contours of San Antonio, Texas, where I live and work?

[1]The concept of religious contours is borrowed from the forthcoming series "Religious Contours of California," edited by Phillip Hammond and Ninian Smart, University of California Press.

3

Conveniently for this example, the city is named for a Catholic saint, Anthony, who was patron of the eighteenth-century Franciscan Mission San Antonio de Valero (better known today as the Alamo). Catholicism has always been a central religion in San Antonio, but there is also a long history of religious and ethnic diversity within local Catholic religious expression. Large clusters of immigrant peoples brought their Catholic faith from other places, like Spain's Canary Islands, Germany, Ireland, France, and Czechoslovakia, and these groups expressed their faith differently. Not only ethnic differences but also class and status differences shaped local Catholicism. In the nineteenth century, the economically comfortable children attending the convent school encountered a different version of their religion than did the poor laborers whose religion was mediated by folk saints, mystery plays, pilgrimages, and a rich popular lore.

Many decades later, despite homogenizing experiences of education, economic mobility, intermarriage, and changes within their church internationally, many of these Catholics still participate in different styles of worship and religiosity, in different languages, gestures and idioms, often celebrating very different aspects of the same world religion. A sociologist would want to know more: What does it mean to believers themselves to be Catholics in San Antonio today? How does their religion help them make sense of their lives? How does it influence their everyday behavior, such as decisions in the workplace or use of leisure time? How, too, does their religion influence their political and moral behavior? Do they experience a sense of belonging to a community of fellow believers? Which fellow believers are experienced as "us," and how does that community, in turn, shape religious expression?

The rough contours of religion in San Antonio also show that Catholicism no longer holds the allegiance of all Latinos (mostly Mexican-Americans), who constitute the ethnic majority of the city's populace. The city abounds with numerous small but active Spanish-speaking pentecostal and evangelical Protestant churches. Separate but similar in style of worship are many pentecostal and evangelical groups serving the city's minority populations of Anglos (whites) and African-Americans. Many of these growing congregations are nondenominational, neither affiliated with nor under the supervision of any larger denominational body.

Effervescent religiosity (including such expressions as speaking in tongues, faith healing, and exorcism) has long fascinated sociologists and anthropologists, who would ask questions like: What is the social nature of these extraordinary religious experiences? For example, how does the group mediate healing to the sick person? How does the group generate and channel such fervor? What are some of the effects of this kind of religiosity on members' everyday lives? How do these groups recruit and convert new members? How are some of the social class, educational, and other background characteristics of members linked with these groups' appeal? What are the effects on group structure, beliefs, and practices of being so independent of external denominational affiliation?

Denominational affiliation is precisely one of the most troubling features of

San Antonio's Southern Baptist churches. In recent years, this mainstream denomination has been wracked with disagreement and uncertainty about the proper role of the national denominational body: Does the faction that controls the national Convention have the right to demand member congregations to conform to its particular version of Baptist belief and practice? What seems like an organization struggle, however, involves some of this religion's core values and beliefs. Southern Baptists are the most numerous Protestant denomination in the city, and local leaders are also prominent in the national struggle. Each time the national conference is held, news of Southern Baptists is a front-page story—indeed, often headlines—in the city's major newspapers.

Sociologists would be fascinated to learn more about this denomination, locally and nationally: What happens (at any level of organization) when one segment of the membership tries to establish an orthodoxy (i.e., boundaries around what one should believe or practice in order to be "one of us")? How do religious groups variously respond to societal changes and modernity? Does change within a religious group occur from the top down or from the bottom up? How does the stance of this denomination affect its larger influence in society and its members' lives?

While orthodoxy is historically rather alien to Baptists, it is both familiar and valued among many hierarchical religions. Besides Roman Catholics and Episcopalians, many ancient hierarchical religious groups are nestled into the religious contours of this city. Making news headlines, for example, was the visit of the Patriarch to the local Coptic church, Saint Anthony the Great. Coptic Christianity dates to the first-century evangelization of Egypt, and today the vast majority of adherents of this faith live in the Mideast. The Coptic enclave in this U.S. city is very much a church in diaspora (i.e., a people settled far from their ancestral homelands), as are the congregations of Lebanese Maronites, Greek Orthodox, and other ethnoreligious minorities.

Surveying these small but significant outlines on the religious horizon of the city, the sociologist would ask, among other questions: How is religion linked with ethnic and language identity? How are those identities preserved in the face of pressures for integration into the larger culture, for intermarriage, and for bilingual or English-only education and mass media? How do religious groups so steeped in tradition respond to modern challenges to traditional norms (such as efforts to redefine women's roles or to democratize decision making in the church itself)? Similar questions would apply to non-Christian traditions active in San Antonio, such as the Jews, Hindus, and Muslims.

Developments in San Antonio's American Moslem Mission mirror national changes. A visitor to this highly committed congregation would immediately notice the balance of U.S.-born black members, together with Mideastern Muslims. The Black Muslim movement was once virulently intolerant of other races and religions; it was also relatively unconnected with Islam as a world religion. Malcolm X was one of the first Black Muslim leaders to embrace a larger vision of Islam, and one faction of the movement followed his lead to greater tolerance and intense study of Islam, becoming part of the Sunni branch of the international faith.

Observing these U.S.-born blacks, with their command of Koranic scriptures and their commitment to the strictures of Muslim religious practice, the sociologist might ask: How does this religious commitment constitute a form of protest against the values and attitudes of the dominant U.S. culture? How does religion figure into racism and other intolerance in U.S. society? How do sectarian enclaves, like the Black Muslims, recruit and maintain commitment? How have the beliefs and practices of this group influenced the socioeconomic situation of its members?

Besides representatives of recognizable world religions, the religious contours of San Antonio include many emerging religious movements. Numerous meditation groups gather to practice eclectic combinations of Eastern meditation techniques, body disciplines, and healing rituals. One New Age group, for example, borrows elements from Zen Buddhist meditation, Huna (native Hawaiian) healing, crystal and color healing, occult and Christian symbolism, and modern psychology. Many of its members simultaneously remain active in their Christian denominational churches. Another group is limited to women members, who seek to create and express a nonpatriarchal religion, focusing on goddess imagery derived from ancient Celtic lore but interwoven with their own new symbols and rituals.

Particularly fascinating to the sociologist-observer is that these emerging religions are frequently found among the relatively well-educated, economically comfortable, young adult or middle-aged members of the community—precisely those who were, according to earlier predictions, most likely to become nonreligious altogether. What do these emerging forms of religion suggest about the place of religion in modern society? What is the basis of their appeal to these segments of the community? How is it possible to put together a "new" religion? How potent are changes in ritual and symbolism for producing long-lasting transformations of people's identities and way of life?

Although these emerging religious movements may not advertise themselves as alternatives to traditional religions, the religious quality of their beliefs and practices is clearly one of their functions. Less obviously religious is the vast range of quasi-religious or para-religious groups in the community. To analyze fully the religious contours of San Antonio, one might include a study of some of these groups that function very much like religion, although whether they are actually religion is a matter of definitional debate (described shortly). One of the oldest para-religious groups in the city is the Ku Klux Klan (KKK), a militant and occasionally violent secret society organized against nonwhites and non-Christians. In the early decades of this century, the Klan paraded openly in full hooded regalia in the streets of San Antonio, asserting their local "muscle" and celebrating their group identity in photographs posed before the Alamo. The rituals, symbols, and group identity of para-religions like the Klan are also worth sociological analysis.

And what should the sociologist of religion make of the quasi-religious rituals and symbols by which the nation is celebrated and understood? The religious quality of the awe inspired by the Alamo, for example, is interesting. Visitors to this "shrine of American liberty" are moved by the image of martyrs, "sacred" places, and objects arranged solemnly within the crumbling walls of the old mission to bring to collective memory certain values of the nation. This collective memory is inspired, despite the conveniently forgotten historical likelihood that the heroes of the Alamo were fighting for the status of Texas within Mexico (rather than for union with the United States) and were motivated by economic self-interest as much as (or more than) political freedom.

This kind of "civil religion" (discussed further in Chapter 6) is of considerable interest to sociologists, who ask: How do complex, modern societies generate, maintain, and express cohesion? How do our myths and rituals involve an ongoing selective remembering? How do these beliefs and practices shape what kind of a nation we want to be and serve as a basis for national self-criticism? How are civil religious ideas and fervor sometimes used to legitimate the interests of one societal group over another? How are they used in political manipulation?

Is religion losing its influence? I once challenged a class to use instances from this community to persuade me that the answer was not an unqualified yes. They presented extensive data about the vitality of church-oriented religiosity in the city, such as the size of church membership rolls and number of people attending services regularly. The most impressive evidence, however, was of the influence of religious expressions not overtly identified with religious organizations. One student described a middle-aged Anglo couple who chose to live, work, and raise their children in the barrio, near one of the worst public-housing tracts in the city. Their daily efforts to help organize the poor for better education, health, and community services were expressions of their religious convictions, even though they did not identify their unpaid work with any specific religious organization. Another instance was a young, college-educated Mexican-American, whose religious commitment was his main reason for devotion to the cause of voter registration (in a state that makes exercising the vote extremely difficult for poor citizens). Yet another example was a group of women who have formed a spiritual and psychological support group for victims of incest and childhood sexual abuse. Instances such as these are surely significant, if less visible, parts of the religious contours of the community.

This rough sketch of just a few parts of the religious situation in one city illustrates that a deeper analysis of the role of religion is essential for an understanding of the whole society. Sociological analysis can give us an interpretive grasp of the everyday meanings, potent traditions, dynamism, and change of religion today.

## The Nature of the Sociological Perspective

Because religion is so intensely personal and reverently held, some people have difficulty grasping a **sociological perspective** on it. The sociological perspective is a way of looking at religion that focuses on the human (especially social) aspects of religious belief and practice. Religion is both individual and social. Even the most intensively subjective mystical experience is given meaning through socially available symbols and has value partly because of culturally established interpretations of such experiences. A personal religious experience such as conversion is voluntary and subjective, yet is situated in social circumstances and given meaning by social conventions. When the individual communicates that religious experience, the symbols used to interpret it are socially determined.

Two characteristics of the sociological perspective separate it from nonscientific approaches to religion: It is *empirical* and *objective*. As much as possible, sociologists base their interpretations on empirical evidence. They seek to verify their images and explanations of social reality by experimental or experiential evidence. And they look for generalizations about a larger societal, historical, and cross-cultural picture, continually asking: Of what larger phenomenon is this particular situation an example? Objectivity means that sociological interpretations of religion do not attempt to evaluate, accept, or reject the content of religious belief. Sociological researchers set aside, for purposes of the study, their personal opinions about religion and try to be as objective as possible in observing and interpreting the religious phenomenon under study. From a sociological perspective, one religion is not superior to another. Indeed, the sociological perspective does not even presume the merits of religious over nonreligious approaches.

These aspects of the sociological perspective on religion may discomfort students who find their cherished beliefs and practices dispassionately treated as objects of study. It may be disturbing to have one's own religion treated as comparable to other religions and not as superior or uniquely true. The perspectives of the religious believer and the sociologist are necessarily quite different, but neither perspective represents the whole reality of religion.

Let's use the analogy of another object of attention. In looking at a flower, for example, several people could view the same one, yet each employ a different perspective. A botanist could employ a scientific perspective, analyzing in great detail the flower's physical properties, comparing it with other plants, and classifying the flower accordingly. An artist could interpret the flower onto the canvas, abstracting those visual qualities that convey its image. A mystic could use the flower as a focal point for meditation, perhaps experiencing a sense of oneness with the flower and the rest of the natural environment. And a child could examine the flower to consider its suitability for a daisy chain. The point is that none of these perspectives has a monopoly on the reality of the flower; what is discovered from the perspective of the botanist does not disconfirm what the artist's perspective has revealed.

Similarly, the religious perspective on human life often produces a very different picture of that reality than does a sociological perspective. What is relevant to the religious believer may be irrelevant or inadmissible evidence to the sociologist. What is central to the sociologist may be irrelevant or uninteresting to the religious believer. The reality perceived from the sociological perspective cannot disprove that of the religious believer; it is not possible for a sociologist to prove, for example, whether or not a religious prophecy or vision is from God. Furthermore, the perspective of the religious believer does not disprove that of the sociologist. If, for example, a religious "revelation" contradicts sociological evidence, the fact that the believer considers the "revelation" to be from God does not disprove the sociological observation.

These two different perspectives are, however, often difficult to reconcile. The sociological perspective, by definition, lacks a key religious quality—faith. The believer accepts certain beliefs and meanings on faith. Faith implies taking certain meanings or practices for granted, implicitly trusting, not questioning. By contrast, the sociologist does not take the believer's meanings for granted but takes them as an object of study.

The sociological perspective sometimes implies that people believe for reasons other than the truth value of the belief system. For example, sociologists have observed that upper-class persons are likely to belong to different Christian denominations than lower-class persons. A 1987 sample of U.S. church members showed that 39 percent of Episcopalians had annual incomes of $40,000 or more, compared with 14 percent of Baptists (Gallup and Castelli, 1989:103, 110). These substantial differences suggest that not only the "truth" of the denomination's belief system but also social class variables account for why people belong to one or the other group.

The more causality that believers attribute to supernatural sources, the less their interpretation can be reconciled with a sociological perspective. The very fact of treating certain interactions like conversion as human behavior, as an object of sociological study, is often incompatible with the basic beliefs of some religious groups. Sociology must necessarily "bracket" (i.e., methodologically set aside) the crucial religious question: To what extent is this action *also* from God? This does not mean that sociology treats religious behavior and experience as "merely" human. Important dimensions of religion may not be accessible to sociological interpretation. Nevertheless, whatever else they may be, religious behavior and experience are also human and are therefore proper subjects for sociological research and understanding. With these limitations of the sociological perspective in mind, let us explore some definitions of religion within this perspective.

## Defining Religion

The purpose of a sociological definition is to bring order to a vast array of social phenomena. The definition of any concept establishes (somewhat arbitrarily) boundaries around those phenomena to be considered as instances of that concept. Thus, to focus our attention on phenomena to be considered "religion," we must establish a working definition of religion.

Try writing your own sociological definition of religion. It should be broad enough to include all kinds of religion but narrow enough to exclude what is similar to, yet not the same as, religion. One way to construct your definition is to think of specific instances of "religion" or "nonreligion" and see whether your definition includes or excludes them. For example, does your definition include the religions of Asia or Africa? Does your definition use terms (e.g., "church" or "god") that, unless clarified, may apply only to some religions? Is your definition narrow enough? Does it distinguish between religious commitment and other commitments such as allegiance to a social club, an ethnic group, or family? There are several phenomena that only some sociologists define as religion (e.g., magic, superstition, witchcraft, astrology, spiritualism). Does your definition account for these phenomena? Why? How do the phenomena of nationalism, Marxism, atheism, or humanism fit into your definition? Does your definition include or exclude them? Why? Do you consider psychotherapy, sports, or rock music to be "religion"? Some sociological definitions of religion are so broad as to include all of these phenomena; others exclude some or all of them. We will examine some of these definitions and some arguments for excluding or including certain characteristics within a definition of religion.

Definitions of religion are an issue of serious debate, not just academic wrangling, in sociology. How one defines religion shapes one's explanation of its role in society. Different definitions of religion result in different interpretations of issues such as social change (see Chapter 7), modernity (see Chapter 8), and nonchurch religion (see Chapter 4). Indeed, part of the problem in determining a definition of religion that is satisfactory for analytical purposes is that the issue of what is "properly religious" is a continuing controversy in modern societies. In recent years, for example, the courts have examined several cases in which their verdicts hinged on deciding whether a given phenomenon was or was not a valid instance of religion. If the definitive boundaries of religion are not very firm for everyday social and legal purposes—as they are not—then producing a neat sociological definition may be counterproductive and misleading (Fenn, 1978:29).

### Definition as Strategy

It is useful to approach sociological definitions as strategies rather than as "truths." A definitional strategy narrows the field under consideration and suggests ways of thinking about it. Definitions can be evaluated according to how useful they are for a given task. What interpretive tasks does any given definition help to accomplish? Pragmatically, then, it is sometimes helpful to use different, even opposing definitional strategies to approach a phenomenon.

Two major strategies used by sociologists of religion are **substantive** and **functional definitions.** Substantive definitions try to establish what religion *is;* functional definitions describe what religion *does.* These approaches can be illustrated by analogy with definitions of the concept "chair." A substantive definition might state that a chair is an object of furniture that usually has four legs and a back; the definition might add further physical details to distinguish a chair from a sofa, bench, or toilet. A functional definition of chair might state that it is a seat, usually for one person. This functional definition is somewhat broader, allowing objects that various cultures use as seats but that may have no legs at all. Defining chair is easier than defining religion, however, because one can point to various physical objects called chair and then derive a set of distinguishing characteristics from observing them. Religion does not have clear-cut physical properties, nor are its characteristics readily ascertained and agreed on.

### Substantive Definitions

A substantive definition defines what religion *is*. It attempts to establish categories of religious content that qualify as religion and other categories specified as nonreligion. Spiro offers a straightforward substantive definition of religion: "An institution consisting of culturally patterned interaction with culturally postulated superhuman beings" (1966:96). By "institution," this definition refers to socially shared patterns of behavior and belief. All institutions include beliefs, patterns of actions, and value systems; the critical feature of religion is that the beliefs, patterns of action, and values refer to "superhuman beings" (Spiro, 1966:98).

Spiro's definition of religion is a good example of a sociological definitional strategy because all of the categories in the definition—"institution," "culturally patterned," "culturally postulated"—are sociologically relevant. His explication of the concept of "superhuman beings" is also sociologically significant because it emphasizes the sense of *power*. "Superhuman beings" are defined as those having greater power than humans, beings who can help or hurt humans but can be influenced by human action (Spiro, 1966:98). Power is one of the most important concepts in the sociology of religion, and a definition that emphasizes power can be useful. Other substantive definitions of religion use similar concepts, including "nonhuman agencies," "supernatural realm," "super-empirical reality," "transcendent reality," and "sacred cosmos."

Besides these theoretical approaches to religion, many empirical studies of institutional religion implicitly use substantive definitions for a different reason: They are easier to adapt to survey research. A somewhat oversimplified example is seen in the questionnaire that asks the respondent's religion, then lists the following alternative choices: "Protestant," "Catholic," "Jew," "other," "none." The underlying assumption of such a question is a substantive definition of religion as a specific Western religious institution.

The major advantage of substantive definitions is that they are more specific than functional ones. They are more explicit about the content of religion. Substantive definitions tend to be narrower and neater than functional definitions; using them, one can specify whether a phenomenon is or is not religion. Substantive definitions also tend to correspond more closely than functional definitions to commonsense notions of religion because they are generally based on Western—especially Christian—ideas about reality. For example, the distinction between "natural" and "supernatural" is a product of Western thinking, such as the elaborate medieval cosmographies (i.e., pictures of the universe) that classified natural and supernatural beings on numerous levels. Although hierarchies of archangels and seraphim may not be relevant in contemporary society, the division of natural from supernatural still seems familiar.

Substantive definitions are appropriate for studying religion in relatively stable societies, which present few problems with issues of social change and cross-cultural applicability. Substantive definitions are problematic precisely because they are historically and culturally bound, based on what was considered religion in one place and time. Because of their basis in Western historical experience, substantive definitions are often too narrow to account for non-Western religious phenomena. Substantive definitions are sometimes deceptively neat. Without specifying the functions of supernatural beings, for example, it is impossible to distinguish the gods from ghosts, Santa Claus, and the tooth fairy.

Some sociologists identify religion with church-oriented religiosity; relatively few non-Western societies, however, have formal organizations like churches. Does this mean that such societies therefore have no religion? If supernatural, nonempirical, or nonrational entities do not figure importantly in a society, does that society lack religion? Confucianism (the state religion of China from about 200 B.C. until the early twentieth century A.D.) is essentially a set of principles of order, especially regarding social relationships surrounding authority and kinship. It does not include the worship of any gods, although nature and ancestors receive much ritual reverence. If supernaturalism is a key criterion of a substantive definition, then Confucianism and some strains of Buddhism (in which there is no deity but only reverence and imitation of holy men) would not be considered religions. A current also exists within contemporary Christian thinking toward leveling this-worldly and other-worldly distinctions (e.g, the theologies of Pannenberg, Metz, Solle, and Moltmann). Would Christianity without supernaturalism cease to be a religion?

Substantive definitions have difficulty accounting for religious change. If religion is identified only in terms of religious expression in one historical period, any change from that form of expression looks like nonreligion. Many theories of religious change, for example, start from a notion of a time when people were "really religious." In Christian history, the thirteenth century is usually identified as the period when religion was a powerful force in the entire society and thoroughly interwoven with other aspects of life—work, education, politics, family, and so on. If one equates that image of religion with "real religion," any change from that pattern is viewed as a trend toward nonreligion ("secularization"). A parallel problem occurs in studies of the family. If "family" is defined in terms of its historical manifestations (e.g., concrete blood or marriage ties), many contemporary living arrangements do not qualify as family. Do two unmarried people and their offspring constitute a family? What about a single woman and her adopted children? What about six unmarried persons sharing sexual partners and collectively caring for their offspring? Because cultures constantly and sometimes rapidly change, it is difficult to create a substantive definition that applies through time. On the other hand, it is difficult to document historical changes in religion's place in society if we lack a sufficiently specific definition of religion. If we compare, for example, the situation of religion in the French Revolution with the religious situation of the American Revolution, we must be able to identify exactly which aspects of social life we mean by "religion."

As a definitional strategy, substantive definitions have advantages: They are more specific and amenable to empirical studies of religion. On the other hand, they tend to be more historically and culturally bound than functional definitions. Substantive definitions, in short, produce a very different interpretation of social change than functional definitions (for essays promoting substantive definitional approaches, see Berger, 1967:175–178, 1974; Robertson, 1970:34–51; Spiro, 1966).

### Functional Definitions

A functional definition of religion emphasizes what religion *does* for the individual and social group. Accordingly, religion is defined by the social functions it fulfills. The content of religious belief and practice is less important for this definitional strategy than the consequences of religion.

Geertz's definition of religion is a useful example of a functional definition: "A *religion* is: (1) a system of symbols which acts to (2) establish powerful, pervasive, and long-lasting moods and motivations in men by (3) formulating conceptions of a general order of existence and (4) clothing these conceptions with such an aura of factuality that (5) the moods and motivations seem uniquely realistic" (1966:4).

This definition emphasizes several sociologically important concepts. The most important element is the provision of meaning because the establishing of shared meanings (i.e., symbols) is an essentially social event. The definition also accounts for social structural and social psychological functions through the concepts of "moods," "motivations," and "factuality" (which also relate to the notion of "institution," as Spiro uses it). The distinguishing features of this definition are the "conceptions of a general order of existence" and their realism. According to Geertz, people interpret events and experiences as meaningful by linking them with a larger sense of *order*. This larger sense of order is perceived from a religious perspective as entirely real—even more real than mundane events and experiences. Empirically this distinction means that the content of religious beliefs and practices does not matter so long as it serves to symbolize some transcendent order to believers (Geertz, 1966:12ff.).

Some of the functions identified by various sociological definitions of religion include the provision of ultimate meaning, the attempt to interpret the unknown and control the uncontrollable, personification of human ideals, integration of the culture and legitimation of the social system, projection of human meanings and social patterns onto a superior entity, and the effort to deal with ultimate problems of human existence. Some of these functions are described further in later chapters.

One distinction used in many functional definitions is the social attribution of the **sacred.** Whereas the natural/supernatural distinction of some substantive definitions refers to the intrinsic quality of the *object* of worship, the sacred/profane distinction refers to the *attitude* of worshippers. The realm of the sacred refers to that which a group of believers sets apart as holy and protects from the "profane" by special rites and rules (Durkheim, 1965:62). The sacred is regarded as especially powerful and serious. For example, a communion wafer may be regarded as nothing but a piece of bread by nonbelievers, but Christian worshippers regard it as special and treat it differently from ordinary bread.

Natural objects such as candles, beads, books, water, oil, and wood can be regarded as sacred. Thus, nonsupernatural cultural systems (e.g., nationalism) could be viewed as sacred systems because of the attitude of their followers. Nationalist groups often treat the state and its symbols—flags, national holidays, shrines—as sacred.

Functional definitions of religion include all that substantive definitions identify as religion, but they are usually much broader. Both substantive and functional approaches would define the phenomena of Calvinism, Roman Catholicism, Methodism, Mormonism, and Reform Judaism as "religion." The inclusiveness of some functionalist approaches, however, extends to some phenomena that substantivists identify as nonreligion: ideologies, ethos, value systems, worldviews, interpersonal relations, leisure activities, voluntary associations, and so on. Geertz (1966:13) points out that his definition of religion would include, for example, golf—not merely if a person played it with a passion, but rather if golf were seen as symbolic of some transcendent order.[2] Functional definitions often include as "religion" phenomena such as nationalism, Maoism, Marxism, psychologism, spiritualism, and even atheism. The religious qualities of less comprehensive human activities, such as sports, art, music, and sex, are incorporated into some functionalist definitions as well.

The primary advantage of a functionalist definitional strategy is its breadth. Functional definitions tend to be better than substantive definitions for encompassing cross-cultural, transhistorical, and changing aspects of religion. Functional definitions encourage the observer to be sensitive to the religious quality of many social settings.

A drawback of some functional approaches to religion, however, is their assumption that society has certain functional requisites. This assumption implies that society requires certain social functions, some of which are uniquely fulfilled by religion. Some theories, for example, assume that society requires cultural integration (i.e., a common core of beliefs, values, and commitments). If religion is then defined as that which provides cultural integration, the theory implies that religion is a requisite for society's existence (Parsons, 1944:86). Such an argument is circular, describing religion as that which provides that which is defined as religion. The assumption that religion is necessary is unproven.

The breadth of functional definitions is a mixed blessing. While functional definitions are less culturally and historically bound, this inclusiveness makes it difficult to use them for empirical studies requiring neat, quantifiable categories. Some critics say that functional strategies result in all-inclusive categories, defining virtually everything human as religion.

Sociologists using functional definitions need to show why they include phenomena that participants themselves do not consider to be religious. From a

---

[2]Geertz, 1966:13. Geertz was probably not surprised by the actual development and international expansion of Japanese "golf religion," the Church of Perfect Liberty. One minister of this church explained that golf is "a little miniature of life. We can learn about ourselves through golf. Our teachings say that in life, anger, worry, and sorrow destroy success, just as in golf. Like golfing, life is an individual game, where you seek individuality; in golf, as in life, it's always the next shot, another chance for perfection" (quoted in the *New York Times*, May 30, 1975).

functionalist standpoint, a good case could be made for considering psycho-therapeutic groups as essentially religious phenomena. Members of these groups, however, often do not recognize their own beliefs and behavior as religious; they may even intensely disavow the religious label. On the one hand, it seems fair to accept the participants' notion of what they are doing; on the other, it is an honorable sociological tradition to point out the facades behind which people mask their activities. Few sociologists would accept at face value prison staff members' assertions that prisons are mainly for the purpose of rehabilitating prisoners. Similarly, there may be good sociological reasons for questioning whether vehemently nonreligious groups are actually nonreligious. Representatives of Transcendental Meditation (TM) have argued strongly in court that TM is not religion and therefore should be allowed in the public schools. The religious aspects of TM's beliefs and practices are downplayed because the desired image of TM is as technique rather than religion. The fact that a movement's desired image may be promoted by denying its religious qualities is worth studying, but sociologists need not accept a group's self-definition (for essays promoting functional definitional approaches, see Geertz, 1966; Lemert, 1975; Luckmann, 1967, 1977).

The choice between substantive and functional definitions is finally a matter of strategy. Each approach has advantages that may recommend it to certain sociological tasks. The two strategies, however, result in very different interpretations of various issues such as social change, secularization, the relationship between religion and other institutions in society, and new forms of religion. An awareness of the limitations and scope of each definitional approach will enable us to evaluate these other issues more critically. This book draws largely on examples that would be defined as religion by both substantive and functional definitions. At the same time, however, it explores phenomena such as civil religion and various new forms of religion that only functional approaches would consider as religion. This approach seems fruitful for two reasons: It allows us to consider an interesting range of phenomena from a sociology of religion perspective, and it raises broader theoretical issues.

## Aspects of Religion

There are many facets of religion that are important for the sociologist to recognize. In Western society, much emphasis is placed on formal beliefs. Religious education consists of informing children of what one's group believes; children read religious textbooks of explanations, learn catechisms (i.e., statements of propositions to which a believer should assent), study scripture, hear sermons. Christianity and Judaism place relatively great emphasis on intellectual and formal belief. In other cultures, however, formal beliefs are relatively unimportant. Other important aspects of religion include ritual, religious experience, and community. Children learn all that is relevant to their faith by participating in group ceremonies and imitating their elders' experiences. The following aspects of religion refer not only to the individual's religion but more generally to the ways in which the religious group organizes itself to focus its shared meanings.

### Religious Belief

Every religion has an essential **cognitive aspect.** The religion shapes what the adherent *knows* about the world. This cosmic knowledge organizes the individual's perceptions of the world and serves as a basis for action. If I believe, for example, that active, powerful evil spirits surround me, I will perceive "evidence" of their activity, and I will take actions to protect myself from them. My belief in evil spirits helps me explain other aspects of my life, such as why I cannot get a job and why I feel depressed and anxious. The same belief suggests appropriate actions, such as necessary prayers to ally myself with good spirits or use of amulets to ward off the influence of these evil spirits.

There is a tendency in modern Western societies to treat religious beliefs as "mere opinion" opposed to empirical beliefs, which are treated as "knowledge." This distinction hides the fact that both types of belief are "knowledge" to the individual who holds them. If a person considers evil spirits to be real, they *are* real in their consequences; they shape the person's experience and actions. The individual who believes that evil spirits cause illness and the individual who believes that germs cause illness are both acting according to their "knowledge."

Most of us are quite familiar with formal religious beliefs, the statements to which adherents of a given religion are supposed to assent, such as a catechism or a creed. The entire enterprise of theology out of which formal beliefs are developed represents a highly specialized and intellectualized approach to religion. But religion also includes less formal kinds of beliefs such as myths, images, norms, and values. Myths of creation and rebirth, for example, are told and enacted in dance and song. These other kinds of symbolization are often more potent influences on behavior than intellectual beliefs.

Religious beliefs are not mere abstractions that are irrelevant to everyday life. People use their beliefs to make choices, interpret events, and plan actions. Myths, one form of religious belief, are paradigms of human existence. There are myths about all major aspects of human life: birth and rebirth, creation and transformation, one's people and place, marriage, work, fertility, sterility, and death. Myths are stories that provide a rationale for a group's actions. They can be metaphors for concrete social structure and for real human events. Individuals draw on these interpretations to give meaning and direction to their own actions. Indeed, the very language in which beliefs are expressed structures believers' perceptions of the world.

Religious beliefs also inform the individual what action is good and desirable or bad and to be avoided. They may tell the individual that marriage is good and right because the holy marriage of the gods is to be copied by humans. Religious beliefs may inform the individual that eating other people is wrong because the gods value people and therefore do not define them as appropriate food. Thus, an entire range of values, norms, and attitudes derives from religious beliefs.

### Religious Ritual

**Ritual** consists of symbolic actions that represent religious meanings. Whereas beliefs represent the cognitive aspect of religion, ritual is the enactment of religious meaning. The two are closely intertwined. Beliefs of the religious group give meaning and shape to ritual performances. Ritual enactments strengthen and reaffirm the group's beliefs. They are ways of symbolizing unity of the group and, at the same time, of contributing to it. Ritual helps generate religious conviction. By ritual action, the group collectively remembers its shared meanings and revitalizes its consciousness of itself. This has important consequences for both the group and the individual member. The group renews its fervor and sense of unity, and individual members come to identify with the group and its goals (Durkheim, 1965:420; Geertz, 1966:28).

Ritual is an effective way of transforming space and time. Ritual places (such as a mountain or a shrine) can be transformed into the locus of power and awe. Time, too, can be changed, becoming a metaphor for sacred meanings and a catalyst for religious experience (Kertzer, 1988; Smith, 1987). For example, a pilgrimage is a ritual journey in which the participant enacts (both literally and metaphorically) a transition from one situation and self to another (Turner and Turner, 1978; see also Frankenberg, 1986; Pace, 1989).

Various religious groups place different emphasis on ritual. Eastern Orthodox, Roman Catholic, and Episcopalian Christians emphasize overt ritual more than Baptists or Methodists. The use of symbolism, such as processions, sacraments, candles, icons, and chanting, aids the collective remembering of the group's shared meanings. Even groups that do not consciously use rituals repeatedly symbolize their unifying beliefs. "Revival" meetings often emphasize spontaneity in worship and downplay formal ritual. Nevertheless, though the words of their prayers may not be formally set, members respond especially fervently to prayers that fit a familiar formula, and their responses are often equally stereotyped and expected. These periodic re-enactments are just as much ritual as the "high church" ceremonies of Episcopalians.

As these examples suggest, the *content* of an act is not what makes it a ritual act; rather it is the symbolic meaning attached to the act by participants. This symbolic value is what distinguishes, for example, lighting a candle to beautify one's dinner table and lighting a candle of the menorah to commemorate the Feast of Lights (the Jewish holiday of Hanukkah). There are also ritual acts that have, over time, become remote from their symbolic significance. For example, Roman Catholic priests wear different colored vestments to celebrate Mass on various holidays, but many people in the congregation may not even know what the symbolism of pink or green vestments means, and thus the ritual significance of the vestments is empty for them. Ritual performed for its own sake, empty of meaning for participants, had led to the notion that ritual itself is deadening.

It is not that ritual per se is dead but rather that the relationship between the symbolism and the group's shared meanings is weak or severed. Events or beliefs that a given ritual has symbolized may, over time, become less important or even forgotten. Perhaps the group may have moved on to new ways of symbolizing itself and its beliefs. Or perhaps the empty ritual may be symptomatic of weak ties in the group itself, that it has no essential unity to celebrate. By contrast, ritual that is a vital symbolic expression of important meanings for the group can be a sustaining, strengthening, and enlivening experience of unity.

The dynamic potential of religious ritual suggests its link not only with religious belief but also with religious experience. Religious symbols, expressed in beliefs and rituals, have *real* power, which can be experienced personally by the individual. Ritual words and ceremonies can evoke experience of awe, mystery, wonder, and delight. Religions often emphasize the power of ritual words, as exemplified by the seriousness surrounding the pronouncement of the words "This is my body" in Christian communion services or by the expectation surrounding the exclamation "Heal!" in a faith-healing service. Ritual has the potential to produce special religious experiences for the group and its individual members.

### Religious Experience

Religious experience refers to all of the individual's **subjective involvement** with the sacred. Although such experience is essentially private, people try to communicate it through expression of beliefs and in rituals. A communal ritual may be the setting for a personal religious experience. Thus, a person receiving communion (i.e., a communal ritual) in a Christian worship service may also experience an intensely subjective awareness of God. Prayer, meditation, dancing, and singing are other common settings for personal religious experience. Similarly, even a private experience has a social element because socially acquired beliefs shape the individual's interpretation of religious experience. The symbolism of various religious traditions shapes the interpretations of even highly mystical experiences through such images as the pilgrimage, perfect love and marriage, and rebirth or transformation (Underhill, 1961:125–148).

Individual religious experiences vary considerably in intensity. They range from momentary senses of peace and awe to extraordinary mystical experiences. Different religions place different emphases on religious experience. Most Christian denominations do not actively encourage highly emotive religious experiences; whereas in some Pentecostal groups, these experiences are central and eagerly sought. In many religions, extraordinary and intense experiences are segregated, appropriate only for certain members or on certain occasions. Thus, among the peoples of the northern Asian arctic region (e.g., Eskimos), extraordinary religious experiences are expected of certain members called shamans. *Shamans* are religious specialists who have undergone an intense encounter with sacred forces, emerged with special powers to effect good or evil on behalf of the rest of the group (see Eliade, 1964). Another example is the segregation of occasions for special religious experiences (e.g., initiation rites).

The content of religious experience varies. It may include pleasurable aspects such as a sense of peace, harmony, joy, well-being, and security. Religious experience may also produce terror; anxiety, and fear. While the content of the experience partly depends on the group's beliefs about what is being encountered, both the pleasurable and frightening experiences are related to the sense of power or force with which the sacred is believed to be endowed. The individual who experiences a sense of security does so because of the power of the sacred to protect from harm; the individual who experiences great fear does so because of the power of the sacred to cause grave harm. The notion of the sacred thus entails both harmful and helpful aspects. Personal experiences of this power can be overwhelming.

William James (1958:67) quotes an account of a religious experience that illustrates the intensity of some of these aspects:

> I remember the night, and almost the very spot on the hilltop, where my soul opened out, as it were, into the Infinite, and there was a rushing together of the two worlds, the inner and the outer. It was deep calling unto deep,—the deep that my own struggle had opened up within being answered by the unfathomable deep without, reaching beyond the stars. I stood alone with Him who had made me, and all the beauty of the world, and love, and sorrow, and even temptation. I did not seek Him, but felt the perfect unison of my spirit with His. The ordinary sense of things around me faded. For the moment nothing but an ineffable joy and exaltation remained. It is impossible fully to describe the experience. It was like the effect of some great orchestra when all the separate notes have melted into one swelling harmony that leaves the listener conscious of nothing save that his soul is being wafted upwards, and almost bursting with its own emotion. The perfect stillness of the night was thrilled by a more solemn silence. The darkness held a presence that was all the more felt because it was not seen. I could not anymore have doubted that *He* was there than that I was. Indeed, I felt myself to be, if possible, the less real of the two.

This kind of religious experience represents an "alternate state of consciousness"—a situation in which the individual's consciousness is relatively remote from the sphere of everyday reality. The person may experience being out of one's body or being one with something or someone else. For example, one person described such an experience as follows: "I came to a point where time and motion ceased. . . . I am absorbed in the light of the Universe, in Reality, glowing like fire with the knowledge of itself, without ceasing to be one and myself, merged like a drop of quick-silver in the Whole, yet still separate as a grain of sand in the desert" (quoted in Happold, 1970:133).

Alternate states of consciousness are not necessarily religious. Substances such as peyote, for example, can produce alternate states of consciousness, but the context in which they are taken (e.g., ritual) and the meanings attached to taking them (i.e., symbolism) determine whether they produce religious experiences. "Alternate states of consciousness" or "peak experiences" are only one extreme of a continuum of religious experiences (see, e.g., Maslow, 1964). Less dramatic religious experiences include a sense of the presence of God, a moving conversion experience, or a deeply absorbing ritual experience.

Contemporary society often ignores or discourages such experiences. Peak experiences such as those described, however, may be more common than generally acknowledged. Approximately 35 percent of a U.S. sample ($N = 1,467$)

had at some time experienced feeling "very close to a powerful, spiritual force that seemed to lift you out of yourself" (McCready and Greeley, 1976:132; see also Greeley, 1987; Hood, 1985; Hood et al., 1989; Spilka et al., 1985). Perhaps a self-fulfilling prophecy may operate in a culture's encouragement or discouragement of such experiences. In those cultures that value and encourage special religious experiences, members have felt and acknowledged these experiences; in cultures that devalue or discourage these experiences, members neither seek nor recognize them.

While a sizable proportion of the U.S. populace have had some special spiritual experience, the culture in general does not particularly value such experiences. This culture places great emphasis on rational, intellectual, "objective" ways of knowing. Religious experience, by contrast, is a way of knowing by subjectively apprehending a reality. One woman described her own experience as follows: "I felt a deep sense of warmth and security and a startling awareness of how much the Lord cares for me personally." This kind of knowledge, although it may correspond with the official belief system of a religious group, is not simply learned, deduced, theologically debated, or received from church authorities.

**Religious Community**

Religious experience may also include the awareness of belonging to a group of believers. Rituals often remind the individual of this belonging, creating an intense sense of togetherness. Communal religious settings can produce a resonance of several individuals' experiences and thus an even deeper sense of sharing "inner time" (see Schutz, 1964). Like making music together, such shared religious experience can create a sense of sharing vivid present time—an experiential communion (Neitz and Spickard, 1990). For example, participants in the several days and nights of a Navajo "chant" can, through ritual acts, attune their individual experiences toward the culturally valued sense of harmony between the individual, social group, and environment (Spickard, 1988).

The community of believers may be formally or informally organized. Formal specialization of a group into an organization such as a church is a relatively recent historical development. The religious group—formal or informal—is essential for supporting the individual's beliefs and norms. Coming together with fellow believers reminds members of what they collectively believe and value. It can also impart a sense of empowerment to accomplish their religious and everyday goals. And the nature of the religious community illustrates the social context of religious meaning and experience. We will explore this aspect further in the next two chapters.

# The Development of the Sociology of Religion

The very development of sociology and the sociology of religion in particular is rooted in some of the social processes we will examine (especially the process of secularization). Only as beliefs and social practices were perceived as something other than taken-for-granted realities could a scientific study of these "social facts" become possible. The development of Enlightenment rationality and paradigms of scientific thought encouraged examination of religion as human action (see Robertson, 1970:7–33).

Ironically, these same developments also acted on religion, possibly changing it fundamentally. Thus, sociology has some complicity in this alteration; to a certain extent, popularization of the sociological perspective on religion has the effect of altering the nature of religion itself. The social sciences, as heirs to Enlightenment humanism, have interpreted religion as a human construction, undermining its claims to transcendence. Some recent sociologists have called attention to this perspective and questioned whether the social sciences can validly claim to be value free if their very choice of an interpretive stance implies certain value judgments. They ask whether it is possible to have a social science of religion that allows for the truth of religion (see Anthony and Robbins, 1975; Bellah, 1970; Greeley, 1972b; and especially Johnson, 1977).

An important strand of classical sociology of religion has a prophetic ring. Weber's doubts about the future of a fully "rationalized" society, Durkheim's concern about the impact of social change upon the cohesion and moral unity of society, Simmel's insights into the threat to the individual of increasing societal controls, and Marx's questioning the continued legitimacy of capitalistic modes of organization all contain strong critical themes (see Durkheim, 1965; Marx, 1963; Simmel, 1971; Weber, 1958a). These theorists saw their contribution to a sociology of religion as part of a larger examination of the nature of modern society. They asked questions such as: In what direction is modern society going? How did it develop to this point, and what factors influence its further development? What is the place of the individual in modern society, and what is the impact of these broader social changes on the individual? These classical contributions are emphasized in several chapters of this book.

Later developments in sociology, especially in the United States, did not continue these themes of concern. The sociology of religion was largely ignored until the late 1940s and 1950s, when an upswing in church membership and participation, combined with increasing interest of religious leaders in institutional research, provided an impetus for sociologists to focus again on religion. The dominant theoretical perspectives on religion then current in sociology were, however, very narrow (see Berger and Luckmann, 1963). Religion was defined strictly in terms of its formal organizational setting: the institutional churches. Much of the research was circumscribed by the needs and interests of these organizations. Demographic features of parishes were studied, for example, in order to predict future building and staffing needs. The standard methodological approaches of that period were similarly restricting; survey research techniques were best suited for measuring clearly definable institutional behavior or opinions. A survey interviewer could more readily ask a respondent to name a denominational preference than to describe fully an entire personal meaning system.

This stage in the development of sociology of religion was characterized by several assumptions, notably the identification of "church" as "religion." Another general assumption of this phase was that individual religiosity could be equated with individual participation in religious institutions. A corollary assumption was that the subjective aspects of religion could be tapped by identifying the individual's religious opinions or attitudes (see Luckmann, 1967:20–23; some criticisms of these assumptions are developed further in Chapter 4). Perhaps as a by-product of this narrow definition of the field's scope, the sociology of religion has been rather isolated from other sociological knowledge and subdisciplines. It was not until the 1980s that sociology of religion seriously expanded its horizons to explore the important connections with other subfields such as women's studies, sociology of health, and mass media (Beckford, 1990).

Beginning in the mid-1960s, the sociology of religion greatly expanded its theoretical focus and research interests. New developments in Western religion spurred this expansion. Following publication of Anglican Bishop John Robinson's provocative book *Honest to God* (1963), theologians debated the viability of culturally bound Christian theology, and a school of theology proclaiming the "Death of God" developed. Simultaneously, Catholic theologians explored new interpretations of their tradition, and the Second Ecumenical Council (Vatican II) established far-reaching structural changes in that church. Religious participation in various human rights movements, the articulation of various national "civil religions," and the emergence of numerous new religious movements raised important sociological questions that pressed the discipline to expand its interpretive scope. Political events have brought religious issues and groups to the fore, stimulating research and analysis on such themes as the prominence of religion in conflicts in Iran, Lebanon, Sri Lanka, and Northern Ireland.

Recent sociology of religion is characterized by its broader emphasis. It encompasses not only Western but also non-Western religious expression; it sees parallels between religious behavior in this society and simpler societies; it is interested in both institutional and noninstitutional religious behavior. Its theoretical concerns reflect both a thoughtful return to classical themes and an ability to go beyond the classics. This development is part of a generally larger interest within sociology in building critically on classical theory. Recent sociology of religion is more central to the larger sociological enterprise than before, asking what can we understand about the nature of society through examination of religion? This focus leads current sociology of religion to themes that are central in contemporary theory, such as the legitimation of society, the individual-to-society link, and the impact of social change (Robertson, 1977).

## Summary

This chapter has examined key features of the sociological perspective on religion: its objectivity and empirical focus. Definitions of religion are matters of serious academic debate, and the choice of definitional strategies (i.e., substantive or functional) influences the observer's conclusions about the place of religion in contemporary society. To understand how religious groups organize themselves around shared meanings, it is necessary to examine several aspects of religion: religious belief, ritual, experience, and community.

The development of the sociology of religion is itself rooted in some of the social processes that have influenced religion's role in society. Only as beliefs and social practices lost their taken-for-granted quality could a scientific study of them proceed. Sociology of religion was central to the focus of classical sociology—its concern about the emerging forms of societal legitimacy, social controls, and the place of the individual in modern society. Recent sociological thought examines these issues with renewed fervor.

## The Focus of This Book

These theoretical themes are the basis of the organization of this book. Chapter 1 has described the scope of the field and illustrates how different strategies for defining religion result in different interpretations of the location of religion in contemporary society. Chapter 2 discusses the ways in which religion provides meaning and belonging for the organization of individual lives and social groups. The capacity of religion to provide meaning and belonging is further illustrated in Chapter 3, which explores how the individual's religion is socially acquired, maintained, or changed. The individual's religion cannot be narrowly equated with participation in specifically religious organizations, as Chapter 4 shows; rather, it comprises a complex array of beliefs and practices organized (sometimes not very coherently) into a personally meaningful and useful way of thinking and doing—a worldview. People do, however, affiliate themselves with groups to express and maintain their religions, and specifically religious organizations have identifiable characteristic patterns, as described in Chapter 5 on religious collectivities. These patterns are dynamic, resulting from tensions within the religious group and between the religious collectivity and the larger society. Chapter 6 explains the dual potential of religion to promote social conflict and to enhance social cohesion; these aspects of religion are also related to religion's role in social change. Chapter 7 examines the conditions under which religion is likely to promote or inhibit social change. Various social changes also influence religion and its social location. Chapter 8 discusses several interpretations of such changes connected with modernization and the structure of the modern world. These theories are directly related to some of the broader issues of contemporary sociology.

Because of this focus, this book de-emphasizes much of the respectable but narrow material from the sociology of institutionally specialized religion. There is relatively little discussion of the clergy and its roles or of congregational polity. For the same reason, methodological issues discussed in this volume suggest ways of exploring nonchurch religiosity and other aspects of noninstitutional religion. Yet much of this emphasis is relatively new, so not many research results exemplify this broader focus, and whole areas of relevant information remain untapped. The sociology of religion has produced much excellent work, but some of the weaknesses of data, methodology, and theory pointed out in this book suggest areas for further development. The implications of contemporary sociology are far-reaching, and I hope to share with the reader the intellectual excitement of pursuing some of these issues.

## Recommended Readings

Robert N. Bellah. "Christianity and Symbolic Realism." *Journal for the Scientific Study of Religion* 9(2), 1970:89–96.

Clifford Geertz. "Religion as a Cultural System." *Anthropological Approaches to the Study of Religion*, Michael Banton, ed. London: Tavistock, 1966:1–46.

Melford Spiro. "Religion: Problems of Definition and Explanation." *Anthropological Approaches to the Study of Religion*, Michael Banton, ed. London: Tavistock, 1966:85–126.

CHAPTER TWO

# The Provision of Meaning and Belonging

Religion represents an important tie between the individual and the larger social group, both as a **basis of association** and as an **expression of shared meanings** (cf. Greeley, 1972a). In this chapter, we examine the provision of meaning for the individual and the larger social group. This provision of meaning is linked with the communal aspect of religion; a community of believers maintains a meaning system and mediates it to the individual. The religious processes described in this chapter are interesting in themselves, but we must ask the broader question: What does our understanding of religion tell us about society itself? Comprehending how the individual and the larger social group are linked in a community of shared meanings is important for the larger issue of how society is possible.

## Religion and Systems of Meaning

The capacity of religion to provide meaning for human experience has been a major theme in the sociology of religion since its early emphasis in the works of Weber (1958a). **Meaning** refers to the interpretation of situations and events in terms of some broader frame of reference. For example, the experience of losing one's job is given meaning when it is interpreted as "bad luck," "market forces," "the boss is trying to get rid of pro-union workers," "ethnic discrimination," or "God's will." As this example shows, meaning here refers to ordinary, everyday interpretations of experience. Sometimes meaning is expressed in grand, theoretical terms, such as elaborate theories of meaning formulated by philosophers and theologians. Most of what is meaningful to people in their everyday lives, however, is less complex.

Have you ever tried to write your autobiography? The ways in which you interpret your experiences and the events, persons, and experiences you choose to remember, as well as how you explain them, reflect the meanings you apply to events in your own life. This chapter discusses how that kind of meaning is attached to individual and social life.

People typically choose their interpretations from a larger meaning system, a broad, interpretive framework. Although the meanings that constitute this personal system are all interrelated, they are not necessarily coherent or internally consistent. People may believe in the idea of germs and in the seemingly contradictory idea of evil spirits as explanations for illness. They may simply use the concept of germs to interpret some illnesses and evil spirits to explain others.

Most historical religions are comprehensive meaning systems that locate all experiences of the individual and social group in a single general explanatory arrangement. A comprehensive meaning system such as these is called a **worldview** (Berger and Luckmann, 1966). We will use the concepts of "religious worldview" and "religious meaning system" more or less interchangeably because most historical religious meaning systems have been comprehensive. Nevertheless in modern society, religious meaning systems compete with many other worldviews. Individuals are less likely to use any single comprehensive meaning system but may apply religious meanings to segments of their lives. For example, some meaning systems (e.g., astrology) explain only certain spheres of their adherents' lives, and believers find other meanings elsewhere.

According to some sociological definitions of religion (especially functional definitions, described in Chapter 1), any comprehensive meaning system is fundamentally religious, regardless of its content. For the purposes of this discussion, however, it is sufficient to say that comprehensive meaning systems that are not overtly religious (e.g., psychotherapeutic or political meaning systems) share most characteristics of specifically religious meaning systems. The process of conversion from one comprehensive meaning system to another is comparable, regardless of the belief content of either meaning system. We must also keep in mind that meaning systems are not abstract but are created and held by people.

According to Berger (1967), the provision of meaning is particularly important for an understanding of religion because of the ways in which meaning links the individual with the larger social group. *Meaning is not inherent in a situation but is bestowed.* The example of how one interprets losing one's job shows the wide variety of possible meanings that could be attached to the event. Although the individual may examine the event itself for clues about which meaning "fits" best, the final choice of meaning is *applied to* the event. The person fired might conclude that, although the boss cited uncooperative attitudes, the "real" meaning of the event points to a greater mission as an organizer of workers elsewhere. The experience is given meaning by this choice of interpretations. Attaching meaning to events is a *human* process (Berger, 1967:19).

Geertz (1966:40) suggests that religion serves as a template in establishing meaning. It not only interprets reality but also shapes it. The template of religion "fits" experiences of everyday life and "makes sense" of them; in turn, this meaning shapes the experiences themselves and orients the individual's actions. A meaning system in which Satan is prominent can explain past events and experiences (e.g., temptation or illness) as evidence of Satan's effects and can shape future experiences (e.g., by warning believers to avoid certain places or activities where Satan is likely to ensnare them).

By examining how this meaning-giving process occurs, we can understand some of the ways religion links the individual with the larger social group.

### Meaning for the Social Group

Meaning systems interpret an entire group's existence. Thus, the group's ways of doing things and its very existence are assigned meaning. The group, for example, may explain its moral norms as instituted by God; its pattern of family life may be interpreted as copying the family of the gods; and its history may become meaningful as the story of the gods' relationship with their people. As Berger (1967:29–33) points out, the meaning system is both *explanatory* and *normative;* that is, it explains why things are the way they are and prescribes how things should be. The meaning system of the group "makes sense" of its *social order*—the present/existing and future/desired social arrangements of a group, such as its form of authority and power, its stratification system and allocation of roles, its distribution of resources and rewards, and so on. These qualities make the meaning system a strong legitimation for the social order of the group.

A **legitimation** is any form of socially established explanation that is given to justify a course of action. Legitimations include any explanation of social practices: Why do we do things this way? Why should we behave according to that norm? Why do we have this position in society? Particularly important are those legitimations that establish authority in the group (Berger, 1967:29ff.).

Legitimations are expressed in a wide variety of forms. Myths, legends, proverbs, folktales, and history are all invoked to justify certain social arrangements. Taken together, these legitimations may be viewed as the "story" out of which a group lives. These examples show that legitimations are seldom intellectual in form. Intellectual legitimations, such as those developed in philosophy, theology, and political and economic theory, are a highly specialized form of legitimation. Nonintellectual forms are far more common.

In India, the ancient Hindu myth of four *varnas* legitimated the division of labor and subsequent rigid social class system developed by the thirteenth century (and continued well into the twentieth century). According to this myth, the highest social group, the Brahmins, issued from the mouth of the Creator. Duties assigned them by the Creator included various religious functions, study, teaching, sacrifice, almsgiving, and the perpetuation of the sacred literature. The next group were Kshatryas, who were born from the arms of the Creator. Their special gift was power, and their duties included fighting and governing. From the thighs of the Creator came the Vaishyas, who were given the gift of strength; they were expected to cultivate the land and engage in other productive labor. The lowest social group, the Sudras, came from the Creator's feet. Theirs was the duty of serving the other three varnas (Lemercinier, 1981:164). This creation myth used potent imagery of the functions and value of parts of the Creator's body to explain the human social structure, to justify the privileges of the higher classes, and to prescribe appropriate behavior of all social groups. The myth itself exemplifies a nonintellectual form of legitimation, but complex intellectual ideologies and legal codes were also eventually developed to justify the Hindu caste system, as further described in Chapters 4 and 7.

Legitimations explain the ways in which the social group has behaved, but they also shape future action by justifying the norms of appropriate or desirable action. Religious legitimations make especially strong claims for the bases of order and authority and for the specific arrangements of the social order by their references to a higher authority. Thus, the social order is represented as more than human convention. The European idea of the "divine right of kings," for example, implied that the king ruled not merely as a political arrangement but as a permanent, God-given right. Contemporary examples of social arrangements that are religiously legitimated include the sanctity of marriage, the right to own and defend property, and the moral rightness of waging certain kinds of war.

The effectiveness of these legitimations often involves a certain amount of *mystification* (Berger, 1967:90). Anything suggesting that the social arrangements are purely human convention is allowed to be forgotten or is deliberately masked. The present way of doing things is promoted as the "true tradition," and contrary evidence is not emphasized in the group's tale of its own history. For example, the Christian ideal of the institution of marriage is often represented as being exactly the ideal held during the whole history of Christianity. Actually, however, both the ideals and practices of marriage have changed considerably, and current religiously legitimated arrangements, such as the practice of church weddings, are relatively recent.

The early Christian churches were ambivalent about marriage; some early church leaders thought that the ideal for committed Christians was celibacy or even castration, with a chaste marriage (i.e., without any sexual activity) as an accommodation for those who could not meet the difficult ideal (Bullough, 1982; McNamara, 1982). Until the sixteenth and seventeenth centuries, ordinary people typically wed by private self-marriage (later formalized as "common-law" marriage). Not until the sixteenth century was there a church prohibition of the practice of concubinage (Brundage, 1982). For most people, self-marriage was an acceptable pattern through the Middle Ages, though couples sometimes went to the church steps for a blessing. Only in the sixteenth century was church marriage declared a sacrament, but during the subsequent Reformation, many Protestant groups denied that marriage was a sacrament. Most Protestant groups did, however, keep the practice of church weddings, and they developed their own rituals for these ceremonies. (Noonan's 1965 analysis of Roman Catholic marriage and family norms and Nelson, 1949, provide further examples of how Christian ideals of marriage and other social institutions have changed considerably over the years.) Thus, we see how the effectiveness of legitimations of present-day marriage practices depends on treating these practices as traditional and on forgetting or de-emphasizing those parts of history that are inconsistent with present norms and practices.

Although religious legitimations are generally used to justify existing social arrangements, they are also potent forms of criticism of the existing social order, as Chapter 7 shows in some detail. The Old Testament prophet Amos legitimated his denunciation of people's ways by reference to their God. Religious legitimations may be invoked to justify even revolutionary action. Thus, religious legitimations are not solely a tool of the dominant group but may justify actions of subordinate and dissenting groups as well.

### The Individual's Meaning System

The individual does not construct a personal meaning system from nothing. An individual's meaning system is learned, for the most part, during the process of socialization. The interpretations that seem most plausible to a person are likely to be those that are familiar and held by others who are important to that individual. So although each individual operates with a highly personal meaning system, that set of meanings is greatly influenced by family, friends, institutions (e.g., education), and the larger society. While the individual comes to accept some of the meanings communicated during socialization, the resulting meaning system is not, however, an inevitable product of socialization. The individual can reject or modify meanings communicated by others. When several competing meaning systems are presented as alternatives (which happens especially in modern society), the individual can choose which meanings to accept. The individual may accept the meaning system presented by a subgroup in society and not that of the larger society. Yet all personal meaning systems gain effectiveness by their link with some community in which they are shared.

Meaning and individual identity are intertwined. People locate themselves and their personal actions in a larger social order by means of their meaning systems. The individual selects an interpretation of events and experiences from the larger interpretive scheme provided by the meaning system, and that personally accepted meaning system becomes the individual's point of reference. The applied meaning system informs the individual what "kind" of person one is, the importance of the roles that one performs, the purpose of the events in which one participates, and the significance of being who one is. A meaning system, in other words, *makes sense of* one's identity and social being.

People can bestow meaning on a situation in various ways. Beliefs are important in this process. Ideas can help locate an experience in a meaning system, but a meaning system cannot be reduced to its formal belief content. Miracles, magic, ritual, and symbols also contribute to a pervasive sense of meaning. The individual can apply meaning to a situation by performing an appropriate ritual for the event. Similarly, by interpreting an event as miraculous or magical, the individual places special meaning on it. The application of meaning to human experience entails social processes. Through everyday conversation, the individual tries out interpretations of experiences. Interaction with others—especially with persons whose response "counts" to the individual—is a significant part of the process of bestowing meaning on a situation.

At the same time, the meaning system informs the individual of the values and norms of the larger group. The following interpretive statements illustrate the variety of ways in which norms and values can be embodied in a meaning system:

"Cleanliness is next to godliness."

"Masturbation is likely to lead to insanity."

"Most of the problems with youth nowadays are due to bad mothers who go out to work instead of taking care of their children."

"Doctors should make a lot more money than others because they had to work so hard and long to become qualified, and their skills are more important to society."

"People who commit adultery will be punished by God."

"He brought his illness on himself by his bitter and hateful outlook on life."

"Hard work and patience are always rewarded."

Even though you may personally disagree with some or all of these statements, you can see how each statement can be used to *interpret* situations or experiences and to *evaluate* general kinds of behavior. Meaning systems embody norms, social evaluations of behavior. To the extent that one applies these normative interpretations to one's own behavior, the meaning system brings an evaluative element to one's identity. Thus, in terms of the interpretive framework, one may conclude: "I am a sinner," "I am a good mother," "I am a successful warrior," "I am ill-fated," or "I am a virtuous person."

The individual's meaning system makes possible the evaluation of past actions and the motivation of future actions. The ability to perceive events as ordered in some way enables one to plan and orient one's actions. If things are experienced as "just happening" in a chaotic, meaningless string of events, the person literally does not know what to do. If, however, these events are given meaning, their interpretation implies an appropriate course of action (even if that action consists in seemingly passive responses such as "suffering through it" or "praying over it"). By relating mundane social life to the realm of the transcendent, religion is particularly effective in motivating the individual's participation in the larger social group. The classic example of such motivation is the concept of "vocation," the idea that one is "called" to an occupational or economic status by God. This idea gives special meaning to everyday work. By understanding such work in terms of "vocation," individuals gain a sense of purpose and value to their labor, and society gains the willing contribution of its members (see Weber, 1958a: Chapter 3).

Both the individual and the social group draw on religion for meaning of their existences. The meaning system provides interpretation for their experiences, locating human lives and events in terms of a larger framework. Religion also serves as an important form of legitimation, or justification, for both the individual and the social order. Religion interprets and evaluates the "way things are to be done" in the social group. These legitimations provide meaning for individual members of the group who accept these explanations and incorporate them into their ways of thinking about the world.

## Crisis of Meaning

The meaning system of the individual or group can integrate most routine events into an understandable pattern, a meaningful whole. Some events and experiences, however, are not so easily interpreted within the existing meaning system. The individual who experiences the death of a loved one, a painful illness, or serious economic misfortune may not be helped by the existing personal meaning system. An entire group can undergo similar meaning-threatening experiences: oppression by an enemy, famine, earthquake, or economic depression. Such events are particularly meaning-threatening if they appear to contradict important aspects of the existing meaning system. Groups who believe in a loving, personal god have more difficulty reconciling disastrous events with their meaning system than do groups whose god is remote and capricious.

Another situation that creates a problem of meaning is serious discrepancy between the ideal that a group promulgates and the actual practice. Inequality and injustice can be especially meaning-threatening when they are inconsistent with the group's ideals. Because the threat of the meaninglessness of such events is so great, individuals and groups try to build special legitimations into their meaning systems to justify these apparent contradictions or discrepancies. In other words, part of the meaning system itself interprets events and experiences that would seem to contradict the meaning system. These meanings are responses to the *problem of theodicy* (Weber, 1963: Chapter 9).

### Theodicies

**Theodicies** are religious explanations that provide meaning for meaning-threatening experiences. Most religions, for example, offer theodicies of suffering and death. The content of these explanations differs among the various religions, but the desire to find meaning for such experiences appears virtually universal. Disaster and death create a problem of theodicy not because they are unpleasant, but because they threaten the fundamental assumptions of order underlying society itself (Berger, 1967:24). Theodicies tell the individual or group that the experience is not meaningless but is rather part of a larger system of order. Some successful theodicies are, in fact, nothing but assertions of order. A woman discussing her personal meaning crisis after her husband's premature death said: "I finally came to understand that it didn't matter whether *I* understood why he died when he did, but that God had a reason for it, and that was all that mattered." For this believer, knowing that an order exists behind events was more important than knowing *what* that order is.

Similarly, theodicies do not necessarily make the believer happy or even promise future happiness. A person suffering poverty and disease may be satisfied with the explanation that the situation has resulted from sins committed by the ancestors. Such an interpretation offers little hope for overcoming the poverty or disease, but it does offer meaning. It answers the question, "Why do I suffer?"

The capability of religion to provide meaning can be illustrated by how a society handles the especially meaning-threatening situation of dying and death. There is an inherent problem of meaning in the prospect of one's own or a loved one's death. Death seems to negate the individual's or group's sense of order. For this reason, the way in which a society handles the process of dying is revealing of its larger meaning system. Long before most individuals face death, their religion has been providing meaning for their various life stages (e.g., by rites of passage into adulthood or by legitimating the norms and prerogatives of old age). In many societies, the meaning system somewhat normalizes dying as a further stage of human development. Some meaning systems affirm an after-life or a rebirth; others emphasize that people live on through their offspring or tribe.

In American society, dying and death have become particularly meaning-threatening. This problem of meaning partly results from the undermining of traditional theodicies; formerly used explanations have become less satisfying to many persons. The problem of the meaning of death is especially acute in this society because its value system assigns comparatively great worth to individual lives. Numerous social arrangements contribute to the problem. The dying person is typically segregated from one of the foremost meaning-providing social supports—the family. Furthermore, the specialized roles of those tending the dying seldom include the provision of meaning. Unlike their counterparts in traditional medical systems, modern doctors do not generally consider as part of their role helping the patient assign meaning to illness or death (Kleinman, 1988). Death and dying have become generally secularized processes, and the problem of their meaning has become acute.

### Anomie

Sometimes a meaning system is completely unable to absorb a crisis experi-ence. The theodicies applied may not be effective in reintegrating the experience into the meaning system, or the group supports of the entire meaning system may be so weakened by the crisis that people are unable to restore a sense of order and meaning. When an individual or group has lost its fundamental sense of meaning and order, ongoing social life becomes virtually impossible. Why should one want to do anything if everything seems senseless? When a meaning system is so weakened, the moral norms it supports are also undermined. And with no underlying order to existence, there is no apparent basis for distinguish-ing between "right" and "wrong" ways of behaving.

It becomes necessary, if social life is to continue, to establish a new basis of order and a new meaning system. If a group's way of life has been thoroughly disrupted by military conquest, for example, the group might reorganize itself around a social movement (e.g., followers of a new prophet) that offers a new basis of order and meaning for group members. The new meaning system sometimes consists of embracing new meanings (e.g., those of the conquerors), a rearrangement and reaffirmation of old meanings (e.g., in nativistic move-ments), or syncretism—the interweaving of new meanings into the traditional meaning system. In colonial Brazil, for example, uprooted, enslaved Africans were allowed to celebrate the *congada*—a symbolic coronation of the King of the Congo, absorbed into the annual Catholic festival of Our Lady of the Rosary, who became patroness of blacks. Several symbolic homologies between Western African religion and Brazilian Catholicism made fertile ground for syncretism (Camara, 1988). Regardless of the source of elements for the new meaning system, its importance lies in its ability to reorganize the basis of order for the group. Social life is given new meaning.

The situation of crisis in the group's meaning and order is described in Durkheim's concept of **anomie** (or "anomy"). The word *anomie* means literally "without order." Durkheim applied this concept to social situations in which there was a deregulation of the public conscience. Anomie means a crisis in the moral order of a social group (Durkheim, 1951:246–257; see also Berger, 1967:22ff.). Although anomie has serious consequences for the individual (e.g., it might lead one to suicide), it is primarily a social situation referring to the inability of the social group to provide order and normative regulation for individual members. Religion's capability of providing meaning and order suggests that religion functions as a protection from anomie in two ways: A firm religious basis of order is a buffer against the occurrence of anomie in the first place; and if the group does experience an anomic situation, religion can potently respond to the crisis of moral meanings. Meaning is fundamental to a sense of order—without it, there is chaos. Indeed, Berger suggests that the opposite of "sacred" *is* "chaos." He states, "The sacred cosmos, which transcends and includes man in its ordering of reality, thus provides man's ultimate shield against the terror of anomy" (Berger, 1967:27; see also Geertz, 1966:14).

### Mazeway Disintegration

Another way of understanding social life in such crises is by examining the parallels between peoples' reactions to major physical disasters and their responses to cultural crises. Wallace (1956, 1957) calls both of these responses **mazeway disintegration.** The cultural mazeway is the socially constructed and learned patterns and rules for interaction that a given group has developed over many generations as "the way" to achieve their wants. Wallace suggests that people in such crises are initially unable to act because their cultural mazeway has fallen apart. The crisis of mazeway disintegration is resolved by reaffirming their identification with some cultural system—traditional, new, or syncretic. For this reason, religious movements that arise to accomplish this reaffirmation may be considered "revitalization movements"; that is, they enable the social group to come alive again.

Although massive disruptions in a group's way of life (e.g., conquest and famine) are obvious starting points of such crises, less dramatic change is more common, and the sense of crisis is less profound. Members of the group may feel only a vague sense of malaise, a discomfort with the way things are going and a general ambiguity about what they as individuals should do. Rapid social change often leaves people unsure about where they "stand." Norms become open; although the old regulations no longer hold, new ones have not been crystallized. Examples of such ambiguities are found in the whole area of gender role expectations in contemporary societies.

An important reason why rapid social change is so threatening to a group's meaning system is that the very *fact* of change undermines many legitimations of the existing social order. It temporarily humanizes the established rules and patterns of interaction by showing them to be changeable products of human convention. Before such change, the group's ways of doing things may have appeared immutable, but the very fact of change belies this perception. In this context, it is possible to understand, for example, the exasperation of a Roman Catholic who, during the period immediately after Vatican II, exclaimed, "How can it be that last year eating meat on Friday was a terrible and grave sin, and this year it is merely an optional and private sacrifice?" Even more threatening are the perceived changes in a social group's fundamental norms and values, such as patterns of family life. Periods of dramatic social change seriously challenge the existing meaning system of a group. Often these periods are marked by the rise of new religious movements that affirm an "improved" meaning system, typically claiming that its own ways are not changeable because they stem from a higher authority than that of the old meaning system.

To summarize then, crises of meaning may occur on the personal level or for an entire social group. The impact of crises of meaning on groups and individuals highlights the significance of the meaning-providing capabilities of religion. Nevertheless, we should not assume an overly intellectualized image of humans. Although a strong case can be made that meaning is a fundamental social requirement (Luckmann, 1967; see also Berger, 1967), we must keep in mind that all people do not necessarily equally desire to find meaning for all aspects of their lives. Similarly, some people are more tolerant than others of inconsistencies in their meaning systems. Some people desire more intellectually elaborate systems of meaning; others are satisfied with quite simple views. Yet historical religions and quasi-religions address similar problems of meaning: discrepancies between the ideal and actual practices in the society; personal suffering and death; and group crises such as war, conquest, and natural disasters. By providing meaning in the face of meaning-threatening situations, religion enables the individual or group to cope with the situation. If the meaning-threatening situation can be successfully reintegrated into the group's sense of order, social life can continue.

## Community and Religious Belonging

A direct relationship exists between the community of believers and the strength of the meaning system. The unity of the group is expressed and enhanced by its shared meanings. The group's meaning system, in turn, depends on the group as its social base for its continued existence and importance. The idea of the "church" (i.e., community of believers) is not merely an organizational feature of religions but expresses a fundamental link between the meaning system and the community that holds it.

### Religion as the Expression of Social Unity

Durkheim treated the essentially communal quality of religion as definitive. The collective nature of religion was central to Durkheim's analysis. He concluded that religious rituals and symbols are, at root, representations of the social group. These *collective representations* are the ways by which the group expresses something important about itself to its own members. Thus, by participating in group rituals, individual members renew their link with the group, and they learn and reaffirm shared meanings (Durkheim, 1965:257). We will explore Durkheim's theories of religious belonging further in Chapter 6. It is sufficient for our immediate purposes to note his emphasis on religion's dual role: It both expresses and contributes to social unity.

### Plausibility Structures of Meaning

Berger emphasizes that meaning systems require social "bases" for their continued existence. These social bases are called **plausibility structures**—specific social processes or interactions within a network of persons sharing a meaning system. As the term implies, the meaning system continues to be plausible (i.e., believable) within these social structures. Berger asserts that *all* religious traditions require specific communities of believers for their continuing plausibility (1967:46). A sound plausibility structure allows the meaning system to be held as a common, taken-for-granted entity. Similarly, it strengthens the ability of individuals to believe, for the group gives them social support and reinforcement in their worldview.

Historically, most religious worldviews were coextensive with the societies that held them. Relatively simple societies exemplify especially well the link between the group's meaning system and networks of fellow believers. If one were born and socialized into a remote, mountain tribe, for example, the religious meanings and rituals of the group would be taken for granted. Simply belonging to the tribe would make one a part of the religion, and the whole tribe would serve as social support for one's belief and practice. Indeed, it would be difficult, if not impossible, for a member isolated in such a tribe to conceive of any other meaning system.

A societywide plausibility structure still exists in many relatively complex societies. In many parts of Latin America, religion is pervasive in all aspects of life, and a single, particular worldview dominates the society. Most people spend their daily lives solely in the company of people who share their entire worldview. Such massive social support for shared religious beliefs and practices makes it relatively easy for believers to take them for granted.

The significance of plausibility structures can also be seen in situations where believers of a worldview are separated from the social group that serves as its basis of plausibility. Missionaries typically went out in closely knit bands among the "unbelievers" (i.e., people with a different worldview from their own). This arrangement was not merely for mutual physical protection but—more importantly—for mutual protection of their view of reality (see Cohen, 1990, on how native worldviews influence missionaries). A meaning system shared even with only a few significant others is stronger than one held alone and surrounded by persons who do not share that meaning. In this context, we can appreciate how grave a punishment it was for some societies to exile a member. Exiles were cut off both from other members of the society and from the social support for their meaning system.

### Subsocietal Plausibility Structures

Pluralistic societies, by contrast, characteristically lack a single, comprehensive worldview. Instead, several different meaning systems compete for adherents. Pluralism itself makes the maintenance of a meaning system problematic because it undermines its taken-for-granted quality. In pluralistic situations, the religious group must organize itself to serve as its own plausibility base because the society as a whole does not support its meaning system. Thus, in pluralistic societies such as the United States and Canada, religious groups often emphasize their communal ties. Three common manifestations of emphatic belonging are sectarianism, tribalism, and localism (all of which characterize many religious and nonreligious groups in these societies).

**Sectarianism.** An orientation by which a group tries to maintain its distinctive worldview by emphasizing primary, face-to-face relationships with fellow believers and high levels of commitment and loyalty to the group is **sectarianism.** These close-knit, intense relationships provide a structure within which the group's worldview is plausible; whereas outside the group, that worldview is disconfirmed. Frequent interaction within this primary group, together with a distancing from nonbelievers' views, helps support members' worldviews in the face of real or perceived opposition from the "world." Even the most mundane actions are transformed to support the sectarian worldview. For example, not only do some fundamentalist Christians eschew "worldly" media but also often immerse themselves in "appropriate" religious audio recordings and reading matter, even to the point of cooking with Christian cookbooks (Ammerman, 1987). The more dramatically the group's distinctive worldview differs from those prevalent in the society, the more important becomes huddling in the sectarian enclave.

Sectarian strategies also include physical or symbolic withdrawal from the "world," limiting outside influences (e.g., television or unapproved books) and restricting members' social contacts. Especially important is the socialization of children; the group protects its children from exposure to competing worldviews, perhaps by running its own schools. Also important is the selection of spouses. The plausibility of a person's meaning system is more likely to be undermined by interaction with an unbelieving spouse than with a co-worker. Thus, a group trying to defend its plausibility often prohibits members from marrying outside the group and arranges social events to encourage marriageable members to meet and marry each other. Groups such as the Amish or Hassidim exemplify extreme sectarian plausibility structures; they are entirely self-contained communities, cut off from the "world." Even large established religious groups, such as Roman Catholics, Lutherans, and Baptists, have used some of these sectarian strategies to defend the plausibility of their worldviews in face of a society that does not generally support them. This sectarian quality of religion in contemporary pluralistic societies is developed further in Chapter 5.

**Tribalism.** Another orientation that protects the group's distinctive worldview is **tribalism**—which, in the United States and other ethnically pluralistic societies, refers to locating oneself within a specific ethnic community and its perspective. Several observers have noted the close link between religion and ethnicity among immigrant peoples. Religious groups provided important resources to newly arrived immigrants, especially during the great waves of immigration in the nineteenth and twentieth centuries. They provided informal networks of association through which the immigrant could obtain a job, help with housing, and other mutual assistance. They also provided protection from the dominant society, keeping alive the old ways, educating children in "safe" environments where their backgrounds would not be held against them, and providing mutual protection from the hostility of those who did not accept the immigrants. When the historical need for these functions has become muted or gone, the ethnoreligious community may still be a source of mutual aid, friendship, and a sense of belonging. For these individuals, the ethnoreligious group as a set of relationships—not the fact of ethnicity itself—provides a stable source of belonging.

Herberg analyzed the relationship between religion and ethnicity, specifically asking why people in the United States tended to identify themselves in terms of one of three religious "communities": Protestant, Catholic, or Jew. He suggested that as the immigrant became assimilated into the dominant U.S. culture, ethnic ties superseded localistic ties that characterized one's self-identity and self-location in the old country. The ethnic church was an important expression of these new ties. According to Herberg, however, ethnic differences are residual and disappearing. As assimilation proceeds even further, religious identity assumes even greater significance—not as actual affiliation with a particular religious group, participation in church activities, or even affirmation of the group's belief system—but rather as a basis of identification and social location. According to Herberg, self-identification in ethnic terms was not altogether satisfactory because it implied incomplete integration into U.S. life. By contrast, religion in the United States is an acceptable way for people to differentiate themselves and thus becomes a way for people to define and locate themselves in the larger society. For this reason, according to Herberg's interpretation, people will identify themselves with one of the three legitimate religious "communities"—Protestant, Catholic, or Jewish—even though they do not necessarily practice that religion or believe its tenets (Herberg, 1960:6–64).

Herberg's analysis was based on earlier waves of European immigrants who fit these three "communities." His conclusions are probably far less applicable to immigrants from religious traditions completely foreign to Western societies. Herberg's thesis might thus apply to the high rates of church participation of Korean Christians, most of whom were Christians before emigrating from Korea (Hurh and Kim, 1990). Communities centered on Vietnamese or Hmong folk religions, however, would *not* readily be accepted as legitimate bases of separate identity within U.S. culture. What little evidence we have suggests that Southeast Asian refugees, resettled in the United States between 1975–1983, often change affiliation to Western religions (a Utah study found 36 percent had "converted"). Whether in Western or Asian churches or temples, however, the immigrants remained in ethnic congregations, which appear to be especially important in dealing with the drastic adjustments these refugees must undergo (Lewis et al., 1988). Thus, tribalism appears to be linked with religion for this new wave of immigrants, even though neither their native religions nor their racial-ethnic identities are as likely as those of earlier European immigrants to be readily accepted as legitimate by the larger society.

Greeley emphasizes that American ethnicity is not merely a vestige of old-country ways but is rather a dynamic, special way of negotiating the transition to U.S. culture. Ethnic communities provide a shelter in which immigrants can assimilate at their own pace, unobtrusively. Ethnicity enables them to legitimate their new identities by reference to their traditions (Greeley, 1977). Belonging to the ethnic community thus promotes a smoother transition to a new identity and sense of belonging. Ethnicity is also a useful variable in explaining diversity of belief or behavior within a religious group; there are considerable differences in the religious approaches of, for instance, Roman Catholics of Slavic, Italian, or Irish ethnic backgrounds (Greeley, 1979b:123). This diversity suggests that religious beliefs and practices may have different significance for various ethnic groups.

Herberg's prediction of ethnicity's disappearance as a basis of personal identity may have been premature (Greeley, 1971:82). For example, an analysis of changing patterns in the U.S. Jewish community suggests that four generations from immigration, intrinsic Jewish cultural patterns (e.g., religious beliefs and sense of having a shared history) are, if anything, revitalized; extrinsic cultural patterns (e.g., ethnically identifiable speech and clothing patterns), however, have virtually disappeared (Waxman, 1983). Some evidence from research on marriage and friendship choices suggests that ethnicity continues to be a significant factor. Nevertheless, higher levels of education, urban residence, and longer residence in the United States reduce the significance of ethnic ties (Abramson, 1973:66, 98; Laumann, 1969). Similarly, studies of three generations of five Canadian ethnic groups found that ethnic-identity retention declines consistently with each succeeding generation, but religious factors are minor or negligible in accounting for the retention of ethnic identity (Currie and Isajiw, 1985; see also Coward and Kawamura, 1978). Evidence from these studies suggests that religious and ethnic identifications are, for some North Americans, subtly intertwined.

Hammond (1988) suggests that changes in patterns of both ethnic ties and religious involvement can be understood by clarifying that both are linked with the individual's identity in two contrasting ways. The "collective-expressive" mode applies when ethnicity and/or religion is an expression of intense primary-group ties, especially if several such connections overlap (such as when one's kin, neighbors, and friends are all involved in the local ethnic church). In the "individual-expressive" mode, involvement in religion and/or ethnic group is largely voluntary and not contingent on other ties. This voluntary involvement may be still extremely important to the person and a major component of identity, but it is not taken for granted.

**Localism.** A similar but broader network of social ties supporting belief is **localism,** an orientation to the local residential community as a source of identity and involvement. In contrast to persons with cosmopolitan orientations, locals focus on their immediate community and their attachments to family, local social groups, and community associations. The localist orientation is parochial and narrow; the cosmopolitan is more open, identifying with wider communities of national or international scope (Merton, 1957:387–420). Mobility, advanced education, exposure to mass media, and urban life appear to diminish localism.

Research data corroborate this interpretation of the relationship between localism and religious plausibility. Localistic factors such as length of residence in a community, belonging to community organizations, having most of one's closest friends residing locally, and having one's network of friends attending one's own religious group contribute to the likelihood that a person will belong to and participate in a church or synagogue (see Gallup and Castelli, 1989:132–148). Another study, using an index of localism, found that "declining community attachment, *more so than the erosion of traditional belief per se,* is the critical factor in accounting for the declines in church support and participation (Roof, 1978:136). A study of Jewish congregations confirmed the connections between localism and both religious involvement and political conservatism (Lazerwitz et al., 1988). A study of church members in Holland observed that because localism is parochial and narrow, it is also correlated with racial and ethnic prejudice (Eisinga et al., 1990). Persons with local group identification have a structure of social support to maintain the plausibility of their meaning system. This support is exemplified by the roles of family, congregation, and community for rural southern blacks trying to maintain their religious worldview in a northern city (see Paris, 1982).

When a religious group holds a worldview not supported by the larger society, it must construct its own plausibility structures. Sectarianism, tribalism, and localism are three orientations that serve this purpose. Other processes that illustrate the way in which religious belonging supports a meaning system include socialization, conversion, and commitment, discussed further in Chapter 3.

## Extended Application: Millenarianism and Dualism in Contemporary Social Movements

Many of the concepts presented in this chapter are particularly relevant to an understanding of contemporary meaning systems. The last part of this chapter is an extended example that applies some of these concepts to modern social movements. We begin by noticing certain themes common to many contemporary movements: millenarianism and dualism. What do these themes imply about the movements' responses to problems of meaning? More important, what does the attraction of such movements imply about people's relationship to the existing social order? This essay is one interpretation of the significance of millenarianism and dualism in contemporary social movements.

These two themes appear regularly in the worldviews promulgated by many developing religious movements in America. A **dualistic worldview** holds that reality consists of two irreducible modes—Good versus Evil. The theme of dualism permeates the belief systems of such diverse groups as the Jesus People, the Unification Church, Catholic Pentecostalism, and Satanism. **Millenarianism** is the expectation of an imminent collapse of the entire social order and its replacement with a perfect new order. The millenarian vision is also a theme in many contemporary religious and quasi-religious movements. The occurrence of movements that emphasize these two elements reveals some of the problems of meaning in contemporary society.[1]

These two elements of belief are not merely quaint characteristics of the movement but are central to their appeal. The appeal of the dualistic interpretation of reality and of millennial expectations results from a need for a new, firmer order in a time when the old basis of order appears to be collapsing. The elements of dualism and millenarianism represent the assertion of order in the face of felt disorder.

### Two Opposing Principles: Good and Evil

A dualistic worldview depicts all reality as consisting of two fundamental modes or opposing principles—one Good and the other Evil. This dualistic perspective is especially important in the belief systems of groups growing from Western religious perspectives (e.g., Islam, Judaism, and especially Christianity). The idea of warring forces of Good and Evil has been recurrent in Christian ideologies, and contemporary expressions of this idea can be traced to enthusiastic religious movements of the sixteenth and seventeenth centuries. Some form of dualism is used by many Christian groups to explain the nature of humankind, the presence of evil, temptation, sin, suffering, and the need for God's help.

Many emerging religious movements particularly emphasize dualistic interpretations of the world (see, for example, Barker, 1984; Bird, 1978, 1979; Richardson et al., 1979; Robbins et al., 1976; Tipton, 1982a; Westley, 1983). Through a new emphasis on mystery, magic, and miracle, believers perceive their world as full of evidence of the action of good and evil forces. Believers experience the world as remystified by the presence of God's spirit; it is also remystified by the presence of Satan in numerous forms. Just as God is influential in their world, so too is Satan immediate and active. Believers see the direct influence of God in the beneficial and pleasant events of their lives; at the same time, they see the Evil One as the source of doubts, temptations, dissen-

---

[1]Because this essay is presented as an example, it is necessarily somewhat oversimplified. See referenced sources for some of the complexities of interpreting social movements. The essay suggests only that dualistic and millenarian theodicies are effective responses to problems of meaning posed by anomie. That does not mean that these movements are "caused" by anomie. A full explanation of these phenomena would need to show that recruits had experienced anomie. It would also need to analyze the sources of problems of meaning in contemporary society.

sion, sickness, and other troubles. This remystification of everyday life is in marked contrast to the usual, more secular interpretation of events.

Some dualistic belief systems identify and personify the forces of Evil in movements, issues, or people that threaten the values of the group. A member of one youth-oriented Pentecostal sect stated:

> This conspiracy is what the Bible calls the false church. There's a false church, the Antichrist, and a true church. This elite is the Antichrist. It has the highest religious leaders, too—the World Council of Churches. They've got definite people, like . . . , who are in with this conspiracy that sit on the board of the Council . . . (quoted in Tipton, 1982a:91).

Dualism in the ideology of some Black Muslims identifies the forces of Evil with the white oppressors (Lincoln, 1989). Similarly, various fundamentalist Christian groups identify communism with the forces of Evil (cf. Himmelstein, 1983:17). Another dualistic approach is the Unification Church of the Reverend Sun Myung Moon, whose belief system holds that the fundamental struggle is between communism and the "God-fearing world" of which the United States is the center (Anthony and Robbins, 1990). Dualism is also a prominent feature of many recent Mideastern Islamic political ideologies. In 1977 one Islamic ideologue wrote:

> Some people today divide the world into democratic or fascist [systems], labor or liberal parties. Islam does not recognize any parties except two: the Party of God and the Party of Satan. The Party of God are those who act as His agents on earth and govern by His laws. The others are the Party of Satan—whatever variation there may be in their system of government and regardless of the conflict among them. In the end, they are one coalition formed to oppose God the Mighty, the Omnipotent (quoted in Arjomand, 1989:119–120).

The cosmic opposition implied in dualism provides a ready explanation for felt opposition to one's own belief system. Numerous religious movements identify their own beliefs and practices as on the side of Good and decry competing religious movements as on the side of Evil. Often, however, Evil is identified with vague general forces that seem to work against the commitments of the religious group. Thus, one neo-Pentecostal member stated:

> I think Satan does his best. I think it's time we recognized that the power of evil lives in the world and that we as individuals have to cope with the evil that's in the world today. And again that's another thing, another advantage of having a prayer community. It's because we gather together for strength and you sort of get to be invincible in a group. If you try to hit it out on your own, I mean obviously, you just can't with those devils. You just don't stand a ghost of a chance. You know, I know a lot of people don't believe the devil exists, but I really don't think things would be as bad as they are, the evil in us, the evil in man. The evil that's in man is the evil that's an extension of possession of the devil. The devil is using that person just like God can use us. . . . I do believe that a lot of evil, a lot of sadness and sickness in this world is brought out by Satan.

I think it's in his scheme of things to turn us against God (quoted in McGuire, 1982:150).

The persistence and strength of belief in dualism suggest that it is not merely one characteristic of these movements but may indeed be central to their appeal.

**Dualism as a Response to Anomie.** Dualistic interpretations are useful to both the group and the individual, especially as a response to anomie, which results in a sense of normlessness and powerlessness. Being out of touch with the source of moral power means being powerless in the face of the nameless terrors of the disordered universe. Only a firm reordering can provide an effective protection against such terror (Berger, 1967:27). A new order must be established and maintained continually against the occurrence of further order-threatening phenomena. Even a firmly established new order is threatened by occurrences such as suffering and death, which perpetually raise the problems of "meaning" and "order."

A dualistic worldview provides both a framework for the *construction* of a new order and legitimation for its *maintenance*. Religious dualism is particularly effective in this reordering because it holds that both the sources of all problems and the "real" solutions to these problems derive from a transcendent realm (see Lofland, 1966:36). Dualism enables believers to name the sources of their anxieties, fears, and problems. This identification is in itself an important source of believers' sense of order and control. Identifying certain difficulties as caused by evil forces also implies a clear-cut course of action. Believers locate their personal courses of action within a cosmic struggle, a continual battle between the forces of Good and the forces of Evil. This identification provides explanations of all events—good or bad—that occur in their lives, and it gives meaning to everyday existence. Even trivial aspects of daily life become part of the order implied in the dualistic worldview.

This order-giving potential of dualism is partly responsible for believers' sense of spiritual opposition. The believer perceives frequent, regular evidence of the immediacy and influence of both good and evil spirits. This evidence, it should be emphasized, is usually in the form of actual events in everyday life. A dualistic worldview contains a built-in tendency to ascribe all good events to God and all bad ones to Satan. The tidiness and order of the dualistic interpretation of the world are part of the basic appeal of such movements. Dualistic figures of ultimate Good and Evil simplify the world to people who are overwhelmed by the ambiguity and complexity of modern life.

In facing the perceived opposition, the group gains cohesion and strength. This sense of opposition is also a useful legitimation for maintaining the new order because it enables believers to see significance in subsequent events, even those that would otherwise threaten their new beliefs. The dualistic worldview heightens the sense that the opposition is a conspiracy against believers and their group (cf. Barkun, 1974; Beckford, 1975b; Westin, 1964).

There are two sides to the believer's sense of order. On the "negative" side is a sense of fearsome powers bent on a conspiratorial attack on believers and their values. On the "positive" side, the same sense of order provides the believer with a feeling of harmony and symmetry. Both positive and negative aspects involve the forming of patterns and imposing of a sense of order (Weil, 1973:178–179). The believer's worldview is founded on the expectation that everything *is* ordered; therefore, order and patterns in the world are subsequently perceived. The "positive" side of this spiritual order can be seen in the believer's constant discovery of "holy coincidence," "providence," and other evidences of the pattern of God's work in everyday life. The dualistic worldview gives form and a "cast of characters" to the patterns that believers perceive in the world around them.

**Dualism as Theodicy.** Dualism also offers the believer an explanatory system for reinterpreting order-threatening events. As such, it functions as a theodicy. The theodicy provides meaning but not necessarily happiness, informing the believer, "You may be miserable, but be comforted that there is, at least, meaning to your suffering."

Dualism is a particularly effective form of theodicy because it is a closed system of legitimation; that is, built into the legitimating system itself is an "explanation" of every argument against the system. Thus, the same theodicy can explain both good and bad events, opposition and confirmation. It also provides a basis for self-justification and for the moral condemnation of others (cf. Rokeach, 1960:69).

Religious groups often use the dualistic theodicy to explain the doubts and uncertainties experienced by new believers: "Your doubts and anxieties are because the devil is trying to confuse and tempt you back to your old way of life." This theodicy gives meaning to the natural doubts and sense of uncertainty encountered by the new believer in the resocialization process. Armed with a concrete identification of the source of these difficulties, the new believer has a greater sense of power in facing such problems of belief. The theodicy that fears and doubts are planted by Satan is a welcome explanation.

The foremost function of dualism is the reshaping of believers' interpretations of everyday events. Believers reinterpret even their own roles. Dualism bestows significance on events that formerly seemed meaningless. Human failure, suffering, social problems, personal difficulties, and death are all given meaning. Events that formerly appeared to be random, haphazard, and disorderly come to be perceived as part of a clear pattern. One's sense of ambiguity and insecurity is resolved by perceiving one's personal role as part of "something big"—a great cosmic struggle in which one gains power and purpose by siding with the good forces. By contrast, in the case of Satanism, members side with the evil forces, figuring them to be more powerful; nevertheless, the same sense of power is gained by Satanists' dualistic stance (see Moody, 1974).

Dualism enables the group to create order out of chaos. The source of threat is not only the external chaos of the anomic social world; intragroup problems constitute a threat, too. Internal conflict is especially disruptive because it undermines those interpersonal relationships that support the believers' worldview. Problems of "disorder" are inherent in the establishment of any new group, and much ambiguity arises during the developing stages of a movement. Also, if members perceive their new belief system to be a departure from their previous one, the fact of change may cause some uncertainty. Emphasizing the group's role in a cosmic dualistic confrontation enables the group to resolve its problems of uncertainty and ambiguity (Slater, 1966). Dualistic beliefs affirm the boundaries between the group and "the world," and they provide mechanisms (e.g., exorcism, confession, and healing) for dealing with internal conflict.

The dualistic theodicy enhances the individual believer's sense of security. Recruits to these movements often express a strong desire to know for sure where they "stand" before God. This desire for certainty is fulfilled by a secure sense of order, which affirms a clear duality between the forces of Good and Evil and shows a distinct path for the believer who would side with Good. This simplified course of action informs the member that, having sided with the Good, one is safe and secure.

### The Imminence of the Millennium

A dualistic theodicy is particularly forceful if combined with a millenarian theodicy. **Millenarianism,** the expectation of an imminent disintegration of the entire social order and its replacement with a perfect new order, is a recurring theme in religious movements. The millenarian theodicy informs the believer that the present chaotic social order, with its misery and deprivations, is transient and that in the coming new order the believer will have a better life and no longer experience the turmoil and malaise of the present. Millenarianism offers both an explanation of the wrongness of the present social order and a hope for future change; as such, it is a potent source of change-oriented behavior.

The millennial dream that the perfect new order is imminent is also a response to anomie. When the old order is seen as nearly defunct, the new order must be near (see Barkun, 1974). The perceived disorder of the United States is especially upsetting to some groups who believe that the nation has a special purpose in God's plan. One television preacher said:

> America's star is sinking fast. If Christians don't begin immediately to assert their influence, it may be too late to save America from the destruction toward which it is plunging. And, since America now stands as the key base camp for missions around the globe, to fail to save America now would almost certainly be to miss its last opportunity to save the world (James Robison, from *Save America,* as quoted in Hadden and Swann, 1981:97).

Similarly, the Reverend Moon warned:

> But today America is retreating. It's not just an accident that great tragedy is constantly striking America and the world, such as the assassination of President Kennedy and the sudden death of Secretary-General Hammarskjöld of the United Nations, both in the same decade. The spirit of America has declined since then. Unless this nation, unless the leadership of the nation lives up to the mission ordained by God, many troubles will plague you. God is beginning to leave America. This is God's warning (quoted in Anthony and Robbins, 1978:84).

People who view this society as badly out of order and consider the established religions impotent to do anything about the "problems" are likely to hope for dramatic change. The ultimate religious "solution" is the end of the world. The coming of the millennium will not only destroy the old problematic order but will also vindicate the faith of the believers.

**Religious Millenarianism.** The idea of an apocalypse—a dramatic end of the world—is not a new religious theme. Many religious groups have anticipated the imminent end of the social order they define as ungodly. Some have expected a total physical cataclysm in which the world would end; others have awaited a god-sent revolutionary end in which the existing social order would cease and be replaced by a perfect new order. This latter hope was frequently combined with the anticipation of a messiah who would bring about the new order. Such themes are particularly prominent in Western religions, especially Christianity (see Barkun, 1974; Bettis and Johannesen, 1984; Cohn, 1970; Festinger et al., 1956; Moses, 1982; Talmon, 1966; Wilson, 1970).

Millenarian movements were widespread in the United States in the nineteenth century, sometimes combined with a messianic vision of a special destiny for the nation. The Mormons, Millerites (and their offshoot, Seventh Day Adventists), Christadelphians, and Jehovah's Witnesses are examples of groups in which the millenarian expectation was or is central (see Wilson, 1970). Nativistic religious movements, which envisioned the millennium as the dramatic restoration of a traditional "native" way of life, arose in the nineteenth century among several American Indian tribes, whose social and economic plight was desperate. The Ghost Dance[2], Handsome Lake, Indian Shakers, and the peyote cult exemplify these movements by which Native Americans responded to the profound disruption of their traditional way of life wrought by white settlers, government policy, and the army (see Aberle, 1966; Barnett, 1957; Mooney, 1965; Myerhoff, 1974). A black syncretic movement, the Lost Found Nation of Islam (i.e., Black Muslims) appealed to both the millennial possibilities in Islam and U.S. blacks' familiarity with Christian millenarian themes. Its early teaching was that 1914 ended the 6,000 years of the "white devils'" rule, after which the lost Nation of Islam would finally realize its destiny (Lincoln, 1989). Later, however, the movement emphasized orthodox Sunni versions of Islam and abandoned this prophecy, while a splinter group headed by Louis Farakkhan reasserted it (Mamiya, 1982; Mamiya and Lincoln, 1988).

Belief in the imminence of the millennium is widespread among new religious movements as well as in traditional sects and some of the more fundamentalist denominations. Peculiarly modern forms of millenarianism include groups that base their millennial expectations on UFOs and extraterrestrial communications (see Balch and Taylor, 1976, 1977; Festinger et al., 1956; Kerr and Crow, 1983). Dispensational premillennialism among conservative Christians

[2]The Ghost Dance was a highly traditionalistic movement that foresaw the destruction of the white man. Although the movement itself was pacifist, its appeal and potential for uniting diverse tribes frightened the frontier army, culminating in the Battle of Wounded Knee, in which over 200 Indian men, women, and children were killed.

has been used for political ends. Building on the long history of millenarian thought, dispensational premillennialism proposes that the world's history is divided into periods of "dispensations," each ended by a wrathful judgment by God. Accordingly, the world is now in the final dispensation, which will end in a "great tribulation" marked by horrible wars and the destruction of much of the world's population; subsequently, Christ will return to reign. Many dispensationalists also believe that the "righteous" (i.e., true Christians like themselves) would be spared this tribulation by the "rapture" in which Christ would "catch" them up before the last seven years of the world. Unlike other millenarian movements that held similar beliefs, this movement actively promotes certain political and military actions that are believed to hasten the end time (Harrell, 1984; Robertson, 1988). For example, despite their widespread anti-Semitism, dispensationalists have raised much money to help right-wing Jewish extremists reclaim the Temple Mount in Jerusalem from the Muslims who have one of their holiest, centuries-old shrines on the site. Unlike other millenarians who merely decry the crises and chaos as signs of impending doom, some Christian dispensationalists actively promote crises and strife.

**Social and Political Millenarianism.** Not only specifically religious movements, but also many other social and political movements, have dominant millenarian themes. Several "isms" of the nineteenth and twentieth centuries project a secular millennium. Thus, many forms of nationalism are based on a vision of a people realizing their destiny in the destruction of the old social order. The millenarian and messianic appeal of the Russian Revolution is another example (Murvar, 1971). Indeed, a persuasive case can be made that the major revolutions of this century (Nazism, Soviet and Chinese communism) were millenarian movements. The potential for modern movements to create and sustain a sense of urgency and imminent disaster in a larger mass population can be partially attributed to the impact of mass media (see Barkun, 1974; Cohn, 1970). Even without reference to a supernatural realm, these belief systems are millenarian responses to problems of order and meaning.

Some social historians observe that there appears to be a cyclical pattern to the rise of millenarianism. In U.S. history, for example, the 1840s, 1890s, 1930s, and 1960s have been identified as periods of intense millenarianism in both religious and other social political forms. If there are recurring patterns, do these periods have certain factors (such as economic conditions, regional growth or decline, moral or political crises) in common (Barkun, 1985; Hammond, 1983; McLoughlin, 1983; Smith, 1983)?

### Dualism and Millenarianism in Social Movements

Expectation of the millennium is not merely an incidental item of belief; rather, it is a significant part of the group's response to the problem of anomie. If people have experienced anomie, the solution *par excellence* is not only to affirm a new order for the present but also to believe in a future reordering that will reintegrate all problematic aspects of the present situation into an overall, meaningful perfect order.

The imminence of this dramatic reordering creates a sense of urgency and makes the mundane business of daily life "in the world" seem trivial. Millennial expectations therefore enhance members' commitment to the group through which they will help usher in the new order. The idea that the end of this world is near makes it much easier for members to give up possessions, jobs, education, and other things that have value only in the nearly defunct world. Like the dualistic theodicy, millenarianism encourages pattern forming. Believers interpret seemingly random events as evidence for the imminence of the millennium.

For example, in 1990, some Christian, occult, and psychic groups saw signs of the impending end in that year's events: earthquakes, eco-disasters, bank and insurance company failures, UFO sightings, war in the Mideast, and so on. One right-wing occult group went so far as to build an entire underground city, in a remote part of Montana, to protect its members who gathered from across the continent.[3]

The millenarian vision is likewise a theodicy for the discrepancies within the present order. Even after conversion, members are likely to have difficulties— friends and relatives "not understanding," prayers seemingly not answered, doubts and uncertainties, and the society still not on the "right track." The belief in the coming of the millennium relativizes the problems and opposition of the present by the knowledge that all of these will be overcome in a glorious future. Members can feel that although "things" are really bad now and will probably get worse, they are not personally threatened by the disorder and ambiguity because they know that they are allied in the present with the source of perfect order and will have a privileged position in the unknown glorious future.

Although infinite possible meanings could be attached to any given event, the themes of millenarianism and dualism illustrate how meaning systems interpret events and locate them in some larger framework. We have seen how individuals and groups use their belief systems to bestow meaning on the social world; and how elements of belief such as imminence of the millennium, messianic expectations, and dualism are responses to a social situation that appears chaotic, ambiguous, or hopeless. Individuals who hold dualistic or millenarian worldviews relate to their social world accordingly. Their worldview organizes their experiences and influences their associations and their very identities.

## Summary

This chapter has outlined some of the ways in which the meaning-providing and belonging components of religion are linked. Religion is both a basis of association and an expression of shared meanings, the importance of which depends largely on the social support of a community of believers.

[3]In the first edition of this book, I observed expectations of the world's impending end from 1973 based on similar events, such as the Kohoutek Comet, war in the Mideast, and UFOs. The second edition observed millennial expectations for 1985 based on earthquakes, a catastrophic volcanic eruption, bank and farm failures, Halley's Comet, and war in the Mideast.

Religious meaning systems are ways of interpreting events and experiences. They assign meaning to a group's existence and to an individual's identity. The meaning system of a group interprets its social order, serving as a legitimation of social arrangements. Meaning systems also locate human lives and events in terms of a larger framework, identifying the individual with an overarching order. The significance of meaning in individual and social life is illustrated by situations in which meaning is threatened. Theodicies are religious explanations that offer meaning for meaning-threatening experiences such as suffering and death. Some crises of meaning profoundly disrupt the social order. The concepts of anomie and mazeway failure describe the impact of such crises on the individual and group.

Analysis of the themes of millenarianism and dualism in contemporary social movements illustrates how groups bestow meaning on the social situation. These two themes can be interpreted as responses to problems of meaning and order. They represent an effort to posit a new order in response to the ambiguity and confusion of meanings produced by social change. Millenarianism and dualism are theodicies that locate believers' experiences in terms of a larger framework of order and meaning.

Religious belonging is a natural outcome of shared meaning systems. At the same time, a community of believers (however small) is necessary for maintaining the plausibility of a meaning system. Where a particular worldview has monolithic status in a society, these plausibility structures pervade the entire society. In pluralistic societies, by contrast, worldviews are in competition. Each group of believers must structure itself to maintain the plausibility of its distinctive meaning system. Three such strategies characterize many religious groups in the United States: sectarianism, tribalism, and localism—all forms of protective enclaves. The interrelationship of the meaning-providing and belonging aspects of religion is well illustrated in the social processes of religious socialization, conversion, and commitment, to be explored in Chapter 3.

# Recommended Readings

### Nonfiction

Peter L. Berger. *The Sacred Canopy: Elements of a Sociological Theory of Religion.* Garden City, N.Y.: Doubleday, 1967. Berger's interpretation of the contemporary religious situation is based upon a carefully developed theory of identity.

Thomas Luckmann. *The Invisible Religion: The Problem of Religion in Modern Society.* New York: Macmillan, 1967. Luckmann develops a theory of modern religion, interpreting nonofficial religion, privatization, and individual religious forms.

Wade Clark Roof. *Community and Commitment: Religious Plausibility in a Liberal Protestant Church.* New York: Elsevier, 1978. This empirical study uses sophisticated methodological tools to apply some of Berger's concepts in an analysis of the bases of community and commitment among Episcopalians.

Steven M. Tipton. *Getting Saved from the Sixties: Moral Meaning in Conversion and Cultural Change.* Berkeley: University of California Press, 1982. Tipton explores the meaning systems and ethical styles of 1960s youth participating in a Pentecostal Christian sect, a Zen meditation center, and a human potential organization (est). The book illustrates well how culturally available meaning systems work or fail to work as individuals develop personal meaning systems in a time of rapid social change.

Max Weber. *The Sociology of Religion.* Trans. E. Fischoff. Boston: Beacon, 1963. This book contains Weber's mature thinking on several important aspects of religion, including theodicy, the idea of salvation, asceticism and mysticism, the role of the prophet, and the religion of nonprivileged classes.

### Other Books

Issues of religious meaning and belonging are often well illustrated in fiction and literary accounts. The following sources are useful:

Walter M. Miller, Jr. *A Canticle for Leibowitz.* New York: Harold Matson, 1959. Describes the role of religion in regenerating civilization following nuclear war, how the Albertian Order of Leibowitz was founded, and how it struggled against the Dark Ages that followed the nuclear holocaust. The few artifacts upon which the new knowledge was built included sacred texts written by the Blessed Leibowitz, such as the following fragment: "Pound pastrami, can kraut, six bagels . . . ."

Kurt Vonnegut, Jr. *Cat's Cradle.* Baltimore: Penguin, 1963. A science fiction account of the end of the world and the ultimate religious movement, Bokonism. Serious humor about the construction of sacred texts, symbols, and meaning.

Elie Wiesel. *Night.* New York: Pyramid, 1960. A moving personal journal of the author's experience as a child in a Nazi concentration camp. The issue of meaning is implicit and powerful.

# CHAPTER THREE

# The Individual's Religion

Sociology of religion emphasizes religious groups and social expressions of religion. The individual members of religious groups are, however, social *actors*—that is, persons with motives and meanings of their own. Although the individual actor's attitudes, conceptions, and behavior may be strongly influenced by social groups, there is no neat or deterministic correlation between what the group believes and what the individual member personally holds central. To understand religious behavior, we must know how religion shapes and is expressed by the individual actor. Chapter 2 suggested some of these ways. Religion provides meaning for the individual's life, enabling one to interpret, evaluate, and project experiences in everyday life. Also, social groups are essential in shaping, maintaining, or changing the individual's worldview.

This chapter examines the processes by which the individual adopts and becomes committed to a religious worldview and community. First, following a life-cycle model, we will explore the social forces shaping and maintaining the individual's worldview. Second, we will analyze the process of conversion by which the individual dramatically changes that worldview. Finally, we will examine the process of commitment, by which old and new believers alike commit themselves to the group of fellow believers. These social processes illustrate the link between the meaning-providing and communal features of religion described in Chapter 2. The individual's meaning system is socially acquired, socially maintained, or socially changed.

## Shaping the Individual's Religion

The individual's religion is not a static entity. It develops and changes over the course of one's life. A person is not born with a full-blown set of religious beliefs and practices; religion is developed and nurtured (or ignored) in the socialization of the child. Although one may have reached an identifiable religious perspective or commitment by young adulthood, it would be a mistake to consider this the end of one's religious development. The individual's religion continues to develop and change, perhaps less dramatically, throughout the rest of life. Indeed, the very meaning of "being religious" changes in different periods of life, and the place of religion in the individual's life also changes. In keeping with this dynamic perspective on religious development, this section follows the life cycle of the individual believer. It focuses (as does religious practice itself) on certain critical periods: early childhood, adolescence, marriage and procreation, and old age and dying. One intermediate period, for which there are no special religious rituals or events, is included: middle age.

One of the difficulties in describing an adequately complex picture of religious development in modern societies is that much of the research in this area has focused on narrowly defined church-oriented religion. There are several studies, for example, about the effect of religious affiliation on fertility rates but little information about how religion shapes the individual's idea of parenting. Similarly, we have considerable data about the correlation between religious affiliation and attitudes toward premarital sexual relations; but we know relatively little about what religious meanings people attach to all aspects of human sexuality. The following description, then, suggests a line of reasoning that future studies might pursue in exploring the relationship between religion and family, community, identity, life cycle, and human development.

### Childhood, Family, and Community

Early childhood is a critical period in the development of the individual's religion. The child begins to learn what it means to be "one of us" (our society, our ethnic group, our religion, our family, our tribe, etc.). Specifically in socialization, the individual internalizes the social group's moral norms and basic values. Socialization also accounts for the development of the individual's attitudes and values, such as attitudes about authority or the relative value of acquiring material possessions. These attitudes and values are closely linked with religious belief, and they vary according to the type of religiosity that individuals acquire through socialization into their group.

In relatively simple societies, there is no distinction between socialization into the larger group and religious socialization. Becoming religious—however defined in that society—is simply part of gradually becoming an adult, responsible member of society. Religious roles in simple societies are not differentiated from other roles such as mother, healer, or chieftain. Similarly, the child's religious roles simply involve the child's participation in ongoing social activities. The child might assist the father in his work and observe the father praying over his tools. Or a child might learn, in helping the grandmother, that one should bless the hearth and fire each morning upon rekindling the fire. He or she might learn the central myths of the tribe from sitting beside a storyteller after a meal. The pervasiveness of religion in simple societies enables the religious socialization of children to occur informally and continuously.

The emphasis on intellectual learning and formal religious knowledge, which characterizes religious socialization in many complex societies, is atypical.

The idea of going to Sunday School to learn specialized religious knowledge is, for example, foreign to most religions because their emphasis is on learning by doing. What the child needs to know can be best learned by accompanying adults and imitating them. By contrast, Western religions have emphasized the importance of formal religious knowledge. Their relatively complex belief systems are presented to the child as a body of knowledge to be mastered. Sunday Schools, Vacation Bible Schools, catechism classes, parochial schools, Confraternity of Christian Doctrine (after-school classes), yeshivot (intensive programs of Jewish religious education), and Hebrew schools (usually weekend classes in religion) are outcomes of this interest. The society's differentiation of religious knowledge from other relevant formal knowledge complicates the child's acquisition of religious knowledge.

Learning the content of the meaning system of one's group is, however, not sufficient. Children need to internalize this meaning system, to make it their own. A computer could teach a child an entire catechism, but that knowledge of itself would probably have very little significance in the child's life. Internalization of the group's meaning system occurs through interaction of the child with specific other members of the group who mediate the group's way of thinking and doing to the child. In early socialization, the child's family is the foremost influence.

**Meaning, Belonging, and Identity.** The significance of specific human others in conveying the group's meaning system to the child emphasizes the connection between the meaning-providing and belonging aspects of religion, as described in Chapter 2. The child does not (indeed cannot) internalize the group's meaning system without having a sense of belonging to the group in which that meaning system is grounded. Typically, in societies with relatively homogeneous meaning systems, belonging to the group means coming to take its worldview for granted.

In societies with competing worldviews, socialization into one's own group includes some awareness of the existence and differences of other groups around one's own. The child's identity includes an important sense of belonging to a specific religion, ethnic group, or nation as well as a feeling of contrast with others not of that religion, ethnic group, or nation. More important, individual identity includes a set of symbols by which the person can socially locate and focus feelings toward the self (i.e., "I like myself," "I am a worthless person") and assert personal identity to the self and others (Klapp, 1969:viii).

The individual's identity and social location are closely intertwined. Often the question "Who am I?" is answered in terms of "This is where I belong." Thus, a woman might describe herself as "a mother, a wife, a Catholic, a Polish-American, a member of the town volunteer ambulance squad, in the church choir, and vice president of the PTA." These roles represent not merely formal memberships but—more important—social locations of her identity. Religion pervades all such social roles in relatively undifferentiated societies. In modern, highly differentiated societies, however, religion is only one significant source of the individual's sense of belonging; it is also intertwined with most other important sources of belonging: family, ethnic community, friendship groups, and nationality.

**The Family.** The first, perhaps most enduring source of the individual's sense of belonging is the family. In the child's socialization, the family is something of a "boundary structure," connecting the borderline between the foundations of personality and the child's beginning participation in society (Parsons, 1969:424). The very young child is not *in* society but is a part of the family, only indirectly participating in society. Later full participation in society is organized according to the values, motivations, and attitudes incorporated during socialization into the individual's personality.

Rituals and symbols allocate identity even in early childhood. One of the most significant and widespread rituals is naming the child. The choice of given names, the affirmation of kinship networks (e.g., naming the child after a grandparent), and the bestowing of the family name (or refusal to bestow it) are highly symbolic acts. Many groups have special ritual occasions for giving a new family member this initial identity. Other occasions of childhood (e.g., birthday celebrations) also affirm the child's identity. The status of the child is relatively clear in most societies; thus, major rituals and symbols of identity are often used to make important transitions, such as the change from childhood to adulthood, in identity.

Many parents value some religious socialization for their children, and children's religious training is a major consideration in adults' decisions to re-establish their church attendance (Gallup and Castelli, 1989:144). A substantial drop has occurred in the proportion of children receiving specifically religious training. In 1952 only 6 percent of adults in the United States had received no religious training as a child; this figure had risen to 9 percent in 1965 and to 18 percent in 1988. According to a 1988 survey, about seven of every ten adults with children ages 4 to 18 were providing religious training (Gallup and Castelli, 1989:66–68).

One study found that family roles themselves—actual or fantastic—were an important factor shaping children's images of God. Children identified God with parental (or grandparental) roles and qualities: big, authoritative, nurturing, understanding, forgiving, controlling, wise, angry, and so on. The motifs of these children's images of God varied considerably according to religious affiliation. For example, Catholic children tended to depict a God intimately involved with family life, as well as to identify God with their concerns with guilt, purity, and forgiveness. By contrast, Baptist children held images of God as a source of nurturance, order, and structure (Heller, 1986).

The family is more than an agent of religious socialization. It is often, indeed, a primary religious group, a fundamental unit of the institution of religion. The trite adage "the family that prays together stays together" is probably reversed in its reasoning. It is entirely likely that any cohesive family expresses, rather than originates, its unity in rituals and symbols that, in most cultures, are also fundamentally religious. At the same time, religion has historically legitimated the family, established rituals to celebrate family unity, and provided norms and social controls to protect the institution of the family. The scope of the family unit in modern society is greatly narrowed from the extended family and large kinship networks of earlier periods. As a result, the primary agents of socialization in the family are fewer, usually only the parents and older brothers and sisters.

**The Community.** Another major source of identity is based on the individual's sense of belonging to a distinctive group—tribe, nation, ethnic group. The U.S. and Canadian history of immigration makes ethnicity an important factor in many people's experience. The degree of affiliation with an ethnoreligious group varies considerably. Ethnicity may be an important basis for a person's choices of neighborhood, friends, job, marriage partner, leisure time activities, and organizational memberships. For another person, ethnicity may be irrelevant to such choices.

The process by which the child is socialized into the family and ethnoreligious group is gradual. It begins with the child's simple experiencing of group membership as a taken-for-granted part of life. Later the child may learn terms for identification with the group (i.e., "I'm a Catholic," "My people are Korean"). This simple identification proceeds from a confused usage of these terms to a clearer conceptualization of their meaning and of belonging to one or another group. One study found that young children (ages 5 to 7) had a vague impression of their religious affiliation as a kind of family name. One 6-year-old responded that it was impossible to be a Protestant and an American at the same time because "you can't have two names," unless perhaps you moved. Somewhat older children (ages 7 to 9) understood religious belonging in terms of wearing particular symbols or of characteristic behaviors (e.g., wearing religious medals or having a Christmas tree). Only much later, usually in adolescence, did children have conceptions of religious belonging that included nonobservable qualities such as faith or belief. Also in later stages of development, some children explained their own identification in terms of contrast to other religious groups (Elkind, 1964; on religious development, see Spilka et al., 1985).

Such studies unfortunately focus on too narrow a definition of religious identity and belonging. Although the individual's gradual self-identification in terms of official religious affiliation is one component of religious identity, we need to know more about the development of other aspects, especially commitment and loyalty. The relationship between identity and values, attitudes, and behavior also needs further exploration without narrow restriction to church-oriented religion. The communities of family, neighborhood, friendship, and ethnoreligious group provide an initial sense of belonging and a foundation for identity; subsequent communities provide a base of support for maintaining that identity or for changing it (as our later discussion of conversion will show). They are also the social base for the continued plausibility of the group's entire meaning system (Berger, 1967:45–47). Thus, the meaning-providing and belonging aspects of religion are directly connected.

Social changes and developments such as pluralism (i.e., the coexistence of competing worldviews) seem to have undermined some of the cohesiveness of these communities. As children emerge from the taken-for-granted small communities of family and neighbors, they will probably encounter a bewildering variety of other groups with other belief systems and ways of life. Some individuals remain immersed in the initial network that supported the plausibility of early beliefs and identity; others detach themselves from these bonds and seek elsewhere a social base for their beliefs and sense of who they are. The very existence of this element of choice in identity and belief system characterizes only certain kinds of society. The possibility of detaching oneself from the taken-for-granted beliefs and social groups into which one has been socialized provides greater freedom—and simultaneously makes both belief and identity problematic (Berger, 1967:137, 151).

**Problems of Family and Community in Contemporary Society.** Many processes in contemporary society have seriously weakened the ability of family, community, and religion to offer the individual a stable source of belonging and identity. Neighborhood communities are decreasingly characterized by face-to-face intimacy and personal, affective bonds. Furthermore, family, neighborhood, and religion alike have been undermined by **privatization**—the process by which certain institutional spheres become removed from effective roles in the public sphere (especially in the arena of economic production and power). Chapter 8 will examine the impact of privatization on religion in greater detail, but it is important here to note the link between privatization of the family and religion.

The power of religion to legitimate other institutions such as the family has been undermined. There is little support for the family from public institutions. What support exists is largely due to the family's significance as an economic unit of consumption (witness the sales appeal of television shows such as "The Cosby Show" and "Little House on the Prairie") and to the family's symbolic importance (e.g., in the rhetoric of politicians). These processes, together with the pluralism of society, have weakened the norms governing family interaction. At the same time, however, greater demands are made on the family. The weaknesses of the neighborhood and community in providing support for members' identities result in greater pressure on the family to be a complete repository of identity (Brittan, 1977:58, 59).

As the individual's control over life in the public sphere diminishes—and as the institutions of the public sphere become less and less concerned with providing meaning for their participants—the individual seeks more fulfillment, security, control, and meaning in the private sphere. The family and other institutions of the private sphere are expected to provide a haven from the public sphere and to compensate for what their members are deprived of in the public sphere (Berger, 1967; Luckmann, 1967:106–114). Privatization thrusts the search for personal identity out of the public sphere. The nature of work and division of labor in modern societies makes it difficult for people to derive their primary sense of identity from their work role. Yet the family cannot sustain the burden of providing all personal identity for its members. One critic of contemporary modes of familism states, "This new emphasis on one's personal feelings and inner needs, one's 'head' or 'life-style', to use contemporary formulations, gives a continued meaning to family life and at the same time threatens to blow it apart" (Zaretsky, 1976:34).

The local community (as a social unit, not just a political or geographical one) has also been undermined. Religion has traditionally been inextricably intertwined with community; the boundaries of one were frequently coextensive with the boundaries of the other. Social and geographical mobility, together with the general acculturation of waves of immigrants, have changed the patterns of community in U.S. society. Local communities are increasingly voluntaristic, segmented, and irrelevant to institutions of the public sphere. The voluntary nature of community has the advantage of freeing individuals to choose their community of identification, but it also reduces their power relative to institutions of the public sphere and their ability to provide stable sources of identity for members. The process of privatization is not inexorable, but its impact on family and community is already evident.

**Alternative Modes of Family and Community.** One historical alternative to the dominant mode of family and community has been the ethnoreligious group that withdraws from the society in order to preserve its distinctiveness. Examples of this kind of community include the Jewish Hassidim, the Amish, Doukhobors, and Hutterites. The Hassidim are particularly interesting because their withdrawal is not to a rural refuge; instead, they have organized their life within urban areas to sustain their religious identity.

The Hassidim are an ultraorthodox group of Jews, mostly first- and second-generation immigrants who fled persecution in central and eastern Europe (see Kephart and Zellner, 1991; Mintz, 1977; Poll, 1969; Shaffir, 1974). There are relatively few U.S. and Canadian converts to Hassidism, although the largest branch (Lubavitcher) proselytes among less Orthodox Jews. More than 100,000 Hassidim live in the United States, and large Hassidic communities also exist in Canada. The Hassidic community is carefully organized to enable members to follow the 613 rules regulating everyday life for Orthodox Jews. Members must live close enough to the synagogue to be able to walk there for services on the Sabbath. Food that passes the highest levels of kosher regulations must be available. Outside the home, men and women are segregated at all times, partly because of religious rules designed to protect men from defilement by contact with a menstruating woman. Thus, the men of the community arrange their work so that they have no interaction with women (e.g., by working in the neighborhood or in an all-male trade). The men further arrange their work time to be free for frequent daytime religious services.

The Hassidic community has its own system of law and its own shops, trades, and schools. The rebbe (i.e., chief rabbi) is the final authority in both spiritual and secular matters. Indeed, there is no distinction between spiritual and secular issues; religion pervades all spheres of life. The socialization of children into the group's approved way of life is relatively unproblematic because of the group's success in limiting outside influences. Children attend religious schools, are heavily occupied with their studies and (in adolescence) with their own religious duties, and are seldom exposed to outside media such as television or movies. The community enclave makes the Hassidic way of life a taken-for-granted reality to the child. Personal identity and a sense of belonging are similarly unproblematic. Roles and norms for appropriate behavior are fixed, with roles available to males varying considerably from those for females. Boys clearly learn their roles as men, and girls learn clear-cut expectations of women; both are rewarded by the close-knit community for meeting these role expectations.

A related kind of local community, the **intentional community**, is one in which members choose to live together communally. This arrangement turns the voluntary nature of its commitment into a virtue. Religious communalism has a long tradition in Western society; monastic communalism dates back to the sixth century A.D. A wide variety of groups dissatisfied with the values and structures of contemporary society have adopted a communitarian form of organization. Intentional communities are potential alternatives to the modes of both family and community life of the larger society. The intentional community allows the group to alter all aspects of social interaction—work, leisure, marriage, family, decision making, education, and so on. Thus, a group can integrate the spheres of work and religion for its members because, in the separate community, these spheres need not be segregated as they are for most members of society. The intentional community also provides a potentially stronger base of personal identity and a stronger sense of belonging than the privatized family and community can. Communitarian groups often resemble traditional communities with their close-knit, affective interpersonal relationships; they differ, however, in being essentially voluntary. Members cannot take their belonging for granted, although certain practices of the group may make it difficult for the member to consider leaving. The group's continual point of reference is the dominant society from which its way of life departs.

The commune movements in the United States, Canada, and parts of Europe illustrate some of these themes. Communes such as Twin Oaks, The Farm, Ananda, Brotherhood of the Spirit, and the Lama Foundation organized themselves as alternatives to the dominant mode of personal, family, and community life. Other intentional communities were outgrowths of specifically religious movements; twentieth-century examples include the Jesus People, Hare Krishnas (Krsṅa Consciousness), Unification Church, Catholic Pentecostalism (later called Catholic Charismatic Renewal), and the Bruderhof (Society of Brothers). Most Catholic Pentecostal prayer groups, for example, consist of persons who live in typical middle-class neighborhoods and maintain ordinary middle-class family styles. Their prayer group attempts to provide additional sources of community support, bolstering families and individual members in their distinctive beliefs and practices. This movement also spawned a number of "covenant communities"—intensive groups whose living arrangements, finances, child care, and other family functions are communally shared. Such communities enable members to have a much stronger base of social support for their way of life than would be otherwise possible.

Contemporary religious communalism can be interpreted both as a celebration of and protest against privatization of religion, family, and community. On the one hand, communitarian groups attempt to restructure social organization to allow religion, family, and community to influence other spheres of their lives such as work, education, decision making, and so on. Their protest is against the fragmentation of social life in society. At the same time, however, they are a celebration of privatization because they have removed themselves from society, presenting little challenge to the institutions of the public sphere. Privatization robs communitarian groups of their impact. Establishing a commune, no matter how dissident, is a somewhat expanded way of "doing your own thing" in the private sphere of family, religion, and community.

### Adolescence

Whereas childhood is a period of initial socialization into a group's way of life and meaning system, the transition to adulthood is a critical transformation in most cultures. Childhood is important in establishing identity and group belonging in a general way, but the transition to adulthood means passage to a new identity involving responsibility, knowledge, ritual and symbolic roles, and acceptance into adult circles. The dramatic quality of this transition is ritually expressed in many cultures by **rites of passage** for adolescents.

Rites of passage are rituals that accompany a change of place, state, social position, and age. One interpretation suggests that rites of passage enable societies to effect an orderly, meaningful transition for individuals and groups who move from one socially recognized stable state (e.g., childhood) into another (e.g., adulthood). The stable statuses have clearly defined rights, obligations, and roles. The transition between them, however, is dangerous because rights, obligations, and roles are temporarily ambiguous and disordered. Both the social group and the individual member are thus served by these rituals that express, yet circumscribe, the ambiguity and disorder of this period.

These rites consist of three phases: separation, marginality, and aggregation. In the separation phase, ritual actions symbolize detachment of the individual from the previous stable state; the individual metaphorically dies to the old self. The phase of marginality symbolizes the ambiguity of the transition. It represents a structureless realm in which previously taken-for-granted roles and relationships are brought into question. The phase of aggregation unites the individual with others in the new status group. It includes the transmission of knowledge needed by the individual in the new role together with the symbolic features of the new role (e.g., clothing representing the new self). Rites of initiation in simple societies illustrate these processes (Van Gennep, 1960; see also Eliade, 1958; Turner, 1974a, 1974b, 1979).

Contemporary religious groups have rituals that are remnants of earlier rites of passage; baptism, for example, has many features of initiation rites. Confirmation among Christians and bar (bat) mitzvahs among Jews also represent the transition from childhood to adulthood in the religious group. These rituals are, however, relatively weak in effecting such transition in modern society. One study noted that, ironically, decline in adolescent religious observance and adherence in their groups' traditional beliefs generally began soon after confirmation, bar or bat mitzvah, or other rituals for adolescents—for most, the last obligatory religious training (Ozorak, 1989). The ineffectiveness of contemporary rituals of transition is partly because religious groups allow ritual adulthood to members whom they do not consider fully socially adult. A 13-year-old boy who had recently become a full member of the Presbyterian church (having attended communicants' classes, made a profession of faith, and received his first communion) wrote to the elders (governing board) of the church suggesting changes in church activities. His letter was received with mirth: "Isn't that cute that he thinks he's old enough now to be telling us what to do." A stronger example of the ambiguity of adult status is the use of forcible "deprogramming," in which parents of legally adult "children" refuse to accept their offsprings' choice of religious identities.

Another more serious basis of this weakness is the quality of adolescence in this society. There is no clear-cut event that confirms adult status. Puberty marks the end of biological childhood, but social adulthood is many years away. The economic structure requires years of preparation for adulthood. A 20-year-old student who is not fully self-sufficient and independent of parental and school control is no longer a child but is not socially recognized as fully adult. Demarcation events such as leaving home, a first self-supporting job, and marriage are recognized as the beginning of adulthood for some individuals but not for others. Some observers suggest that the forms taken by adolescent rebellion against adult authority are themselves attempts to symbolize a transition (Berger, 1969).

Rather than the *medium* by which transition to adulthood is structured, religion in this society often ends up as the *content* over which adolescent rebellion is staged. Adolescence is often a time of identity crisis because the ambiguity is not culturally resolved. The assertion of one's identity (separate from the child's identity, which is defined by family and community values) sometimes takes the form of religious rebellion. The young person tries to assert an adult self by denying those aspects of life perceived as symbolic of the childhood self and tie to parents or family. An obvious example of this rebellion is the youth who joins a religious group that appears diametrically opposed to his or her childhood religion. A very different pattern—the young person who becomes superreligious in the childhood religion—is another way of expressing rejection of the family's "low level" of religiosity and their "hypocrisy." A holier-than-thou stance is comparable to changing religions as a way of asserting an independent identity and rejecting parental values (Greeley, 1979a). These seemingly dissimilar adolescent strategies may account for a number of contemporary religious expressions: the attraction of youth to countercultural religious movements, their seemingly fanatical adherence to both old and emerging religious groups, and the agonizing, often bitter struggles between parents and youth over a newly asserted religious identity.

During the period of transition from childhood to adulthood, the direct influence of parents on their adolescents' beliefs and practices often diminishes, whereas peer influences may increase (Potvin and Lee, 1982). Nevertheless, the parents' beliefs and practices are reasonably good predictors of certain aspects of their adolescents' religiosity. One study of a conservative religious group found that parents with relatively traditional religious beliefs and practices were likely to produce adolescents with relatively traditional religiosity, whereas less traditional parents were likely to raise less traditional adolescents. Both sets of adolescent children, however, were collectively less traditional than their parents (Dudley and Dudley, 1986).

Another study surveyed second-year high school students and later restudied them when they were approximately 27 years old. During high school, students exhibited a high degree of conformity to their parents' religious orientation, beliefs about God, and levels of church attendance. A decade later, the influence of parents' religious practices and beliefs was, at most, indirect. The religious involvement of one's spouse became far more salient for these young adults (Willits and Crider, 1989). Similarly, a study among Mormons found that one of the main enduring effects of youth religious socialization was channeling members into a social network of friends—a personal community of social support—which also served to sustain their religious worldview (Cornwall, 1987; see Benson et al., 1989, for a review of literature on adolescence and religion). Thus, both the original family and, later, personal community and new family through marriage served as significant plausibility structures (as described in Chapter 2), but none of these sources had fixed or deterministic influence.

### Marriage, Sexuality, and Procreation

Marriage marks a status passage and the beginning of a new family of procreation. Especially for women in most cultures, marriage is the entry to full adulthood. A wealth of ritual and symbolism accompanies and defines this status passage. Religion often directly promotes the ideal of marriage by setting the norms for marriage and establishing the appropriate behavior of members before, during, and after the marriage ceremony. In societies where several religious or ethnoreligious groups coexist, religious groups also delimit the pool of acceptable spouses; members are not supposed to marry "outside the fold."

Images evoked in the marriage rituals of historic religions symbolize this status passage. The traditional transition of the woman from her father's possession to her husband's is symbolized by the father "giving away" the bride. Some denominations have attempted to change this symbolism to fit the realities of contemporary family life—for example, by having both parents escort the bride to the groom or by allowing the bride to present herself. Nevertheless, the symbolism represents a fundamental transition of status for both bride and groom. Other symbols (e.g., wedding bands, ceremonial binding of hands, exchange of ceremonial crowns) represent the unity of the new social group—husband and wife.

Although historically religion has been closely linked with family life and the regulation of marriage and sexual behavior, its impact in contemporary family life appears to be more indirect and perhaps weakened. Unfortunately, the data available are narrowly limited to the impact of religion on marriage in four particular areas: endogamy (i.e., marrying within one's ethnoreligious group), sexual norms (e.g., restrictions of premarital, extramarital, homosexual, or "deviant" sexual activities), divorce, and reproduction. Furthermore, these data describe only relatively recent attitudes and behavior. We have no comparable information about the impact of religion on family life in, say, 1830 or 1600. These data are also generally based on a narrow consideration of the term *religion;* correlations usually refer only to "religious affiliation."

Religious endogamy is decreasingly common in the United States and Canada, and it is of less importance in mate selection than even a few decades ago (Glenn, 1982; Larson and Munro, 1985). In this century, U.S. Catholics, Jews, and Protestants (except the most conservative denominations) have all experienced steady increases in the rate of exogamous marriage (McCutcheon, 1988). One Canadian study found that religious homogamy (in which both partners are of the same religion) was not a significant predictor of marital commitment (Larson and Goltz, 1989). Religious exogamy may, however, be correlated with denomination switching because often one spouse changes to the religion of the other (Bahr and Albrecht, 1989; Hoge, 1981).

Although all Jewish and Christian groups emphasize the ideal of permanent marriage, divorce rates and attitudes toward divorce have been changing. Assessing the impact of religion in these changes is difficult. Although lack of religious affiliation is correlated with marital disruption (i.e., divorce and separation), this correlation results largely from differences in socioeconomic status between religiously affiliated and nonaffiliated persons. Considerable diversity also exists among various denominations and ethnic groups in rates of marital disruption (e.g., the divorce rate among Baptists is higher than among persons with no religious affiliation). The overall rate of marital disruption for Roman Catholics and Protestants is roughly the same (although Roman Catholic divorces may be somewhat underreported because of the church's restriction on formal divorce). Jewish marriages are the least likely to end in divorce (Bumpass and Sweet, 1970; see also, McCarthy, 1979). The rate of marital disruption of all three groups is rising, as is the proportion of members who approve of divorce (and remarriage of divorced persons). Although the official teachings of the Roman Catholic church are strongly against divorce and remarriage, fewer than 20 percent of Roman Catholics disapprove of divorce (*N.Y. Times*/CBS Poll, reported in Berger, 1985).

One study observed that religion's influence on marital commitment is often indirect. For example, it shapes the degree of emphasis on caring for one another and the couple's definition of gender roles, which subsequently have an impact on the marriage relationship and partners' commitment to that relationship (Scanzoni and Arnett, 1987). Another study found, however, that the linkage between religious involvement and marital commitment held mainly for male, religiously involved, conservatives; by contrast, women's marital commitment was influenced far more by their degree of satisfaction with marriage and family life than by their religious involvement (Larson and Goltz, 1989).

Considering the importance of religion in fostering love and mutual commitment, social science's emphasis on religion's social control functions in marriage is imbalanced. It would be useful to know how religion (defined more broadly

than religious affiliation) influences the extent to which people perceive themselves as having choices in life (e.g., the choice not to stay married). How does
religion influence the individual's general orientation toward life (e.g., fatalism
or hopefulness) and toward others (e.g., are others perceived as basically good
or bad)? What impact, if any, does religion have on the individual's sense of
self-worth and independence; does the individual's religion contribute to the
sense of self outside the marriage or family? These broader influences of religion
are probably more critical for our understanding of marriage and divorce than
are formal moral norms of specific religious organizations (see D'Antonio and
Aldous, 1983).

Similarly, it would be fruitful to explore what kinds of marriages result from
different kinds of religious orientations. For example, are the religious orientations and perspectives of the spouses important in determining the power
relationships between them? Their ideals of married love? The degree of interpersonal communication in their marriage? The allocation of tasks, resources,
decision making, and rewards in the marriage and family? Probably, religion is
one important factor in shaping the quality of marriage relationships, but the
precise nature of its influence has yet to be documented.

A related correlation is the influence of religion on gender roles. Various
religious perspectives lead to different images of maleness and femaleness.
Gender roles are extremely important in marriage, as in other institutional
settings. What a man expects of himself and what others expect of him as a man
directly influence the kind of activities and relationships in which he engages. If
it is considered unmanly to care for babies, a man is likely to avoid situations in
which he would do (or be seen doing) this. Religious beliefs, myths, images, and
symbols have been important forces in shaping gender roles, as illustrated in the
Extended Application section of Chapter 4. Religious groups also maintain a
system of social support for appropriate gender-role behavior and social control
for "deviant" gender-role behavior. In recent years, for example, a Mormon
feminist was taken before a church court for her open support of ERA (i.e., the
Equal Rights Amendment to the U.S. Constitution).

Similarly, a broader approach to the influence of religion on childbearing,
childbirth, and family life is needed. Numerous studies have correlated religious
affiliation and fertility. In general, Roman Catholics tend to have more children
than Protestants, who have more children than Jews. Most of these differences,
however, also partly result from differences in social class, education, and
ethnicity. Roman Catholics increasingly approve of "artificial" birth control; in
1985, 68 percent of all U.S. Catholics (compared with 71 percent of all non-
Catholics) favored use of artificial birth control, and 83 percent of Catholics in
the main childbearing years (18 to 39) approved (N.Y. Times/CBS Poll, reported
in Berger, 1985). Already, there is considerable convergence between Catholic
and non-Catholic contraceptive behavior and fertility (Eckhardt and Hendershot, 1984; Westoff, 1979).

Studies of comparative fertility rates, however, have focused too much on
the differences between religious organizations in relating their norms of
assumed control over the reproductive behavior of members. Far more interest-

ing would be information about how religion shapes the attitudes of men and women toward children. The meaning of having children is different for men than for women, for older couples than for younger ones, for Italians than for Germans, for Jews than for Hindus, and so on. What exactly is the impact of the individual's religion on his or her attitude toward children—especially one's own children? Also, what is the impact of religion on attitudes toward reproductive events per se? How does religion shape the individual's ideas of and feelings about menstruation, intercourse, pregnancy, and childbirth? How does religion influence the person's experience of these events?

One important step in this direction is a recent study of women's positions on the abortion debate. Female "pro-life" activists and female "pro-choice" activists were found to operate from totally different worldviews, which entailed radically different meanings of motherhood, family, women's roles, children, and the importance of the individual. Although religious affiliations informed these opposing worldviews, another explanatory factor was the difference in the two groups' socioeconomic situations. Pro-choice activists tended to have experience and educational credentials to pursue careers and higher-paid jobs; ability to control their reproductive functions was crucial in being able to achieve their career goals. By contrast, pro-life activists had chosen life patterns generally out of the paid workforce (and most were ill-prepared to succeed in that sphere); for them, childbearing and childrearing were central parts of their sense of purpose and self-esteem. Indeed, because of their values about children and motherhood, these women were typically opposed to the use of "artificial" birth control, even though its use would obviate the need for many abortions. From their respective worldviews, both groups of women had made life commitments that limited their ability to change their minds and life directions (Luker, 1984:158–215). Some of the fervor of the abortion debate is fueled by the stakes that persons on both sides have invested in their worldviews.

Finally, the narrow focus of research on attitudes toward sexual norms has failed to tap the broader issue of sexuality itself. All religions have attempted to interpret sexual themes or experiences. Religious symbolism frequently deals directly with themes of sexuality. Important parallels exist between spiritual and sexual ecstasy; sexual images are linked with religious images in the writings of many great mystics such as Saint John of the Cross and Saint Theresa of Avila.

Religion's link with human sexuality is understandable because both are direct, personal experiences of *power,* which evokes a sense of chaos and need for control, pollution and need for purification, danger and need for protection (Douglas, 1966). Religion, with its capacity to give meaning and order, offers control, purification, and protection from the chaotic power of sexuality. The establishment of moral norms to regulate sexual behavior is only one aspect of this control. Other important aspects include religious interpretation of sexual experiences and events (e.g., interpreting marital intercourse as imitative of the marriage of the gods). Religion also provides rituals and symbols to deal with some of the more potent and awesome aspects of sexuality. Many religions have a ritual of purification for women after childbirth (e.g., the "churching" of women in some Christian groups). Indeed, the human body and its functions are potent symbols. Corporality (the quality of being a body) is a common metaphor for social and religious cohesion (see Turner, 1980, 1983). Religion also offers ways of channeling sexual energies into spiritual energies. Many religious groups believe that sexual activity reduces one's spiritual energies; thus, norms of abstinence from intercourse are based not on notions of right and wrong but on the idea of heightening one's spiritual powers.

These reflections about the interrelationship of religion and sexuality suggest some of the directions that further study could take. Existing research on religion and marriage, sexuality, and procreation is much too narrow in both its use of the concept "religion" and its definition of what is relevant in the study of these important human experiences.

### Middle Age
Significantly, there are no rites of passage to middle age. In earlier times, fewer people lived in relatively good health much beyond the time their children became adults. Thus, there was no major gap between the years of active childrearing and support and the years of relative retirement from work and old age. Now, however, with increased longevity and smaller families, a gap of twenty or more years between the time children leave home and parents reach retirement age is not uncommon.

Middle age is often a period of "identity crisis" for both men and women. Many women discover that they have invested their entire sense of purpose in serving their families, but their status as "mother" means little in middle age. The correspondence of this social loss with the time of biological change (i.e., menopause) can be doubly traumatic because some women consider their reproductive capacity a basic part of their sense of self and personal worth. Although the existence of a male biological equivalent to menopause is debated, the social transitions of men in middle age are often as pronounced as those of women. A man may have invested much energy and time into his career and, in middle age, begins to realize that his career goals are unlikely to be achieved. Or he may come to doubt the goals and values that had motivated him in earlier years of his career. Precisely at the stage when his children are becoming more independent of their parents, he often begins to want a closer relationship with them. The structure of family and work in this society makes middle age a problematic time: One's chances of achieving socially desirable goals in family and job are considered past or severely reduced, but one is not yet free to quit striving for those goals.

In Western societies, chronological age is a key criterion by which social roles are linked. It becomes an evaluative criterion when the individual or social group measures the age appropriateness of the person's achievements and activities. A 25-year-old student who returns to college after several years of apparent disinterest in educational goals is called a "late bloomer"; and statements such as "He is too young to be married" or "She is too old to be working full-time" exemplify the social basis of these definitions of age appropriateness.

The person may be physically capable of full-time work or marriage and childbearing, but the society has established its criteria of the appropriate ages for these roles (Kearl, 1980; see also Berger et al., 1973:73).

Age is also a key criterion in comparing oneself with others. Many social institutions encourage and even formally organize these comparisons. Schools compare the individual's achievements with others of the same age or grade; businesses compare the achievements of age cohorts of workers. Self-evaluation is also often based on the individual's comparison of self with others of the same age. As age progresses, however, opportunities for comparing favorably with others of one's cohort decrease. By middle age, many people sense that their number of remaining years and resources for competition severely limit their likelihood of "getting ahead." Middle age brings the peak of socioeconomic mobility; for many individuals, it may mark the beginning of the downward mobility characteristic of old age. Institutions, especially in the sphere of work, further solidify this sense; often workers' job statuses remain relatively static in later years of employment. Retirement marks the worker's entry into "old age"—again, a socially defined threshold of age-appropriate behavior.

Contemporary religious groups do little to smooth this status passage with ritual or symbolism. Some groups give some special status to middle-aged persons (e.g., by making them elders of the church). Overall, however, the main impact of religion on this status passage is not its interpretation of the meaning of the status through a symbol system but its offering of potential significance to the middle aged through participation in the religious community. Middle age is generally a period of increasing church involvement for many people. The overall age pattern in adult church attendance shows a steady increase from late teens to a peak in the late 50s to early 60s, followed by a slight decline in later old age (Atchley, 1980:330–340). These figures do not necessarily reflect greater religiosity of middle-aged and old-aged persons but should be viewed as changing patterns of involvement in voluntary associations (involvement in work-related, sport-related, and school service associations drops dramatically during these same years). This associational involvement may indirectly influence the members' beliefs and practices.

Generally, however, middle age is not a valued life period in our culture, the dominant values of which emphasize youthfulness: youthful standards of beauty, energy, bodily functioning, and carefree life-style. Middle age is, by contrast, a time of increased awareness that one has begun aging, both biologically and socially, and is on the verge of an even less valued period of life in this culture—old age.

### Old Age

Popular imagery holds that religion is more important in the lives of old people. Thus, one study asserts that the holding of orthodox beliefs (such as certainty in the existence of God), the evaluation of oneself as religious, and the degree of satisfaction received from religion all increase in later years of life (Gray and Moberg, 1977). Other, possibly contradictory evidence shows that, on measures of well-being used in the NORC General Social Survey, 1972–1977, religious older people do not have consistently greater feelings of well-being than less-religious older people (Steinitz, 1980). The elderly tend to become disaffiliated from religious and other voluntary organizations (Bahr, 1970). Most of these data refer only to church-oriented religiosity, and little information exists about the place of individuals' personal religion in their older years. Although physical limitations might prevent an older person from attending church-related activities, it is possible that the person might pray more frequently, remember religious experiences or events more intensely, or base more everyday activity on religious values (cf. Mindel and Vaughan, 1978).

Problems of meaning and belonging may be particularly acute for older persons. Difficulties of financial and physical limitations may compound the broader problem of the society's general devaluation of old age. A 1974 survey found that only 6 percent of U.S. adults consider the "best years of life" to be one's 50s, 60s, or 70s. By contrast, 34 percent felt that one's 60s and 70s were the "worst years of life" (Harris and Associates, 1975). At the same time, elderly persons become increasingly aware of the imminence of their own death and must often cope with the reality of the deaths of their spouse and friends. To the extent that important aspects of the individual's identity are supported by relationships with such relatives and friends, these deaths may mean real losses in the individual's own life (see Hall, 1985).

Popular notions also suggest that the greater religiosity of old age is something of a last-chance "cramming for the final exam." Although there may be some validity to this conception, the religiosity of the elderly is probably more of a reflection of their social situation. Old people may emphasize spiritual and human relationship values in their lives because the society has relegated them to the private sphere—out of the world of work. Our society places great value on one's work role (especially for men, but increasingly also for women) throughout the individual's adult life, and old age brings the often abrupt end of that role. Retirement effectively means, for many people, leaving the public sphere. Thus, the elderly person must find all bases of identity and self-worth in the private sphere—family, leisure-time activities, religion, neighborhood. Perhaps because they have been characteristically confined to the private sphere, women often adjust better than men to old age. They have already developed more social roles in the private sphere, and society has not expected them to invest themselves in their employment roles (if any) as heavily as men (Myerhoff, 1978).

Increased emphasis on religious bases of individual identity may thus represent an attempt to transform the previous (i.e., culturally encouraged) valuation based on work and parenthood to a new valuation based on spiritual status. Because this represents a fairly dramatic change in self-evaluation, however, it probably does not succeed for many people in our society.

By contrast, some societies provide recognizable spiritual roles for elderly persons to assume. Hindu men who have raised and supported their children to adulthood are allowed to retire in honor to a life of contemplation and spiritual exercises. Upper-class, married Hindu women are allowed a similar freedom after menopause to perform *habisha*—rituals to protect their husbands and develop their own spirituality. Their freedom is temporary, however, ending when they become widows and must observe the social and ritual prohibitions for that status (Freeman, 1980). Elderly women in other cultures often have spiritual roles as healers and midwives—positions of culturally recognized spiritual power. Urban Western cultures, characteristically, do not provide such roles. Individuals may be recognized and honored for their "holiness" or "goodness" within their own immediate religious group, but such roles are privatized. The honor an elderly woman may receive at her Wednesday night prayer meeting does not carry over to her treatment by the Social Security bureaucracy, hospital clinic, or other tenants in her apartment building.

One of the critical problems of meaning in middle and old age is modern society's sense of time. Primitive religions integrate all human action into cosmic time; the events of one's life can be interpreted as part of a larger cosmic drama. The individual's passage through the life cycle repeats and imitates the deities' birth, adolescence, marriage, childbirth, parenting, work, play, fighting, aging, and death. In this religious perspective, time has sacred significance. It collapses past and future into an eternal present.

Historic religions such as Christianity and Islam also give sacred significance to time. The past is full of the deities' self-revelation to humans; the present is important in the working out of the deities' will for humans; the future will bring the full realization of that will and celebration of a glorious reward. In this perspective, time promises immortality. Old age has sacred meaning, both as a fulfillment of divine will and as a threshold to higher levels of spiritual rewards. Modern society, by contrast, encourages a profane image of time. Time "passes," and its passage signifies decay or entropy. Time is a resource to be used but contains no special meaning; when the resource is depleted, life ends. In this context, old age has no special significance with reference to past accomplishments (e.g., social rewards for living a good life) or to future rewards (e.g., heaven, nirvana). Old age means merely the end of full life opportunities (Kearl, 1980).

U.S. society values economic roles especially highly. The individual's occupational role shapes the society's evaluation of that person and his or her own identity. Consumer roles are equally important. Social status is based partly on evaluations of the individual's ability to maintain a certain standard of consumption (e.g., quality of house, car, neighborhood, clothing). Elderly persons are often deprived of valued statuses in both kinds of economic roles; they are retired from their work roles, and, simultaneously, their fixed incomes leave them unable to maintain valued standards of consumption.

The legitimation offered for this loss of valued statuses is the idea of "retirement." Retirement is supposed to be an economic and moral vindication for growing old. It is described as a time for individualism (e.g., the freedom to move away to a retirement resort, play golf, lounge around, putter in the garden, and escape social obligations). The concept of retirement implies that the individual has earned this escape by having fulfilled life's social obligations. This individualism (if, indeed, the retired person could afford such luxuries in retirement) does epitomize what younger members consider to be freedom and a desirable reward. It cannot, however, provide meaning to life and death, a sense of belonging or self-worth, for the retired person (Kearl, 1980).

This life is made even more problematic in its ending. Often biological death follows the individual's social death by many years; the individual may become physically, financially, or mentally unable to sustain interactions that the society considers "alive." Yet the person is kept biologically alive, often by extreme measures of medical intervention. It becomes difficult to die "on time." The medical supervisors of death not only treat it as meaningless but also often segregate dying persons from family, neighbors, or friends who could support their personal meaning system. Religion has traditionally given meaning and dignity to old age and dying. Attempts to retain these values are not supported, however, by the structure of modern society. Old age and dying are generally perceived as times to fear.

## Conversion

The capabilities of religion for providing the individual with a sense of both meaning and belonging are especially evident in the process of conversion. **Conversion** means a transformation of one's *self* concurrent with a transformation of one's basic *meaning system*.[1] It changes the sense of who one is and how one belongs in the social situation. Conversion transforms the way the individual perceives the rest of society and his or her personal place in it, altering one's view of the world.

This definition of conversion distinguishes simple changes in institutional affiliation from more fundamental alterations in the individual's meaning system. An Episcopalian who marries a Roman Catholic may join the Catholic church to accommodate the spouse's wishes. Such a change of affiliation is not necessarily a conversion. Similarly, a Presbyterian who moves to a new town and, finding no local church of that denomination, joins a Congregational church has probably not—strictly speaking—converted. Such denomination switching is relatively common in the United States and is strongly correlated with socioeconomic mobility (Stark and Glock, 1968:183–203). Several studies indicate that some 40 percent of the U.S. populace have switched denominations at least once (Roof and McKinney, 1987:165), and about one-third of those have switched more than once, mainly due to factors such as marriage, friendship, and family (Roof, 1989). Sometimes switching denominational affiliation symbolizes a more dramatic transformation of self and meaning system; typically, however, it is not a conversion but simply a change of affiliation from one organization to another. Yet we will see how important a new group affiliation *can* be in making the change of self and meaning system possible.

[1]Throughout this discussion we will emphasize the broad concept of *meaning system* more than the specific term *religion*. This usage is helpful because the processes described here apply to other comprehensive meaning systems as well as to specifically religious ones. The processes of conversion and commitment can apply not only to religious changes but also to psychotherapeutic and political transformation.

### Kinds of Conversion

There is considerable diversity even among "real conversions." One distinction is the degree of transformation that takes place. How different are the new meaning system and self from the former ones?

**Radical Transformation.** The extreme case is a radical transformation of self and meaning system such as when a highly committed Conservative Jew converts to a fundamentalist Christian worldview. Not only are such extreme conversions relatively uncommon, but they rarely occur as dramatically as popular imagery implies. The processes by which such radical transformations occur are actually similar in kind, though usually not in degree, to less extreme conversions (see Berger and Luckmann, 1966:157–163). These processes are described in greater detail in the next section.

**Consolidation.** Less extreme cases include conversions in which the new meaning system and self represent a consolidation of previous identities. Some young men who became ba'ale teshuvah—members of strict Orthodox Jewish yeshivot (i.e., commune-schools)—had come from non-Orthodox Jewish homes but rejected their Jewish way of life and had then tried various alternative worldviews (especially countercultural "trips" such as meditation, communes, and drugs). Becoming Orthodox was not a return to their former Jewishness because their new meaning system and selves were dramatically different from the old ones. Yet the way of life of the ba'ale teshuvah enabled these members to consolidate elements of both their former identities into a new, "superior" self (Glanz and Harrison, 1977; see also Danzger, 1989).

**Reaffirmation.** Another less extreme type of conversion is an alteration of self and meaning system representing a reaffirmation of elements of one's previous identity. Many adolescent "conversions" fit this model. It is difficult to specify how much change such conversions really entail. Often they involve no change in one's religious affiliation, yet exhibit real changes in the individual's personal meaning system and sense of identity. Roman Catholics who become Pentecostal (i.e., in the Catholic Charismatic Renewal) typically undergo a conversion experience, yet remain Catholics and often become even more active in parish life. They change their personal meaning systems and selves, but they interpret these changes as consistent with their former meaning system. They view their former religiosity as a vague groping for the truth, which they now have found. This type of conversion does not necessarily entail a total rejection of the previous meaning system.

Some identity-consolidating conversion experiences involve little or no change in meaning system and sense of self. Many religious groups expect young members to make a personal faith decision and to undergo a conversion experience as they approach adulthood. Such groups provide opportunities, such as retreats or revivals, in which the necessary conversion experience is more likely to occur. These experiences, although very real and meaningful to participants, are often not conversions (as defined) but rather rituals of reaffirmation of the person's existing identity and meaning system (Wimberley et al., 1975). Even these quasi-conversions, however, occur by some of the same social processes as more radical transformations of self; both are forms of religious socialization and are thus comparable in many respects. Some experiences are part of a process of commitment in which the individual's self-concept as a religious person becomes more central (see, for example, Staples and Mauss, 1987).

**Conversion Accounts and Rhetorics.** The main difficulty in distinguishing the degree of change that occurs in any given conversion is that the individual who converts reinterprets past experiences in relationship to the new meaning system. Therefore, it becomes difficult to determine what amount of the convert's description of the changes experienced represents the objective process of conversion and how much expresses the convert's subjective reinterpretation of those events. The convert constructs the story of conversion, drawing on a socially available set of plausible explanations or "rhetoric." Several rhetorics may be used to "explain" conversion: rhetorics of choice, rhetorics of change, and rhetorics of continuity.

**Rhetorics of choice** emphasize how much the change resulted from a personal, often agonizing decision. Our society places much value on individual decision, so these rhetorics are prominent in explanations of conversion. In a culture where personal decision was less valued or discouraged, rhetorics of choice would not be emphasized; indeed, the convert might not even experience "making a personal decision" (Tippett, 1973).

**Rhetorics of change** emphasize the dramatic nature of personal change in the conversion. Converts may compare the evil or unhappiness of their previous way of life with how wonderful their new way is.

**Rhetorics of continuity** focus on the extent to which one's new meaning system and self are the logical extension of earlier beliefs and experiences. The convert might remember important past experiences as tentative steps toward the newfound truth.

Religious groups often encourage the application of one type of rhetoric over another. Thus, Catholic Pentecostals generally encourage new members to interpret their conversion as continuous with their former way of life, whereas Jesus People typically encourage new members to interpret their experience as a dramatic change. Because the main source of information about conversion is the converts themselves—and because their explanations of events surrounding their conversions are reinterpretations consistent with their new meaning systems—evaluating evidence about conversions is difficult. Some sociological theories of conversion have mistaken these interpretations and the rhetorics that express them for the objective events of the conversion (Beckford, 1978a; Burke, 1953).

Another problem in interpreting conversion accounts is the point in the convert's life history at which the account is given and the social context in which it is given. A psychotherapist hired by "anticult" parents to help their son or daughter to overcome a disvalued conversion experience would almost certainly hear a different story about that conversion than would a sociologist participant-observer talking to a satisfied group member. Even though both "stories" might be truthful, they would probably differ dramatically in what events were remembered and how those events were interpreted (Richardson, 1985a; Robbins, 1984). Just as there is some pressure from fellow members on believers to witness appropriately to their conversion experience (McGuire, 1982), there is also a set of expectations, especially from the family, for ex-members about how they should account for their experiences (Beckford, 1985). Thus, believers' accounts of conversion, although methodologically important and useful, cannot be taken at face value.

### Explaining Conversion

A theoretical understanding of how conversion occurs is nevertheless worthwhile because it reveals much about the connection between the individual's meaning system, social relationships, and very identity. Because conversion consists in a change of the individual's meaning system and self, it has social, psychological, and ideational components. The social component consists of the interaction between the recruit and other circles of associates (e.g., parents, friends, co-workers). The psychological component refers to emotional and affective aspects of conversion as well as to changes in values and attitudes. The ideational component includes the actual ideas the convert embraces or rejects during the process. These ideas are rarely very philosophical or theological; they are simply a set of beliefs that both justify the new meaning system and negate the former one.

**Factors in Conversion.** An adequate theory of conversion must take all the aspects mentioned into account without overemphasizing any single component. Some theories give too much weight to social factors by creating the image of a passive person being pushed and pulled by various social forces. Although very real social pressures are exerted on the potential convert, the person who converts is not a passive object of these pressures. Conversion entails an interaction during which the recruit constructs or negotiates a new personal identity (Beckford, 1978a; Kilbourne and Richardson, 1989; Straus, 1979). Furthermore, only some of those exposed to such social pressures do decide to convert (Barker, 1983).

Some theories of conversion overemphasize ideational components of the process. These theories are consistent with the ideological claims of the religious groups themselves. Religious groups like to believe that the truth value of their beliefs alone is sufficient to compel a person to convert. The content of the belief system *is* a factor in conversion. Some beliefs are more appealing than others to people in certain circumstances; indeed, the potential convert will likely be recruited to a group whose perspective is consistent with that person's previous outlook, even though the specific content of the group's beliefs may be unfamiliar (Greil, 1977; Harrison, 1974). Also, we must acknowledge people's religious reasons for their religious behavior and not try to reduce every motive to some psychological function.

Nevertheless, ideas alone do not persuade a person to convert. Even in the scientific community where objective facts and the truth value of interpretations are supposed to be paramount, there is considerable resistance to change from an established interpretive paradigm to a new one—even when the old paradigm is inconsistent with the "facts" (Festinger, 1957; Kuhn, 1970). How much more are religious believers, with their emphasis on supraempirical reality, likely to resist changing their ideas? Thus, although the ideational component is important in the appeal of a new belief system, it is not sufficient to bring about conversion.

Other theories place too great an emphasis on psychological factors in conversion, explaining the change entirely in terms of the individual's personality, biography, and personal problems. Psychologistic explanations are attractive because they mesh with many of our individualistic cultural values. Nevertheless, they are too one-sided, leaving out social situational factors and other important components. Also, some of these theories tend to assume that conversion to unusual religious groups (e.g., "cults") entails "sick" behavior. Yet adherents' behavior is quite understandable and rational within their alternate meaning systems. If one believes that astrological forces influence human events, it is perfectly rational to act in accordance with those forces.

**"Brainwashing."** One particularly misleading psychologistic model of conversion is the "brainwashing" metaphor. This model is based on studies during the 1950s of the processes by which certain U.S. military personnel in the Orient were pressed to convert to Chinese communism. The popular image applied to this process was "brainwashing," conveying the idea that the convert's mind was cleansed of prior beliefs, values, and commitment, then filled with a new belief system. Psychological studies of this process identified several factors producing conversion against the wills of the converts (Lifton, 1963; Sargant, 1957). Various social scientists have subsequently generalized the interpretations of this drastic type of political conversion to other forms of conversion. Some accurate parallels do exist between forcible "brainwashing" and conversion, but these characteristics apply to all forms of resocialization. Thus, the training of soldiers for combat and rehabilitating juvenile delinquents also involve these processes. To say that conversion is a form of resocialization does not mean that it is therefore an extreme, involuntary form of resocialization.

The key problem with the "brainwashing" metaphor is its ideological use and potential application for abuses of civil liberties. Nonconverts often feel threatened by the conversion of someone close to them. The convert has rejected their own dearly held views and norms and has indirectly threatened the nonconverts' own meaning system. When people cannot understand why an individual would *want* to convert to an unfamiliar religious perspective, they find "brainwashing" an attractive explanation. This metaphor implies that the converting individual did not change voluntarily. The metaphor also allows people to negate the ideational component of the convert's new meaning system. A convert's parents can feel, "He doesn't believe those ideas because they are meaningful to him but because his mind has been manipulated." In its extreme form, the "brainwashing" metaphor has been recently used to justify the denial of converts' religious liberty on the ground that they do not know their own minds (Anthony, 1990; Bromley and Richardson, 1983; Richardson, 1990a, 1991; Robbins, 1979/80; Shupe and Bromley, 1980).

An interesting parallel with the current anticult charge of "brainwashing" is the nineteenth-century anti-Mason movement. Freemasonry is now a legitimate, middle-class form of fraternal organization, but it was severely attacked in the nineteenth-century United States as subversive to democracy. Other, now respectable groups that were attacked (often violently) were Roman Catholics and Mormons. The key themes of the movements against Masonry, Roman Catholicism, and Mormonism emphasized that, unlike conventional denominations that claimed only partial loyalty of their members, these groups allegedly dominated their members' lives, demanded unlimited allegiance, and conducted some activities in secrecy (Holt, 1973; Vaughn, 1983).

This parallel suggests that the "brainwashing" controversy is an ideological issue at another level (see Richardson, 1981; Robbins and Anthony, 1979). The society defines as "deviant" one who is *too* committed to religion, especially authoritarian religion. The resocialization processes themselves are less of an issue than the legitimacy of the group's religion itself. To illustrate this discrepancy, two researchers compared conversion and commitment processes of the Unification Church of Sun Myung Moon and similar contemporary sects with the nineteenth- and early twentieth-century practices of the now socially acceptable Tnevnoc "cult." The Tnevnoc practices were essentially comparable and seem bizarre until we discover that the authors were actually referring to life in the convent—which, spelled backward, is Tnevnoc (Bromley and Shupe, 1979b). The chief difference between many modern "cults" and groups such as Roman Catholicism, Freemasonry, and Mormonism is that the latter groups have now achieved social legitimacy.[2]

Keeping these cautions in mind, we can examine some of the factors in conversion. By emphasizing conversion as a *process* rather than an event, we take into account the fact that the convert has both a history and a future. Although an individual may experience conversion as a discrete event, numerous other experiences lead up to and follow that event that are also parts of the conversion. The following description examines the sequence of events in the process of conversion; but let us remember that no single step in the sequence is itself sufficient to "cause" conversion (Beckford, 1978a; Heirich, 1977; Richardson, 1985a).

### Predisposition to Conversion
Several personal and situational factors can predispose people to conversion by making them aware of the extent to which their prior meaning system seems

---

[2]To reduce the impression that conversion and commitment processes characterize only "weird" religions, I have drawn examples from both traditional and "new" religions.

inadequate to explain or give meaning to experiences and events. By contrast, if individuals can satisfactorily "handle" experiences and events within the framework of their meaning system, they have no desire to seek alternative meanings for their lives. Sometimes the individual who acutely feels the need for a new set of meanings becomes a *seeker*—that is, a person who actively looks for a satisfactory alternative belief system. A seeker often tries many different alternative beliefs and practices (Balch and Taylor, 1977). For example, many eventual converts to Krishna Consciousness had gradually been drawn to Eastern thought, then actively sought a satisfactory guru or path to spiritual enlightenment—often in the course of several years of seekership (cf. Shinn, 1989). Often, however, the individual experiences the desire for a more satisfactory meaning system as a vague tension, a malaise.

Many converts describe a crisis that they felt was a turning point in their lives. It is very difficult to evaluate the extent to which such crises actually precipitate the individual's conversion. Some crises may disrupt a person's life so completely that the individual has difficulty integrating them into the previously held meaning system. Natural disasters, war, and personal tragedy are particularly acute challenges. Serious illness or unemployment may be experienced as turning points. Social events such as an economic depression or anxiety over social conditions (e.g., crime or erosion of morals) may predispose people to conversion. Nevertheless, such crises do not *cause* religious conversion. Religious conversion is one among several possible resolutions of tensions and problems created by the crisis. Thus, serious illness might predispose one person to convert to a new meaning system; another person with a similar illness might find great meaning and comfort in his or her existing meaning system and have that belief confirmed by the crisis experience. Diverse other responses are possible, including alcoholism, political conversion, psychotherapy, suicide, and so on. Individuals converted to religious meaning systems are typically people whom previous socialization has predisposed to a religious perspective (see Greil, 1977; Lofland, 1966, 1977).

Determining to what extent the convert's description of a crisis experience is the result of after-the-fact interpretation of events is also difficult. Some religious groups encourage their members to witness about their conversion experience by telling the group how they came to "see the light." Whether the events thus described were really critical when they occurred is therefore difficult to reconstruct. Often the new group itself promotes the recruit's experience of a crisis. Much of the initial interaction between group members and the potential convert is devoted to making the recruit dissatisfied with the prior meaning system. Similarly, many groups raise the recruit's anxiety over social and personal problems. Jehovah's Witnesses, for example, often approach strangers with a message about common worries—war, inflation, crime. By "mediating" anomie, the group encourages the individual to convert. The group encourages an experiential crisis of meaning by emphasizing order-threatening conditions and magnifying the potential convert's feelings of dissatisfaction, fear, and anxiety, which the person may have previously felt only vaguely (Beckford, 1975b:174).

**Initial Interaction**

Most recruits are drawn to the group by friends or relatives. Besides introducing the newcomer to the group and its beliefs, these pre-existing networks of friendship account for the plausibility of the beliefs and the attractiveness of belonging. Thus, a person might be impressed by a roommate's happiness in a religious group and be curious enough to "check it out." The fact that a person whom one knows and likes belongs to the group attests to the normalcy or desirability of the group's way of life (Gerlach and Hine, 1970; Lofland, 1966). Even groups, such as the Nichiren Shoshu (a Japanese movement brought to the United States in 1960), which conspicuously proselyte in public and anonymous settings typically recruit most of their members through pre-existing networks of friends and relatives (Snow et al., 1980).

Through interaction with members of the group, the recruit is gradually resocialized into that group's way of life. This resocialization consists of the individual's reshaping of identity and worldview to become consistent with those considered appropriate by the group. Several social processes enable the individual to make this transformation. Group support is particularly important. The recruit enjoys warm, affective relationships with the new group. Members of the Unification Church, for example, shower the potential member with attention and affection (sometimes called "love bombing"). These bonds affirm the new self and meaning system. As the recruit gradually withdraws from competing social relationships, the new group's opinions become increasingly important (Berger, 1967:50–51). Intensive interaction and close affective bonds with group members are central to the conversion process because they link the individual's new identity with the organization's perspective and goals (Greil, 1984).

At the same time, the recruit also weakens or severs those relationships that support the old self. Former attachments compete with new commitments, symbolize a worldview that the recruit wishes to reject, and are based on an identity that the recruit wishes to change. Imagine, for example, that through two good friends you are introduced to a group of U.S. Sufis (Sufism is a mystical sect within Islam). Suppose that you learned enough about the group that you decided you would like to join them. How would your non-Sufi friends and relatives react? Your roommate might say, "Oh, come off that mysticism kick. I liked you better when you were a drinking buddy." Imagine your parents' reaction when you announce at supper that weekend, "Guess what, Mom and Dad, I've decided to join this fantastic bunch of Sufis, and I'm going out to the West Coast to live in a commune with some of them!" Most converts find their former set of friends less than supportive of their newfound truth and new self. Relationships with the new group therefore become even more important to counteract opposition from others.

During this resocialization, recruits learn to redefine their social world. Relationships once valued become devalued, and patterns of behavior once undesirable become desirable. New believers may redefine their families to exclude the biological family and to include the new family of fellow believers. A Shaker hymn thus celebrates the severance of old family ties:

*Of all the relations that ever I see*
*My old fleshly kindred are furthest from me*
*So bad and so ugly, so hateful they feel*
*To see them and hate them increases my zeal . . .*

*My gospel relations are dearer to me*
*Than all the flesh kindred that ever I see . . .*
(quoted in Kanter, 1972:90).

This process results in a whole new way of experiencing the world and oneself. The individual comes to "see" the world with an entirely different perspective; indeed, the new believer may say, "I once was blind, but now I see." This phrase is not merely metaphorical because the new perspective actually causes the individual to *perceive* the world differently. That which was marginal to consciousness becomes central, and that which once was focal becomes peripheral. Every worldview entails the selective perception and interpretation of events and objects according to its meaning system. Conversion means adopting new criteria for selecting (Jones, 1978; Snow and Machalek, 1983).

Recruits also redefine their own biographies. They remember episodes that appear consistent with the newfound perspective and interpret them as "part of what led me to the truth." Events are reinterpreted in terms of new beliefs and values. Remembering how proud she was to have achieved scholastic honors in school, one young woman said, "What a fool I was back then to have put so much store on worldly achievements."

The actual interaction between the recruit and the group is especially important in bringing about this transformation of worldviews. The most obvious action of religious groups is **proselyting** potential converts. Proselyting means that an individual or group actively tries to persuade nonbelievers to become believers. Some religious groups (e.g., Jehovah's Witnesses) do much proselyting among nonbelievers; others (e.g., Conservative Jews) do virtually none. Although proselyting activities are relatively conspicuous to outsiders, they are important as a commitment mechanism for already converted members (Beckford, 1975b; Festinger et al., 1956; Shaffir, 1978). One study of a growing Pentecostal denomination found that members with extensive religious experience (e.g., speaking in tongues, divine healing) were especially vigorous evangelizers of potential recruits (Poloma and Pendleton, 1989). Thus, a mutually reinforcing relationship developed between members' involvement in the group and their attempts to share their religious faith and experience.

Mutual witnessing within existing friendships appears to be especially effective in bringing about the recruit's conversion to a new meaning system. Thus, the newcomer may mention an apparent coincidence that had recently happened, and a group member may respond, "That was no coincidence. That was God trying to show concern for you so that you will change your life." Or a member might say, "I used to be just like you. I had my doubts and didn't know what to believe, but now it all fits together. Now I can see what I was missing."

Through these informal interactions, the recruit may gradually "try on" the interpretations suggested by members and apply their meaning systems to personal experiences. Thus, the new believer comes to share their distinctive worldview.

### Symbolizing the Conversion

The part of the conversion process typically identified as "the conversion" is essentially some form of *symbolizing* the transformation that has already been occurring. The convert affirms the new identity by some symbolic means considered appropriate in that group. In many Christian groups, baptism is meaningful as a symbol of conversion. Other ritual expressions of the new self include speaking in tongues (i.e., glossolalia) and witnessing. Some groups have very formal means for symbolizing transformation; others have more informal symbols that are not obvious to the nonbeliever (Gerlach and Hine, 1970; Lebra, 1972).

Although conversion is a gradual process, many recruits who have decided to convert adopt some of the symbols of conversion rather dramatically. Part of the resocialization itself is learning to act, look, and talk like other members of the group. Indeed, groups such as the Unification Church may focus on teaching recruits to display signs of commitment to the neglect of socializing them to actual commitment (Long and Hadden, 1983:9). The imitation of signs of commitment—which might be called "doing being converted"—is an important part of the resocialization process. It is, however, often mistaken by observers as evidence that a dramatic conversion has been accomplished.

Some groups may encourage the new member to seek a conversion experience. This special emotional and spiritual event thus symbolizes the person's full conversion. Such experiences, however, are only part of the larger conversion process but are valued ways of symbolizing the transformation in some groups. Often these conversion experiences are brought about in carefully orchestrated settings. In a revival meeting, the timing of the altar call is synchronized with music and spoken message to proclaim, in essence, "Now is the appropriate time to have that special experience you came for" (W. Johnson, 1971; Walker and Atherton, 1971).

Another symbolic expression of the new self is changing one's name. Nuns traditionally changed their names upon taking their vows. The new name both symbolized the person's new identity and helped confirm that identity every time she was addressed. Some groups also encourage converts to confess the wrongness of their previous way of life. Among the Society of Brothers (Bruderhof), such a confession event symbolizes the conversion and demonstrates how new members' views of themselves have been transformed (see Zablocki, 1971:239–285). In the confession, converts affirm their new selves by derogating the behavior of their old selves. These symbols of conversion illustrate the complexity of the larger process of resocialization (see Goffman, 1961, especially his comparison of the "mortification of self" in several different institutional settings).

## Commitment

The process of conversion does not end when the recruit formally joins the group and symbolically affirms the conversion. Rather, the conversion process is continued in the **commitment** process, by which the individual increasingly identifies with the group, its meaning system, and its goals. Groups that have highly effective recruitment strategies may not be able to keep and effectively deploy their members if their commitment processes are not also effective (Long and Hadden, 1983). Popular imagery somehow envisions the converted recruit as permanently changed. The turnover of members in contemporary religious movements belies this notion. A British study found that, in the highly organized Unification Church, the majority of converts leave the group, of their own free will, within two years (Barker, 1983). Similarly, a Canadian study of numerous new religious and para-religious movements found that, while participation rates were relatively high, the proportion of adherents staying in these movements was extremely low (Bird and Reimer, 1982).

Commitment is a problem not only for contemporary movements. George Whitefield, an eighteenth-century Calvinist preacher, observed that his efforts were not as successful as those of John Wesley, the principal founder of Methodism. The key difference was that, although both preachers achieved many converts, Wesley insisted that the localities where he preached should establish "classes" to ensure the commitment of new converts (Snow and Machalek, 1983). Relative commitment of members is no indicator of the truth or "deviance" of a religious group. Maintaining commitment to *any* group is always problematic, and it is especially difficult in a modern, pluralistic, mobile, individualistic society.

Commitment means the willingness of members to contribute in maintaining the group because the group provides what they want and need. Commitment therefore implies a reciprocal relationship. The group achieves its goals by fulfilling the needs of its members, and the members satisfy their desires by helping to maintain the group. Persons who are totally committed to a group have fully invested themselves in it and fully identify with it. Commitment is the link between the individual and the larger social group. A person cannot be coerced into commitment but decides to identify with the interests of the group because of personal values, material interests, or affective ties (Kanter, 1972: 65–70).

Conversion is a resolution of the individual's problems with former meaning systems and former self, but conversion alone is not sufficient to resolve new problems. The group's commitment processes help prevent the individual's doubts and new problems from undermining the conversion. The final result of the entire conversion process is not merely creating new members but creating members who will invest themselves in what the group is believing and doing. The same process also ensures the commitment of all members, new and old, to the group's values and objectives. *Commitment processes build plausibility structures for the group's worldview and way of life.*

The level of commitment that a group expects varies. Most major denominations in the United States do not expect intense commitment from their members or for that commitment to influence all aspects of their lives. Other religious groups (e.g., Pentecostal and Holiness groups, as well as many communal groups ranging from Hassidic Jews to the Bruderhof, Trappist monks, and the Unification Church) expect members to demonstrate intense religious commitment in all spheres of daily life. Commitment mechanisms in dissenting or deviant groups are especially important because of their difficulty in maintaining their worldview in face of opposition both from established religious groups and the larger society. All social groups, however, need some commitment from their members in order to maintain the group and achieve their goals (see Gerlach and Hine, 1970). And all social groups (including nonreligious groups such as the army) utilize commitment measures similar to those used by religious groups.

The processes by which the group fosters commitment are similar to processes of conversion. Both processes urge members to *withdraw* from competing allegiances and alternate ways of life, and both processes encourage members to *involve* themselves more deeply in the life of the group, its values and goals. These commitment mechanisms are used to some degree by all social groups. Groups desiring more intense or total commitment of members, however, are likely to use more extreme commitment processes. Groups gain greater commitment of members by asking them to sacrifice something for the group, but the degree of required sacrifice varies widely. Most religious bodies ask their members to give up some of their money and time for the group's goals and projects; and some groups expect their members to tithe a specific, substantial percentage of their income. Still other groups ask members to give up all belongings to the group and to live communally.

### Withdrawal from Competing Allegiances

Any degree of sacrifice enhances the individual's commitment because giving up something makes the goal seem more valuable. Sacrifice gives observable evidence to the group that the member is committed, and it "weeds out" members who are not sufficiently committed. Religious groups further encourage sacrifice by signifying it as a consecration, so the act of sacrifice gains sacred status. Some groups ask members to sacrifice time and energy (e.g., devoting a certain number of hours each week to proselyting new members). Some groups expect members to abstain regularly or periodically from certain foods or from alcohol, tobacco, drugs, or sexual relationships. Several Christian groups encourage or require their members to fast during Lent. Either for all members or for an elite core group, many religious groups place special value on celibacy. Other groups may expect members to do without "worldly" pursuits (e.g., dancing, going to movies, wearing makeup or stylish clothing). Such sacrifices are demanded by most contemporary religious groups, especially marginal ones. Celibacy, for example, figures in the commitment process of such diverse groups as Roman Catholic clergy, some neo-Pentecostals, the Divine Light Mission, and the Unification Church. Vegetarianism and abstinence from smoking, drinking, and drugs are commonly required sacrifices (see Kanter, 1972, for examples from nineteenth-century communal groups; for contemporary examples, see Gardner, 1978).

Some sacrifices may also be interpreted as forms of **mortification**, the process of stripping the individual of vestiges of the "old self." Groups seeking to resocialize their members into a new identity consistent with the group's beliefs and values often encourage mortification. Members are asked to let go of those areas of life that compete with the new, desired self. They may have to wear prescribed dress and hairstyle, do without makeup or jewelry, and give up certain prized possessions. They are asked to sacrifice not because these things are wrong in themselves but because using them supports the "old self." Other forms of mortification include public confession, giving up control over one's time and personal space, and relinquishing personal choice in a wide range of matters (Goffman, 1961; see also Chidester, 1988; Zablocki, 1971).

The group sometimes promotes further withdrawal of members from their former way of life by asking them to renounce competing relationships. Many sects and cults discourage members from interacting with the "outside world" and may adopt special social arrangements to insulate members from outside influences. Groups such as some Mennonites, Jesus communes, and Christian monks geographically separate themselves from the rest of society. Other groups insulate their members by operating their own schools, places of work, and social clubs. For instance, numerous conservative evangelical groups have instituted home schooling or separate sectarian schools (see Peshkin, 1986; Rose, 1988; Wagner, 1990). The group may also exercise control over the communication media to which members are exposed or may limit interaction with outsiders.

More important than physical withdrawal from "the world" is the creation of psychic boundaries between the group and the outside. By use of these boundaries, members come to think of the group as "we" and the rest of society as "they." Furthermore, members perceive their in-group as good or superior and the outside as evil or degraded. Thus, the individual member's withdrawal from competing activities is motivated not only by controls that the group exercises but especially by the wish to identify with the in-group and to avoid the negative influences of the outside. Even groups that have not withdrawn fully from the larger society often create these kinds of psychical boundaries, for example, by urging members to join parallel religious organizations rather than secular organizations. Thus, groups like Christian Veterinarian's Fellowship or Fellowship of Christian Athletes provide some of the prestige and business networks of professional associations, while protecting members from secular society (Elzey, 1988).

Withdrawal from competing relationships often entails changes in the member's relationships with parents, spouse, and close friends. Typically, the individual identifies with outside relationships less and less, while simultaneously drawing closer to fellow group members. As we have seen, this disengagement from relationships that do not support the member's new identity and beliefs is partly necessary to maintain the individual's new worldview.

Many religious groups try to exert some control over the member's choice of a marriage partner. Marriage to someone who does not support (or who even opposes) one's worldview can undermine the believer's meaning system. Also critical is competition between religious groups over the socialization of children from a "mixed" marriage.

The group frequently tries to guide or control even its members' relationships within the group. Close relationships among a small part of the group may detract from commitment to the group as a whole. The attachment of a married couple to each other or of parents to their children can compete with their involvement in the larger group. Groups that seek intense commitment from their members often have special structural arrangements to reduce this competition. The Oneida commune of the nineteenth century had a form of "open marriage" that diminished the pairing off of couples. Other groups (e.g., the Israeli kibbutzim) consider children to belong to the whole community rather than solely to the biological parents.

These processes all promote commitment to the group by encouraging individual members to withdraw themselves from those aspects of their former life that prevent them from being fully a "new self." The degree to which any given religious group asks its members to withdraw from nongroup loyalties depends largely on the type of group. The more marginal, sectlike groups typically expect high levels of attachment. Their commitment processes are therefore more intense and extreme than those of ordinary denominations.

### Attachment

At the same time that groups encourage members to withdraw from other allegiances, they also urge members to become more and more involved in the group itself, drawing them into greater oneness with the group. This sense of unity is clearly related to the concept of belonging, discussed in Chapter 2. Some studies of fundamentalist and charismatic congregations suggest that the provision of a sense of belonging is as important, if not more important, than the specific belief content of the group in maintaining high levels of member commitment (Ammerman, 1987; McGaw, 1980).

Activities that draw the member into the fellowship and consciousness of the larger group promote both the cohesion of the group and the commitment of individual members. These commitment mechanisms make belonging to the group an emotionally satisfying experience. Commitment mechanisms for attachment are also likely to differ according to the intensity of the commitment desired by the group. Groups that expect intense commitment of members use stronger measures to promote attachment to the group.

The "we feeling" of group consciousness is promoted by homogeneity of membership. The more alike members feel, the easier it is for the group to gain a sense of unity. Established religious bodies typically achieve member homogeneity by self-selection. Individuals choose to join a church or synagogue with membership characteristics comparable to their own social status, racial, ethnic, or language group and educational and religious background. Sectlike groups, by contrast, have more selective memberships, screening out or discouraging unacceptable members. Sectlike groups put greater emphasis on resocializing new members, thereby creating more homogeneity (Kanter, 1972:93, 94).

Group unity is also enhanced by the sharing of work and possessions. The extreme form of such sharing is full communal living, in which all possessions are held in common and all work is performed together. At the opposite pole are nominal forms of sharing, such as gathering Thanksgiving baskets for the needy or a painting party to decorate Sunday School rooms. Even these minimally demanding kinds of sharing promote a sense of unity in the group. Much sharing in religious groups consists of people taking care of each other. Group members may aid the family of a hospitalized member by caring for the children, preparing meals, and comforting the worried spouse. This kind of sharing promotes the commitment not only of the family receiving care but also especially of those giving the care.

Regular group gatherings also bring about greater commitment of members, and such gatherings need not be for overtly religious purposes. A church supper helps increase members' feeling of belonging to the group. For example, ethnographies of African-American churches note the centrality of preparing church suppers and of eating food together as an expression of group solidarity and sharing (Goldsmith, 1989; Williams, 1974). Communal groups meet very frequently, sometimes each day. Other sectlike groups also urge their members to meet often. Some groups hold prayer meetings three nights a week in addition to Sunday services and church socials. Not all religious groups, however, identify commitment to the congregation as critical. Some groups emphasize both family-level religiosity and supracongregational commitment as very important. Religious gatherings of the family (e.g., the family saying the prayers of the rosary together or praying special Sabbath blessings, Shabbat b'rachot) serve similar functions of commitment but not necessarily to the congregation.

The content of group gatherings can also promote the commitment of individual members. **Ritual** is one particularly important aspect of a group gathering. By ritual, the group symbolizes meanings significant to itself. Ritual gives symbolic form to group unity, and participating individuals symbolically affirm their commitment. Ritual both reflects and acts on the group's meaning system. Too often we tend to think of ritual as being empty and a matter of "going through the motions." Even going through the motions can promote a sense of unity, but in many groups the content of ritual is highly meaningful and especially successful in creating a sense of oneness. Rituals important in many religious groups include communion and other ritual meals, healing services, symbols of deference, embraces, special prayer postures, hymns, and rituals of purification.

Mutual witnessing continues to be as important in the commitment process as in the initial socialization or conversion of the believer. Through witnessing, members show themselves and others how their daily lives can be interpreted in terms of the group's meaning system. This kind of witnessing is prominent in Pentecostal and Jesus movement prayer meetings and in the satsang of the Divine Light Mission. Witnessing is a transformative process. All events, thoughts, and experiences are transformed into significant events, meaningful thoughts, and special religious experiences. Everyday and nonbelievers' interpretations of events are devalued during witnessing and replaced with religious interpretations. Witnessing can be relatively public or can occur in the setting of a small fellowship group or family. The public proselyting of Jehovah's Witnesses, Mormons, and Hare Krishnas is different from the relatively private witnessing of the evangelical Women Aglow movement or the pentecostal Full Gospel Businessmen's Association. Groups that consider themselves in opposition to the rest of "the world" are more likely to emphasize witnessing as a commitment mechanism (cf. similar functions in Alcoholics Anonymous, Weight Watchers, and psychotherapeutic groups; see Allon, 1973; Jones, 1975; Rudy and Greil, 1987, 1989; on witnessing in religious groups, see Ammerman, 1987; Kroll-Smith, 1980; McGuire, 1977; Shaffir, 1978; Tyson, 1988).

Commitment to a group can be strengthened if the group convinces the member that the group itself is extraordinary. If, for example, members become convinced that the group is in itself the exclusive path to salvation, they are more likely to remain with the group. Groups that expect the imminent end of the world typically portray themselves as the elect who will be saved. Other groups represent their rituals and practices as necessary for salvation in the next life. Many of these groups teach that fallen-away members will be even worse off than people who never knew the "right" way.

Group practices that promote a sense of awe further emphasize the significance of the group itself. These practices make the actions of the group appear more than mundane; mystery, magic, and miracles surround the group actions. When the leaders "receive directions" from God, as among Mormons and Pentecostals, the directions seem far more awesome than if members had voted on them. Social and symbolic distance also promote a sense of awe. Thus, medieval churches used physical barriers (e.g., rood screens) and space to separate the body of the congregation from the central ritual performance. Even today many religious groups have certain sacred spaces in their places of worship where ordinary members cannot routinely go. These practices may not have been deliberately created to generate commitment, but the production of a sense of awe does result in enhancing members' commitment.

## Disengagement

Although many members continue or increase their commitment to their religious group, others become less committed or drop out of the group. In many respects, the process of **disengagement** is the reverse of conversion. Like conversion, disengagement typically entails a transformation of self and basic meaning system. When a religious meaning system and identity have been especially central in a believer's life, the process of altering self and meaning system may be a wrenching transformation.

Both conversion and disengagement are forms of "status passage" in which the individual leaves one role and enters another (Glazer and Strauss, 1971). Persons in modern, mobile societies may go through such role-exit experiences, such as moving out of one's parents' home, divorcing a spouse, discharge from the armed services, moving away from a community, leaving a hospital or prison, changing jobs or occupations or political parties, retiring, or quitting a social club. Some of these roles are more central to the individual's identity, and thus role exit may be more complex and emotion-laden than for other secondary roles.

### Stages of Disaffiliation

Like the commitment process, the disengagement process involves the pushes and pulls of various social influences. Individuals must weigh what they are accomplishing by being members against opportunities to reach those (or other) goals elsewhere. Typically, members reach the decision to leave the group only gradually, but just as members' retrospective accounts of their conversions are transformed to fit their new beliefs and image of self, so too are ex-members' accounts of the events and decisions that led up to leaving the group trans-formed. Four stages characteristic of role exit include (1) first doubts, (2) seeking and weighing role alternatives, (3) a turning point, and (4) establishing an ex-role identity (Ebaugh, 1988).

Disengagement is often the result of a breakdown or diminished effective-ness of the plausibility structure (described in Chapter 2) that supports the religious group's beliefs and practices. One study found that this breakdown of plausibility resulted from such factors as reduced isolation from the outside world, competing commitments (such as intimate dyadic relationships within the group or family links outside the group), lack of movement success, and apparent discrepancies between leaders' words and actions (Wright, 1987). Particularly if the social-emotional climate of a religious group was important in a recruit's initial conversion or commitment, those interpersonal experiences that fail to fulfill the believer's affective needs can be disconfirming (cf. Jacobs, 1989a).

Just as conversion involves coming to take a new member role for granted as part of the recruits' identity, so too the "first doubts" stage of role exit typically involves experiences that call that taken-for-granted reality into question. Often the doubts are not about the group's beliefs; for example, many nuns who eventually left the convent were not doubting their Christian faith but rather their role commitments as nuns (cf. Ebaugh, 1988). Like the conversion experi-ence, disengagement often entails a turning point, which is remembered more vividly in retrospect than many other parts of the gradual change. Just as the researcher must be careful when interpreting retrospective conversion stories, so too must tales of disaffiliation be treated with caution (Richardson et al., 1986).

Ex-members' recollections of their experiences with various contemporary religious movements were reconstructed differently, depending on whether they left the religious group voluntarily and on their degree of contact with deprogrammers and anti-"cult"[3] counselors (Solomon, 1981). Persons who had left voluntarily were more likely to remember their affiliation with the group as a growth experience, even though they had later changed directions. By contrast, people who were deprogrammed were more likely to recall their experience as one of deception and brainwashing. If they subsequently became involved in the anti-"cult" movement (which actively sponsored much deprogramming), their new movement encouraged telling atrocity tales about their former groups (Lewis, 1989). By invoking the brainwashing explanation of their own conversion, these ex-members also could disclaim personal responsibility for whatever unhappiness their previous religious commitment may have caused family and friends.

The believer might respond to first doubts in a number of ways, such as seeking advice from other believers, trying to relieve doubts, reaffirming belief and allegiance to the group, or reconsidering commitment. Furthermore, persons who are disengaging from a group typically try to identify and weigh their alternatives to the member role. Whatever the outcome, neither conversion nor disaffiliation are passive, mindless processes.

### Coerced and Voluntary Exit

The inadequacies of the "brainwashing" explanation of conversion become particularly acute in interpreting disengagement. If, as the "brainwashing" thesis holds, converts are coercively persuaded to belong to a religious group, then once converted and held in the group, they would be unable to exercise the will to change beliefs and leave the group. This interpretation became the justification for the practice of forcible "deprogramming," by which believers were kidnapped, held against their wills, and subjected to a barrage of tactics designed to turn them against their former religious group and to convert them to an alternative perspective (often a fundamentalist Christian, anti-"cult" worldview).

Sociological studies of disengagement from religious groups show, however, that the brainwashing/deprogramming conceptualization is a grossly inaccurate portrayal of these processes. Using this model, sensationalistic media convey the impression that converts are trapped, indeed "lost" indefinitely or at least for a long time. In fact, much conversion (especially to demanding religious groups) is temporary (see Wright, 1988).

One indicator of the inadequacy of the "brainwashing" thesis is how rapidly new members of many religious groups decide, of their own accord, to disengage (see Bird and Reimer, 1982). For example, even socially "acceptable" conversions to well-established religious movements (like a Billy Graham cru-

---

[3]The use of the pejorative term "cult" for these movements is sociologically incorrect (as discussed in Chapter 5); most of them are more properly identified as sects because they tend to be more authoritarian and consider their religious path to be uniquely legitimate. Because the mass media have accepted the incorrect label of "cult" and used it widely and uncritically, the movement against these religious groups generates considerable legitimacy by identifying itself as against the so-called cults. We retain quotation marks on the word "cult," therefore, to show that it is not a sociological label, nor even the correct usage of the sociological concept of cult. The concept of "cult" has become so laden with negative value connotations that it biases judgments regarding the groups and their members (Pfieffer, 1989; Richardson, 1990).

sade) are often very tenuous and short-lived (W. Johnson, 1971). A study of recruits to the Unification Church (one of the religious movements frequently accused of brainwashing) found, during the movement's growth period in the 1970s, that of some 1,000 recruits who were interested enough to attend a residential workshop, only 8 percent joined and remained as full-time members for more than one week. And of those few who did join, only about 5 percent remained full-time members for a year. Those recruits most likely to have joined with enthusiasm and later disengaged were those who had converted with idealistic expectations that the movement would make the world a better place, whereas those who continued in the movement were more likely to have joined for personal spiritual goals. The study concluded that it was "perfectly plausible" for the action-oriented idealists to decide to disaffiliate from the Unification Church when it appeared not to live up to its promise as an agent of social change (Barker, 1988a). Mounting evidence from studies of the many "new religious movements" of the last three decades indicates that, far from being the malleable, passive, gullible dupes portrayed in the media, recruits to religious groups are both open to the spiritual alternatives and yet relatively resistant to conversions that entail high levels of commitment and drastic changes in identity.

Some research suggests that the "brainwashing"/deprogramming approach to conversion and disaffiliation may be actually creating some of the problems it proposes to explain. A study of persons who had left new religious movements found that those who were forcibly removed and resocialized were particularly likely to describe unhappy psychological responses (such as suicidal thoughts) after they left. One respondent said:

> I was kidnapped and deprogrammed, so my whole world was suddenly ripped out from under me . . . before, every single action had cosmic importance, but after, I felt so small and unimportant. Life had no meaning (quoted in Lewis, 1986).

A Dutch study of former members of similar new religious movements found that about half of the respondents had never experienced social emotional problems, before, during, or after conversion and disaffiliation. For those who did experience social-emotional difficulties, the existence of such problems *before* conversion was strongly predictive of subsequent social-psychological problems at the time of exit or later (Derks, 1983). Thus, commitment to the religious group was probably not the source of many ex-members' problems. Rather, it appears that pre-existing social-emotional problems (which members may have sought to resolve by joining the group) resurfaced after leaving.

A similar correlation appears in studies of the so-called cult withdrawal syndrome. Psychologists affiliated with the anti-"cult" movement hypothesized that these religious groups used "mind-control" techniques that produced "a disorder of awareness" producing "long-term mental and emotional effects." Some symptoms of this "syndrome" or "information disease" were nightmares, amnesia, hallucinations, altered states of reality or sense of floating, inability to break mental rhythms of chanting, violent outbursts, and suicidal or other self-destructive tendencies (Conway and Siegelman, 1979, 1982).

Because there were serious methodological and data-analysis flaws with their report, subsequent studies attempted to document the nature of ex-members' emotional problems using more representative samples. A more precise test of the "cult withdrawal syndrome" hypothesis elicited responses from 154 ex-members of groups labeled "cults" by the anti-"cult" movement. Unlike the sample used by the psychologists, this sample included numerous persons who had left their groups voluntarily and had not received anti-"cult" counseling after leaving. This study found that persons who were forcibly "deprogrammed" were significantly more likely to experience troubling symptoms than were persons who left the groups voluntarily and were not subject to any "exit counseling." For example, whereas 58 percent of those receiving involuntary counseling reported nightmares, only 11 percent of those with no coercion and no counseling reported them. Far more persons receiving involuntary counseling struggled with violent outbursts (42 percent) and suicidal tendencies (31 percent) than did those who left voluntarily without counseling (only 9 percent for each symptom). Although all major role transitions are stressful and often emotionally disruptive, these data suggest that much of the reported emotional distress may be a product of the anti-"cult" movement's intervention tactics and that the higher rate of reporting symptoms may be an artifact of anti-"cult" socialization norms (Lewis and Bromley, 1987).

### Collective Forms of Disengagement

Doubts and unsatisfied needs may propel a member to leave a religious group, but individual disaffiliation is not the only response. One possible response to doubts and disillusionment is internal reform of the religious group. For example, during my own research, members of one large meditation group gradually realized that their guru was using his esteem and influence in sexual advances to several women members. Because the disappointed members still very much believed in their group's spiritual practices and ideals, instead of leaving the group, they began discussing how to change the group. They decided to send the guru back to India and continue without him.

Another historically important form of disengagement is for dissatisfied members to break off from the religious group and form a separate group. Rarely do schismatic groups consider themselves to be *leaving* their faith; rather, they view their exit from the group as keeping the true faith. They view their new ideology as more pure and their new practices as more faithful than those of the parent group.

A related response is when a group of dissatisfied members collectively defect over movement politics, often staging their departure to make a strong statement to the leadership of the movement organization. For example, a group of former ISKCON (Hare Krishna) members formed the "Conch Club" to collectively defect from the national organization, which they rejected as unworthy of their commitment due to its direction after the death of its founding guru. They continued to retain their beliefs and identity as Krishna devotees, locating themselves within the larger movement but rejecting the organization of that movement (Rochford, 1989). These collective forms of disaffiliation are also occasions of some role transition and perhaps personal anguish, but they are rarely so wrenching as individual role exit because members have the mutual support and idealism of a group making the exit together.

## Summary

This analysis of the processes of religious socialization, conversion, and commitment illustrates the interrelationship of religious meaning and religious belonging. The individual's meaning system is socially acquired and supported through early socialization and interaction with other believers throughout life. If the individual changes meaning systems, it is through social interaction. And the processes that promote commitment to the meaning system and the group supporting it are fundamentally social processes.

Social factors are important in shaping the individual's religion, and examination of critical periods in the individual's life cycle suggests some of these factors. Early socialization in the context of the family, neighborhood, and ethnoreligious community is particularly important in establishing not only the basic beliefs and values but also the connection between the individual's belief system and very identity. Rites of passage to new statuses are often filled with religious significance. Passage to adulthood and marriage illustrates some of the ways in which religion shapes critical moments as the individual takes on a new social identity. Middle age and old age are, however, problematic in this society, and there are no satisfactory transitions into these devalued statuses. The society's secularized conception of time may be an important cause of some of these problems. Evidence about the nature of the interrelationship between the individual's religion and social factors is, however, generally limited either to nonmodern examples or to studies of narrowly defined, church-oriented religion. Thus, this chapter has only suggested some of the directions that further research into the individual's religion might take.

Conversion is essentially a form of resocialization similar to nonreligious resocialization. Through interaction with believers, the recruit comes to share their worldview and takes on a new self consistent with that meaning system. Conversion includes a range of changes: radical transformation of self and worldview, identity consolidation, and reaffirmation. Some changes are mere transfers of organizational affiliation and not real conversions of worldview and identity. The process of conversion funnels recruits from a general predisposition to conversion, through interaction with group members, to growing identification with the group and its belief system. The conversion process is generally gradual, although it may appear sudden and dramatic because of the way it is symbolized by some individuals or groups.

Commitment mechanisms promote the loyalty and attachment of all members, new converts and old members alike. Groups such as most denominations, which expect only partial commitment of members, typically use less extreme commitment mechanisms than do sectarian groups, which expect members' total commitment and immersion in the life of the group. The process of commitment involves simultaneously the individual's withdrawal from competing allegiances (e.g., by sacrifice) and greater attachment to the group (e.g., by frequent interaction with fellow members). Through these commitment processes, the group builds a firm plausibility structure for its meaning system. Like conversion, disengagement from religious commitment entails a transformation of self and worldview, supported by a changed plausibility structure.

## Recommended Readings

### Articles and Essays

Erving Goffman. "Characteristics of Total Institutions." *Asylums*. Garden City, N.Y.: Doubleday, 1961, pp. 3–74.

James Richardson. "Studies of Conversion: Secularization or Re-enchantment." In *The Sacred in a Secular Age*, P. Hammond, ed. Berkeley: University of California Press, 1985, pp. 104–121.

### Books

Nancy Tatom Ammerman. *Bible Believers: Fundamentalists in the Modern World*. New Brunswick, N.J.: Rutgers University Press, 1987. This highly readable ethnography shows how members of a fundamentalist congregation maintain their worldview and commitment.

Eileen Barker. *The Making of a Moonie: Choice or Brainwashing*. Oxford: Basil Blackwell, 1984. Winner of the 1985 Distinguished Book Award of the Society for the Scientific Study of Religion, this book is a methodologically exemplary and highly readable description of the beliefs, recruitment, conversion, and commitment processes of the Unification Church of the Reverend Sun Myung Moon.

James A. Beckford. *Cult Controversies: The Societal Response to New Religious Movements*. London: Tavistock, 1985. Drawing on data from Europe and North America, this book provides a balanced interpretation of conversion, commitment, and disengagement in "new" religious movements.

Leon Festinger, Henry W. Riecken, and Stanley Schachter. *When Prophecy Fails*. New York: Harper & Row, 1956. A fascinating account of a group that received a prophecy of the imminent end of the world.

Rosabeth Moss Kanter. *Commitment and Community*. Cambridge: Harvard University Press, 1972. Using evidence from nineteenth-century American communitarian ventures, this book presents a well-organized theoretical schema for analyzing commitment.

# CHAPTER FOUR

# Official and Nonofficial Religion

I n the United States, the mention of "religion" or "being religious" typically evokes the image of church-oriented religion. When we think of religion, we generally think of Protestantism, Catholicism, and Judaism. When we think of being religious, we tend to locate that religiosity in the social framework of Old First Church, Saint Mary's Church, or Temple Sholom.[1] Church-oriented religion is a prominent and important social form of religion in Western societies, but some other modes do not conform to the "official" model of religion.

This chapter contrasts official religion and its characteristic expressions of religiosity with "nonofficial" religion and religiosity. Both of these models illustrate the capacity of religion to provide meaning and belonging. A related topic in this chapter is the difficult problem of how to measure or scientifically explore the quality of an individual respondent's religion and religiosity. The Extended Application at the end of this chapter describes the social definitions of women's roles by both official and nonofficial religions and the concurrent shaping of women's religion and religiousness.

## The Official Model of Religion

The development of a specialized official religion distinguishable from popular or folk religion and other nonofficial religious elements is the result of the historical process of institutional differentiation. Earlier religion was relatively diffused throughout all aspects of social life; the practice of religion was relatively unspecialized. Modern religion, by contrast, is characterized by **institutional specialization**: standardization of the worldview in a well-defined doctrine,

---

[1]For sources on official religions in the United States, see the Recommended Readings at the end of this chapter; also see Carroll et al., 1979; Castelli and Gremillion, 1987; Greeley, 1977; Heilman, 1976, 1990; Heilman and Cohen, 1989; Hunter, 1983, 1985, 1987b; Lavender, 1977; Lazerwitz and Harrison, 1979; Liebman, 1988; B. Martin, 1978; Marty, 1980; McKinney and Roof, 1990; Roof, 1982; Rosenberg, 1989; Scherer, 1980; Schoenfeld, 1978; Sharot, 1976; Sklare, 1972.

religious roles performed by specialists, and an organization to control doctrinal and ritual conformity, promulgate group teachings, and promote organizational programs (see Luckmann, 1967:66; also Bellah, 1964). Interestingly, this same process of differentiation created the possibility of organized irreligion—the institutional expression and consolidation of beliefs and practices of hostility or indifference toward religion. Organized irreligion (e.g., the Secularist, Positivist, and Ethical movements) closely parallels official models of religion (see Campbell, 1971).

Specialized institutions for religion thus become the focus for societal attention to religion. Each institutionally specialized religion typically consolidates its beliefs, values, and practices into a coherent model. These official models typically include a prescribed doctrine, set of ethical standards, cultic expression, and institutional organization.

### Doctrine
The official model of a religion is a coherent and consolidated meaning system, articulated by doctrinal experts (e.g., theologians) and typically promulgated only after official approval. Although church doctrine changes over time, a church can state its official model of religion fairly specifically at any one time: "This is what we stand for." Religions vary as to the extent of doctrinal sophistication and the importance of formal, prescribed doctrines. The Roman Catholic church has a vast, complicated doctrinal system, with some teachings having the weight of dogma (i.e., doctrines to which members are obliged to assent). By contrast, Judaism has no dogma but emphasizes the continual unfolding of religious truths through study of sacred scripture and texts; nevertheless, Judaism has some fairly explicit definitions of its shared meaning system.

### Ethics
Official religion also characteristically prescribes a set of norms and regulations consistent with the group's doctrine. These obligations for behavior specify what actions are necessary to be a member in good standing. They include ideals toward which members should strive. Simultaneously, official religion usually uses specialized religious measures for the regulation and control of conduct so that misconduct can be noted and punished. In religions containing normative prescriptions that are complex, the organization sometimes supports specialized religious courts and ethical specialists to interpret the body of regulations. Other religions, whose ethical norms may be just as strong, rely on informal social control measures.

### Cultic Expression
Each religion has a set of observances and devotions by which its meaning system is ritually expressed. Official religion tends to encourage standardized cultic expressions such as a liturgy or formal order of worship. Even less central expressions (e.g., devotions) are often standardized. Such standardized forms of expressing religious beliefs and sentiments include the Roman Catholic "Way of

the Cross" and the Eastern Orthodox ritual of kissing icons. Some religions emphasize spontaneity in cultic expression; even so, however, the religion encourages certain regularized ways of being spontaneously religious. A minister might pray in his or her own words but choose words appropriate to the King James version of the Bible (e.g., "Almighty God, Thou hast showered us with Thy abundance"). Official religion also typically uses ritual specialists—priests, ministers, cantors, music directors, organists, choirs, and liturgical experts. Thus, official models of religion consolidate and standardize the cultic expression of religion.

### Institutional Organization

Official religion typically organizes itself as a specialized association. It develops organizational specialists such as a professional clergy, organizational hierarchies (both lay and clerical), and often a host of auxiliary functionaries (e.g., religious teachers, church librarians, secretaries, program directors, and building service personnel). There are often formal specifications and procedures for membership and jurisdictional boundaries (e.g., parish boundaries). There are programs to be organized—Sunday Schools, "outreach" and "mission" programs, fund raising, youth clubs, study groups, and so on.

Especially important in the institutional organization of religious groups is the form of **polity** adopted. Polity refers to the arrangements for exercise of legitimate authority in the organization. The three characteristic forms found in most Western religious organizations are episcopal, presbyterian, and congregational polity. Episcopal polity refers to an organizational arrangement in which authority is centralized in a hierarchy (i.e., a pyramid of channels of authority). Thus, in the Roman Catholic church, local parishes are served by priests (sometimes hierarchically arranged even in the parish itself) who are responsible to bishops. The bishops place priests in the local parish, and the congregation has no official authority. Similarly, the bishops are responsible to the pope, using his delegated authority in their dioceses. Other examples of the episcopal form of polity include Eastern Orthodox churches and Episcopalianism; Methodism has a modified form of episcopal polity. Presbyterian polity is a form that uses a representative government by both clergy and laity. The local church is run by its clergy and lay-selected representatives. They, in turn, are responsible to a higher authority of elected representatives such as a presbytery, synod, or general assembly (as in Presbyterianism).

The congregational form of polity is characterized by high degrees of autonomy of the local congregation, including the power to call or dismiss clergy. This form is characteristic of Jewish congregations, Baptists, and the United Church. Denominations with congregational polity are less likely to conform with a uniform doctrine and ethical or ritual standards among congregations than denominations with other kinds of polity.[2] Similarly, as developed further

---

[2]It is therefore not surprising that sociologists trying to research members of these religious organizations have greater problems studying, as aggregates, the adherents of congregational polities, such as Southern Baptists, in contrast to hierarchical polities, like Mormons. How do they determine, for example, who is a Southern Baptist (cf. Hadaway, 1989; Cornwall and Cunningham, 1989)?

in Chapter 7, individual congregations in denominations with the congregational form of polity can be highly resistant to external pressures for change, ignoring the urgings of national denomination councils and firing change-oriented clergy. For the same reason, however, congregational and presbyterian forms of polity are more receptive than episcopal polity to the initiatives or needs of local congregations.

## Individual Religiosity in the Official Model

The specialized religious institution mediates the official model of religion to the individual believer. Thus, for example, being a Presbyterian might be experienced in terms of attending Presbyterian Sunday worship services and Sunday Schools in a particular building of one's congregation; receiving communion in the Presbyterian church and being a member of the committee that washes the communion cups; interacting with Presbyterian clergy; singing in the church choir; having a grandfather who is a member of the Session (i.e., elected Board of Elders); having once memorized but now forgotten the Westminster Confession of faith and answers to the catechism; and learning other Presbyterian beliefs and norms in Sunday School and adult discussion groups or from sermons. In other words, the individual member is socialized into the model of his or her church.

For this reason, individual religiosity in official religion can be described in terms of its conformity to the official model. The individual's beliefs are somehow related to church doctrine, and the individual's ethical standards are measured by conformity to the ethical teachings of the official religion. Cultic expressions of the religion are translated in terms of individual observances and devotions. The individual's relationship to the institutional organization is expressed as membership and participation in that organization (see Luckmann, 1967:74).

Thus, it is possible for a sociologist studying an official religion to operationalize—to create a working definition of—individual religiosity in terms of the officially defined model. The Roman Catholic church (especially before Vatican II) used a fairly explicit official model of expected individual religiosity. Thus, in his studies of Roman Catholic parishes, Fichter (1951, 1954) was able to operationalize individual religiosity for research purposes, using criteria such as Mass attendance, reception of communion, sending children to parochial schools, participation in parish activities such as Holy Name Society, and interaction with clergy. This approach to operationalizing the concept of religiosity works, to some extent, *only* because the official model of Roman Catholic religion was precise and mediated to individual members.

Nevertheless, we cannot infer that these aspects of official religion explain *why* the individual member participates in any aspect of it. We cannot infer that the nearly 40 percent of the U.S. populace who attend religious services each week do so in precise conformity with what the official model of that religion puts forth as the purpose and meaning of those services. One researcher suggests that the typical churchgoer is primarily interested in "friendliness and fellowship, the chance to meet 'good' people and associate with them" (Wood, 1970:1066). Thus, although the official model of religion has implications for individual religiosity, there is little indication that all aspects of official religion hold equal importance for each individual believer.

## Differences from the Official Model

Official religion, then, is a set of beliefs and practices prescribed, regulated, and socialized by organized, specifically religious groups. These groups set norms of belief and action for their members, and they establish an official model of what it means to be "one of us." Nevertheless, the actual religion of the individual member may not correspond very closely to the official model. What is the *operative faith* of the individual? What norms, ritual actions, and beliefs hold priority in the individual's life?

A number of factors bring about this discrepancy between the individual's religion and the official model. One factor is how the individual has been taught and socialized into the official model. Some persons have not learned the official model of their group's religion. The official belief systems of Christian and Jewish groups, for example, generally include a set of moral norms called "the Ten Commandments." According to Gallup Poll data, however, a majority of U.S. adults could not name more than half of those commandments (Gallup Poll, 1984:14). Many members may not know items of official belief such as catechisms, creeds, and confessions of faith. Furthermore, members vary in how adequately they have internalized the official model. It is one thing to know that one's group holds certain beliefs; it is another to believe them oneself.

Individual members may have incorrect knowledge as evaluated by the official definitions of the belief system. Although representatives of religious organizations may decry this misinformation or lack of knowledge, it is an understandable product of socialization. In socialization, the official model is mediated (if taught at all) through other persons who are significant in the child's life: parents, friends, teachers, clergy. The child may be exposed to a relatively incoherent assortment of stories, admonitions, images, threats, and examples.

The official model is often overlaid with other religious themes, perhaps drawn from folk religion, mythology, popular culture, or the teacher's own "misinformed" view of the official model. For example, one woman said: "I was in my thirties before I realized that a lot of the stuff I learned about religion in grade school wasn't the official church teaching. I'm sure the church has some theology about sin and grace, but to this day whenever I hear those words I think of what my religion teacher in fifth grade taught us about good angels whispering in our right ears and bad angels urging us in our left ears." Although official religion is characteristically a coherent, consistent body of beliefs and practices, the individual's version of it is not necessarily coherent or consistent. Individuals vary in how much consistency they expect in their own lives, but many people appear to be comfortable with highly incoherent assortments of beliefs and practices (see Bibby, 1983; Hertel, 1980).

Individual members may also deviate from the official model of religion in their priorities, often emphasizing aspects of belief and practice that the official model holds as relatively unimportant. Thus, a religious group may emphasize norms against cheating and stealing, whereas an individual member may downplay those norms while emphasizing norms against drinking and gambling.

Sometimes individual members may disagree with the official beliefs and practices. A sizable proportion of U.S. Roman Catholics do not agree with the official church teaching on birth control, abortion, divorce, women clergy, and clerical celibacy. In a 1987 poll, 77 percent of U.S. Catholics said they would be more likely to follow their own conscience rather than the Pope's teachings on these matters (Gallup and Castelli, 1989:99). Another study found that most Catholics surveyed felt that a person could be a good Catholic without going to church every Sunday (70 percent), obeying church teachings on birth control (66 percent), going to private confession at least once a year (58 percent), or obeying the church's teachings regarding divorce and remarriage (57 percent) (D'Antonio et al., 1989:57).

In the 1950s when official national organizations of major Protestant denominations made strong statements about Christian responsibility in the issue of racism, many individual members chose to disagree with these teachings (Hadden, 1969). Similar recent disagreement has occurred in both Protestantism and Roman Catholicism over teachings on homosexuality, women's roles, abortion, war, and divorce. Even less volatile and less current teachings do not receive the full assent of members. How many Presbyterians fully believe the historically central doctrine of predestination (i.e., the belief that God knows and decides beforehand who is destined to salvation or damnation)? Similarly, though belief in a hell is part of official Roman Catholic and most Protestant teaching, only 70 percent of U.S. Roman Catholics and 68 percent of Protestants polled believed in a hell (Alston, 1972:180). Assertion of an official belief, such as in response to a poll, does not necessarily mean the believer understands that belief with any theological or logical sophistication. Fifteen percent of the U.S. populace polled, for example, expressed a belief in heaven but not in afterlife—a belief that, in most theologies, is necessary for explaining the idea of heaven (Hertel, 1980).

The operative religiosity of individuals may vary considerably in degree. Some people are simply more interested and involved in religion than others. Various religious groups expect different minimal levels of individual religious expression, but even within these groups there is considerable variation. One problem in measuring religiosity is operationalizing the concept to account for all of the different ways of being religious. Just as individuals select various official beliefs and practices with which they agree, so too they vary in their chosen ways of being religious. One expression of religiosity may be church attendance; another, watching a religious TV show at home. For one individual, social activism may be an essential expression of religion; another individual might emphasize interior religiosity to the complete exclusion of social action.

It is therefore difficult to describe individuals' religions merely by the name of the official religion to which they belong. Two Roman Catholics may be as different from each other in their individual operative religions as a Roman Catholic is from a Baptist or Jew.

## Measuring Religiosity

In recent years, sociologists have attempted to understand the impact of religion on society by studying the religious attitudes and beliefs of individuals. They have asked, for example, whether highly religious persons are more or less likely than less religious persons to be prejudiced toward other ethnic, racial, and religious groups. Other studies have correlated individuals' religious attitudes with their attitudes toward work or education. Typically, these studies apply the research methods of large-scale opinion surveys to understand religion.

Because we need to know more about the values, attitudes, feelings, and experiences that motivate the individual believer, this emphasis on understanding individual religiosity is worthwhile. The methods and measures used, however, have serious shortcomings that cannot be overcome by minor refinements. One problematic issue is that most of this research has not been interested in the individual's religion per se but in generalizations about the "sum" religiosity of larger groups (e.g., denominations or ethnic groups). Thus, the research does not really try to grasp the total picture of each individual's religion and commitment to it; instead, it isolates certain readily measurable aspects of each individual's religion, then lumps these into a composite picture of the larger group.

The second major problem with attempts to measure religiosity is in creating measurable operational definitions of "religion" and "religiosity." Proceeding from fairly narrow substantive definitions of religion, most measures of individual religiosity define it in terms of the official models of specific organized religions—particularly traditional Protestant Christianity. In these operating definitions, "religion" refers to the respondent's affiliation with an institutionally identifiable belief system (e.g., Greek Orthodox or Seventh Day Adventist). "Religiosity" refers to the intensity of commitment to this belief system, as expressed in institutionally identifiable attitudes or behaviors (e.g., receiving communion or agreeing with the group's moral condemnation of certain actions). As with all methodological approaches, this "way of seeing" is also a "way of not seeing" (Burke, 1935:70). The perspective of official religion focuses the concepts of religion and religiosity but simultaneously makes other modes of religion and religiosity "invisible." Let us now examine some of the most commonly used measures of religiosity.

### Unidimensional Measures
Many studies have characterized the individual's religion or religiosity by a single measure. One familiar measure is religious affiliation or preference. A

questionnaire may ask the respondent to indicate whether he or she is Protestant, Catholic, Jew, other, or "none." A further refinement of this is specification of the denominational affiliation (e.g., Presbyterian, Southern Baptist, Conservative Jew, Mormon, etc.). These categories then become the basis for broad comparisons of behavior (e.g., college attendance) or attitudes (e.g., approval of desegregation).

Another measure focuses somewhat more on the quality of the individual's religious participation or involvement. The survey may ask "How important is your religion to you in your everyday life?" Or it may seek some quantitative measure of commitment such as frequency of church attendance. Some studies combine several such measures into an index of religiosity; such an index may include frequency of church attendance, frequency of receiving communion, and percentage of income donated to the church.

These definitions of religiosity are severely limited, and not all religious groups emphasize such expressions of religiosity as important for their members. Some groups urge their members to tithe a minimum of 10 percent of their incomes; other groups do not consider tithing a necessary mark of a religious person. Also, individuals differ in their ways of being religious. One person might express religion by meditating regularly, another by attending church, another by reading certain literature, another by participating in a civil rights demonstration. Recognition of these variations led to the construction of multidimensional measures of religiosity.

### Multidimensional Measures

Only relatively recently have research designs recognized variations in religiosity. Some studies as early as the 1950s demonstrated important differences in involvement among members of the same religious organization (Fichter, 1951, 1954). In the 1960s, seeking measures that could be applied in large-scale survey research, researchers began to formulate more formal distinctions of dimensions of religiosity (Fukuyama, 1961; Lenski, 1963). The most influential of these is the classification proposed by Glock (1965). He distinguishes five dimensions: experiential, ritualistic, ideological, intellectual, and consequential.

a. The **experiential dimension** includes feelings or sensations that are considered to involve communication with "divine essence." It refers, for example, to a feeling that one has been saved or healed, a feeling of deep intimacy with the holy, or a sensation of having received a divine revelation of some sort.
b. The **ritualistic dimension** includes religious practices such as worship, prayer, and participation in certain sacraments.
c. The **ideological dimension** refers to the content and scope of beliefs to which members of a religious group are expected to adhere.
d. The **intellectual dimension** encompasses the individual's knowledgeability about the basic tenets of the group's beliefs and sacred scriptures.
e. The **consequential dimension** includes the effects of religious belief, practice, experience, and knowledge upon the individual's behavior in institutional

settings that are not specifically religious (e.g., business, voting, family, and leisure-time activities).

If these dimensions are discrete factors, each dimension must be translated into operational (i.e., working) terms in order to measure it. The most serious shortcoming of this process is the quantitative study of individual religiosity. No research has yet defined these dimensions in terms that are not biased toward the orthodox position of certain historic (i.e., traditional Judeo-Christian) expressions of religiosity. In their study of U.S. church members, Stark and Glock's operational definition of these five dimensions illustrates such a bias (Stark and Glock, 1968:22–80). Thus, a person who assented to a higher proportion of the following beliefs in an index of the ideological dimension was considered more religious than someone who assented to only some:

a. Unqualified certainty in the existence of God
b. Belief in a personal God
c. Belief in miracles as described in the Bible
d. Belief in life after death
e. Belief in the actual existence of the devil
f. Belief in the divinity of Jesus
g. Belief that a child is born into the world already guilty of sin (i.e., original sin)

By these measures, liberal denominations appear less religious than conservative ones, and liberal members of any denomination appear less religious than conservative members. Devout Quakers would, by these measures, probably score very low on religiosity. Subtle cultural and gender biases may also influence these measures. Apparent differences between men's and women's religiosity and between blacks' and whites' religiosity may be, in part, an artifact of how religiosity is operationalized and measured (Feltey and Poloma, 1990; Jacobson et al., 1990).

Many other researchers build uncritically upon these multidimensional indices. Most of their working definitions of dimensions of religiosity result in measurements of beliefs *about* these other dimensions rather than actual behavior. Rather than tapping the meaning of ritual to the individual, such scales often ask questions like "Do you feel it is possible for an individual to develop a well-rounded religious life apart from the institutional church?" (See Faulkner and DeJong, 1966.) Although answers to questions such as this are interesting in themselves, they hardly measure a person's actual ritual experience. Although statistical analyses of these five scales show them to vary independently of each other, a valid criticism has been raised: Are they perhaps only separate aspects of a single dimension, that of *ideological commitment?* Thus, may the belief scales identify some of the content of the individual's ideology, whereas experiential, ritualistic, intellectual, and consequential scales measure expressions of strength of commitment to that ideology? (See Clayton and Gladden, 1974; Weigert and Thomas, 1969.) Some indicators of religious commitment may be appropriate to certain research tasks and not to others (Roof, 1979).

Recognition of the cultural biases of some ideological items tapped in these scales has led Glock and others to formulate parallel measures for emerging religious patterns in the United States. In an ambitious study of nontraditional religious affiliation in a California community, Glock and his co-workers included several non-Christian options such as Eastern mysticism, astrology, Transcendental Meditation, Satanism, and Yoga (Glock and Bellah, 1976; see also Glock and Wuthnow, 1979; Wuthnow, 1976b). This research asked about religious practices other than specifically Christian ones—whether respondents meditated, for example, and what kinds of techniques they used in meditation (e.g., a mantra, drugs, breathing techniques, prayer, etc.). Similarly, a Canadian study tapped several alternative meaning systems beyond the usual official or "visible" religions; researchers identified the "invisible" religious cultural themes as mystical, paranormal, familist, feminist, and positivist (Bibby, 1983). Other researchers have attempted to overcome possible gender biases by eliciting respondents' images of God and by including a measure of perceived closeness to God (see Feltey and Poloma, 1990; Nelsen et al., 1985; Roof and Roof, 1984).

Although such studies greatly expand the range of religious beliefs and practices considered, they are nonetheless limited to discrete, identifiable clusters of beliefs. It may not be possible to devise a research instrument to tap all possible religious beliefs, practices, knowledge, and experiences held by any individual in a culture as diverse as the United States'. Survey research is predicated on measuring the distribution of opinions, attitudes, or attributes that can be specified in advance (i.e., in wording the questions of the survey). Such a specification is only possible by using the criteria of official religion. A number of features of religion and religiosity in contemporary societies make it problematic to assume that an individual's religion can be adequately described by measuring its correspondence to orthodox official religion (Luckmann, 1973).

Let us now consider some of the ways in which the religion of the individual may differ from official religion.

## Nonofficial Religion and Religiosity

Alongside or overlaying official religion is another pattern of religious belief—**nonofficial religion**. Nonofficial religion is a set of religious and quasi-religious beliefs and practices that is not accepted, recognized, or controlled by official religious groups. Whereas official religion is relatively organized and coherent, nonofficial religion includes an assortment of unorganized, inconsistent, heterogeneous, and changeable sets of beliefs and customs (Towler and Chamberlain, 1973). Nonofficial religion is sometimes called "common," "folk," or "popular" religion because it is the religion of ordinary people rather than the product of religious specialists in a separate organizational framework. Only in societies where religion is separated as an institution from other institutional spheres does an official model of religion emerge, leaving elements of belief and custom not accepted as "official" to be carried on outside the religious institution (Luckmann, 1967:66, 67). Individuals often practice both the official and common religion simultaneously; others adhere only to official or only to popular religious patterns. While popular, folk, or occult religious beliefs and practices may be defined as "deviant" by arbiters of official religious or scientific knowledge, in everyday life they are in the range of "normal" and plausible to many people (Campbell and McIver, 1987).

### Popular Religion
Popular religion is no single entity, and its elements are diverse. One aspect is a popular version of official religion but completely uncontrolled by official religious organizations. A large body of inspirational literature exists that is nontheological and unrelated to the official religion of its readers, and the religion promulgated in this literature is largely quasi-magical, emphasizing religious techniques. One author of many such books, Dr. Norman Vincent Peale, urges, "Learn to pray correctly, scientifically. Employ tested and proven methods. Avoid slipshod praying" (quoted in Schneider and Dornbusch, 1957:479; see also Meyer, 1965). Values and ideological perspectives embodied in popular literature, from *Reader's Digest* to *Gentlemen's Quarterly*, find their way into popular religion (Elzey, 1976; 1988).

A parallel and even more popular source of popular religion is inspirational programming on radio and television. Some mass-media preachers gain a sufficient following to build up a large-scale organization. Oral Roberts, for example, began as a popular tent revivalist and faith healer and subsequently developed a large, well-financed sectlike organization. Televangelists (i.e., television preachers) exemplify nonofficial religious independence from doctrinal, organizational, and fiscal controls.

There is a complex relationship between religion as promulgated by popular culture and that of official religious groups. On the one hand, studies have found that the "electronic church" (especially television religious shows with popular preachers) is watched almost exclusively by already converted, relatively committed conservative or fundamentalist Christians (Stacey and Shupe, 1982). According to a CBN/Nielsen study, in 1985 the total audience for the top-ten U.S. religious programs was 67.7 million households, but because many households watched more than one program, the actual unduplicated audience was about 34.1 million households (cited in Hadden and Shupe, 1987). Rather than change viewers' beliefs, these programs may simply affirm them. On the other hand, many organized religious groups—including notably some of the highly conservative fundamentalist denominations—have criticized the "electronic church." Part of their objection may be, perhaps, due to competition for the same audience and its participation and contributions. Another part of their concern, however, may be the freedom enjoyed by independent evangelists from outside evaluation and control, especially regarding the content of their preaching.

Since televangelists are entrepreneurs, their success hinges on achieving an audience response that can be converted into contributions of money. This necessity is often connected to the beliefs they teach: For example, many entrepreneurial preachers tell their viewers that success, prosperity, and answered prayers come to those who contribute, especially to that particular evangelist's organization (Hadden and Swann, 1981). Thus, while popular evangelists present a version of religion, that version often differs in content and emphasis from orthodox teachings of official religions.

The ideas and practices from popular culture religion are widespread. Artifacts of popular religion bear some resemblance to those of official religion: a plastic Jesus on the car dashboard, gold-plated crosses on neckchains, bumper stickers proclaiming "I Found It" or "My God Is Alive; Sorry About Yours," plaques, plates, and posters with messages such as "Smile, God Loves You." One shrine of popular religion is Holyland, U.S.A.—a large sculpture garden where busloads of tourists visit various scenes depicting biblical events such as Noah's Ark and the Crucifixion (cf. Elzey, 1975; 1988). Popular religion also includes religious or quasi-religious themes from patriotism or nationalism (explored in detail in Chapter 6). National symbols such as the flag or Statue of Liberty form parts of some individuals' belief systems and are often indistinguishable from elements of official religion.

Popular religion has political significance. One reason for its political influence is that popular religion is often inextricably linked with a nation's civil religion (see Chapter 6). For example, in parts of Venezuela, the "Liberator" Simon Bolivar has the status of a folk saint, and some rituals of popular religion celebrate his memory (Diaz, 1980). Furthermore, popular religion mediates cultural values (such as kinship and ethnicity) that are linked with public/political life. Some aspects of popular religion indirectly legitimate and reproduce specific sociopolitical arrangements. For example, the relationship between worshipper and patron saint reflects and subtly legitimates the system of reciprocal obligation that underpins clientalistic politics in many countries (Kokosalakis, 1986). At the same time, however, popular religiosity can also be used as a powerful resource for social change (see Chapter 7).

### Superstition and Magic

Another overlay from common religion onto the official model consists of a variety of superstitious and magical beliefs and practices. Some superstitions pertain directly to practices of official religion. Asked why she had all of her babies christened, one woman responded, "Don't know. People have it done. It keeps them safe like wearing a St. Christopher" (quoted in Towler and Chamberlain, 1973:23). Another somewhat magical use of official religion is stichomancy—the practice of divining or getting a "message" from random opening of the Bible. Similarly, quasi-magical attitudes often accompany religious actions. At the end of a healing service observed in a mainline denomination, the minister offered to bless salt and oil for home prayer for healing. With a rustling of brown paper wrappers the congregation opened the spouts of salt boxes and lids of cooking oil bottles so the blessing could "get in." This overlay suggests a difficult methodological problem: how to interpret a situation in which members participate in rituals of official religion with attitudes and motives inconsistent with the official model.

Much superstition is related to more mundane matters and represents an attempt to control or explain one's environment. One study in England found superstitious beliefs and practices to be fairly common. Thus, 22 percent of the sample believed in lucky numbers and 18 percent in lucky charms, 15 percent avoided walking under ladders, almost 50 percent threw spilt salt over their shoulders, and over 75 percent touched wood for protection. Some superstitious actions were not serious, but a core of 6 to 8 percent of the sample said they became uneasy if they did not perform the appropriate actions. This study found persons who characterized themselves as religious more likely to be superstitious than those characterizing themselves as nonreligious; churchgoing "religious" persons, however, were much less superstitious than nonchurchgoing "religious." This evidence suggests that for some religious persons, official and common religious beliefs and practices become indistinguishably merged but that churchgoing may filter out nonofficial elements of belief (Abercrombie et al., 1970:98, 113; see also Jahoda, 1969, regarding the affinity between belief patterns of superstition and official religion).

### The Paranormal and Occult

Other beliefs and practices concern **paranormal occurrences** (i.e., events outside the usual range of experiences). Belief in paranormal occurrences is fairly widespread. Approximately one-fourth of U.S. adults believe they have seen a ghost or spirit of the dead (Greeley, 1975:36). One U.S. national survey (Gallup and Castelli, 1989) reported that 42 percent of respondents felt they had undergone an experience of being really in touch with someone who had died. In the same study, 67 percent reported experiences of déjà vu (i.e., the sense of having already seen a place where they had never been before), and 67 percent reported experiences of ESP (extrasensory perception). Similarly, surveys of European beliefs and values found about 25 percent of British, German, and French respondents believed they had communicated with the dead, and about 35 percent reported experiences of ESP (Gallup Poll/Social Surveys, 1984). Some persons attribute great significance to paranormal experiences; whole belief systems (e.g., Spiritualism) allocate special meaning to such experiences. Other persons who have had such out-of-the-ordinary experiences recognize them as unusual but place no special interpretation on them.

A study in an English industrial city found comparably high rates of extraordinary religious experiences in a random sample in which very few claimed active membership in an official religious institution. Despite the fact that many of these experiences were described as personally very meaningful, the majority of respondents kept the experiences private or altogether secret for fear of ridicule or being labeled mentally imbalanced (Hay and Morisy, 1985). Thus, some elements of nonofficial religion are "invisible," not so much because of their deviance from official religion, but more because the general culture treats them as inappropriate.

Extrascientific explanations and techniques are also part of common religion. These are ways of interpreting and manipulating the natural and social environment that are not accepted by this culture's official religious or scientific groups. Included are beliefs in astrology, hexing, palmistry, numerology, amulets and charms, water divining, UFOs, divination (by Tarot, pendulum, rod,

I Ching, entrails), and witchcraft. Science fiction, for example, deals with a number of quasi-religious themes about humanity and its boundaries, which have been embodied in religious groups such as UFO cults and groups that believe they communicate with nonterrestrial beings (see Mörth, 1987). One study found disproportionate numbers of young, highly educated males among those attracted to deviant (or occult) sciences and occult quasi-religious practices (Ben-Yehuda, 1985). It asks, Why this prevalence of nonofficial religion (and nonofficial science) among precisely those social groups most favored in modern, technological society?

A major focus of extrascientific practices is healing (e.g., warding off the "evil eye," divining causes of illness, conjuring, herbal cures, and enlisting the aid of the spirits for curing or causing illness). Some of these beliefs derive from the folk traditions of various ethnic groups (see, for example, studies reported in Crapanzano and Garrison, 1977; also Roebuck and Quan, 1976). Some evidence exists, however, that marginal medical beliefs and practices are not uncommon among middle-class, relatively well-educated persons. Marginal medicine often overlaps the official model of religion in many religious groups, as exemplified by miraculous and faith healing. Some of these beliefs and practices are organized as groups; others are highly individualistic and are transmitted by separate practitioners or anonymous media such as magazines.

Beliefs and practices pertaining to paranormal occurrences and extrascientific explanations and techniques are part of what is generally termed the "occult." **Occultism** is a set of claims that contradict established (i.e., official) scientific or religious knowledge. This focus on the anomalous makes the occult seem strange and mysterious (Truzzi, 1974a:246). Some observers have suggested that the 1960s and 1970s were periods of occult revival in U.S. popular culture (Marty, 1970; Truzzi, 1972). The esoteric and occult have long been part of countercultural currents in Western societies. Many wide-ranging innovations have begun in the "seedbed" of the esoteric, and modern medicine, chemistry, and psychology owe much to the esoteric sciences of earlier times (Tiryakian, 1974:272, 273).

Although some of the recent interest in esoteric beliefs and practices may be merely countercultural experimentation or fads, many people take their belief in the occult seriously. Witchcraft exemplifies beliefs and practices followed by adherents with varying degrees of seriousness and involvement. At one end of the spectrum are people for whom witchcraft is a fad or form of entertainment. Others believe in witchcraft and therefore seek occasional spells for health, protection, and success; they might also ascribe special importance to astrological events such as solstices and eclipses. At the other end of the spectrum are people who are fully involved in the beliefs and practices of witchcraft, belong to a group (i.e., coven), and organize their lives around witchcraft as a counterofficial religion.

There is a distinction between white and black magic. White witches employ their magic for only beneficial purposes such as healing. They often emphasize the continuity of their tradition with the medieval European white witches— "wise women" and men who combined pre-Christian symbolism and lore. It is not uncommon for white witches also to belong to church groups of official religion. One interesting contemporary development of witchcraft and neopaganism has been the adoption of some of their symbols and organization for the expression of feminism. Some women are proud of the deviant heritage that the white witch represents (Christ, 1987; Culpepper, 1978; Neitz, 1990). By contrast, practitioners of black magic use their powers for evil as well as good ends. They believe they have allied themselves with the forces of evil (i.e., Satan), from which they obtain power and direction (Alfred, 1976; Bainbridge, 1978; Moody, 1971, 1974; Truzzi, 1974b).

Secret societies have long been a part of nonofficial religion. Mystery cults, kabbalistic groups, Freemasonry, and Rosicrucianism exemplify their variety. Mystery and awe surround the esoteric knowledge protected by the secret society, and initiation rituals symbolize admission of the member to each level of secret knowledge. The protection of this secrecy promotes the solidarity of the group (Carnes, 1989; Kertzer, 1988; Simmel, 1906; Tiryakian, 1974:266, 267). Seldom is there any clear border between secret societies that are overtly religious and those that are not specifically religious. Similarly, some religious movements (e.g., the Divine Light Mission) are comparable to secret societies because they are organized around a *gnosis* (i.e., special knowledge).

Another strand of esoteric culture that is a much more familiar part of common religion is **astrology**, a set of beliefs and practices predicated upon the idea that impersonal forces in the universe influence social life on earth. Believers try to read and interpret zodiacal signs (e.g., the position of the stars at the time of one's birth) in order to predict outcomes of the influence of those forces, to choose the best moment for certain actions or the best probable course of action, and to discern when humans bear responsibility or blame for certain outcomes. Under the general category of astrology fall a wide variety of beliefs and practices ranging from highly sophisticated, complex speculations to pop versions such as newspaper horoscope columns (Fischler, 1974).

Astrological beliefs and practices are widespread in modern Western societies. According to a 1988 survey, 15 percent of the U.S. population believe in astrology; 25 percent follow their horoscope, and 85 percent know their zodiacal signs (Gallup and Castelli, 1989:75–76). Similarly, a Canadian survey found that about 35 percent of respondents believe in astrology and 75 percent consult their horoscope at least occasionally. The study noted that many conventionally religious people supplemented their beliefs and practices with nonofficial religion such as astrology, ESP, and psychic beliefs (Bibby, 1987). Although some of the symbols and imagery of astrology were used by the 1960s youth culture (e.g., the song "Age of Aquarius" in the musical *Hair*), belief in astrology is spread across all age categories. One study found that although young people were somewhat more likely to know about astrology and to be interested in it, a higher percentage of older persons were actually committed to astrology (Wuthnow, 1976a:159).

The nature of belief in astrology illustrates the methodological difficulty of studying nonofficial religion. Astrological beliefs are widespread, significant parts of many people's worldviews. Some people hold them in combination with official religious beliefs; for others, they are substitutes for official religion. There is no "church" of astrology. Although there are some individual astrological practitioners, most adherents learn their beliefs and practices through relatively anonymous media—books, newspapers, magazines. Some evidence suggests that, like superstition, commitment to astrology is strongest among traditionally "religious" persons who are not involved in official religion (Wuthnow, 1976a:166).

### Interpreting Nonofficial Religion

It would be a mistake to view nonofficial religion as a vestige of rural or folk religiosity. Although some elements of common religion (e.g., root doctors and herbalists) are kept alive by distinct rural or recently rural subcultural groups, other aspects of nonofficial religion are characteristic of urban mass society. Astrology is more prominent in an urban milieu than in a rural one (Fischler, 1974:287). Furthermore, some items of nonofficial religion have been borrowed from earlier folk religions but given new significance (e.g., white witchcraft).

One interpretation suggests that modern (especially urban) society is increasingly characterized by a new mode of religiosity, in which individuals select components of their meaning systems from a wide assortment of religious representations. But while traditional representations from official religion may be among the components selected, other elements are drawn from popular culture—newspaper advice columns, popular inspirational literature, lyrics of popular music, the Playboy philosophy, "women's magazine" versions of popular psychology, horoscopes, night-school courses on meditation techniques, and so on (Luckmann, 1967:103–106). This interpretation may somewhat overstate the contrast of this eclectic religiosity with that of earlier times; it is entirely probable that earlier generations held widely varying versions of official religion, melding together combinations of folk beliefs, superstition, pagan customs, and misinformed interpretations of the official model of religion to form individual meaning systems. The sharp contrast between official and nonofficial religion is probably itself a product of relatively recent social processes by which society has come to define religion as the domain of official religious institutions and by which religious organizations have articulated their distinctiveness from popular religion.

The complex threads woven together to comprise the individual's operative religion make it difficult to describe that religion by reference to any single model such as an official religion to which the person may belong. Thus, the religion of a person who is a member and occasional attender of an Episcopalian church may simultaneously include orthodox Episcopalian beliefs and practices, popular religious items gleaned from inspirational literature and television, patriotic piety, popular psychology from *Reader's Digest*, careful attention to certain superstitious practices, pieces of astrological advice, and beliefs and techniques learned in an adult school course on psychic healing. This amalgam of beliefs and practices is not at all unusual.[3] Evidence on the distribution of

---

[3]This example is, in fact, drawn from my own research, in which many respondents described these and more diverse elements of belief and practices.

nonofficial religious beliefs and practices shows that many adherents are also adherents of official religion. Only a small portion of nonofficial religion is practiced in lieu of official religion.

Merely listing the elements of a person's religion does not adequately describe it. We also need to know which beliefs and practices are central and which are peripheral. Perhaps psychic beliefs are very important to one individual, but astrological beliefs are relatively unimportant. Similarly, we need to know when the individual applies which beliefs and practices. The believer may turn to official religion to help socialize children but seek nonofficial religious help for healing a chronic back problem. We need more information about the effective religion of individuals in its complexity and richness.

## Extended Application: Women's Religion and Gender Roles

As we have seen, the individual's religion is a personally meaningful combination of beliefs, values, and practices that is usually related to the worldview of a larger group into which that individual has been socialized.[4] In socialization, the individual typically receives many of these beliefs, values, and practices from representatives of the larger group such as parents and teachers. Much of this received meaning system becomes internalized—that is, made a part of the individual's own way of thinking about self and others. The meaning system received by the individual includes a number of beliefs, images, and norms about that group's definitions of maleness and femaleness. All religions have addressed the theme of human sexuality and gender roles because sexuality is a potent force in human life and because gender is, in most societies, a major factor in social stratification.

This analysis will focus first on how women's religion influences their gender roles and identities, illustrating some of the concepts in Chapter 3 on the individual's religion. A similar application could be drawn for men as males, showing how their gender roles are linked with religion, although this application is not pursued here.

Second, this analysis will illustrate why focusing only on official religion and religiosity is sometimes misleading. As religion became a differentiated institutional sphere, many of women's religious roles and expressions were excluded from official religion and, if continued at all, were kept alive in nonofficial forms such as healing cults, popular religiosity, mediumship, witchcraft, and spiritual midwifery. We need to examine the structural bases for women's status; official religious institutions have historically epitomized the structural and ideological suppression of women. The religious legitimation of their gender role, then, raises some interesting questions about women's religiosity. Do they come to think of themselves as the official model of their religion portrays them? Impressive historical evidence exists that nonofficial religion is one vehicle for women's assertion of a different religious role. A related question is

---

[4]This extended Application exemplifies concepts in both Chapters 3 and 4.

whether official religions are capable of changing to legitimate and express gender equality even if they were to adopt these goals.

### Gender-Role Definitions

The definitions of maleness and femaleness are culturally established. On the basis of these definitions, a group develops and encourages certain social differences between men and women. U.S. society expects and encourages little boys to be more active and aggressive than little girls; little girls are expected and encouraged to be more polite, passive, and nurturant. In socialization, males and females are taught their culturally assigned **gender roles**—the social group's expectations of behaviors, attitudes, and motivations "appropriate" to males or females. Historically, religion has been one of the most significant sources of these cultural definitions of gender roles; and religion has been a potent legitimation of these distinctions.

Through socialization, the group's definitions of appropriate women's roles become part of the individual woman's self-definition. She evaluates herself in terms of these definitions: "I am a good girl/daughter/woman/mother/wife." The group's definitions of gender roles are subtly interwoven into the individual's learned "knowledge" about the social world. They are embodied in the language and imagery of the group and thus indirectly shape the members' thought patterns. The culture's use of words such as "bitch," "slut," "sweetie," "doll," "housewife," and "mother" has strong evaluative connotations. These words imply qualities that some people attribute to women and for which there are no obvious male equivalents. Numerous words (e.g., "slut") exist for promiscuous women; there are no equivalent words for promiscuous men. Similarly, a man who is said to "father" a child performs a single biological act; a woman who "mothers" a child provides continuing nurturant care. The language embodies the different standards of the society for men's and women's roles (Lakoff, 1975). In learning the group's language, the individual internalizes these images and symbols. Specifically religious symbols and images also shape the individual's gender-role concept.

Although both men and women belong to the same official religions, the individual's religion is not necessarily a small carbon copy of the group's entire official stance. Women's versions of a certain religion are probably very different from men's versions; a woman may focus on those aspects of the group's worldview that speak to her social situation. Thus, an Orthodox Jewish man's personal religion might focus on his public ritual roles (e.g., carrying the Torah and forming part of a minyan, i.e., the ten men necessary for a prayer gathering to be official). He might emphasize the intellectual development, discussion of sacred texts, and careful application of religious law expected of men in the Orthodox community. Women, by contrast, have few ritual duties (e.g., the ritual bath after menstruation, making the Sabbath loaf, and lighting Sabbath candles). Their religion is more likely to be focused on the home, in instructing the very young, and in enabling the men of the home to participate in their public religious roles. The very meaning of being Jewish is thus probably very different for the men than for the women. Central religious activities (e.g., attending Sabbath services at the synagogue and family worship around the Sabbath candles) also probably mean something different to each sex.[5]

A woman's religious experience and what she holds religiously most important are qualitatively different from men's religious experience and focus. Women's religion is nevertheless shaped heavily by the larger religious group because it is not a separate religion. The larger group attempts to form the individual's role through its teachings, symbols, rituals, and traditions. Thus, in the example of the Orthodox community, women's religion is strongly influenced by men's ideas of what a properly religious woman should do and be. Men interpret the Law and run the congregation and the religious courts. In most historic (i.e., official) religions, women have had considerably less power than men in establishing social definitions of appropriate gender roles. In the Roman Catholic church, at least since the time of the early church, men have held all significant positions of authority to set and interpret religious norms, practices, and beliefs. The beliefs, ritual expressions, norms, and organizational structure of official religion, in short, effectively subordinate women.

The religious legitimation of women's roles is very similar to the religious legitimation of other caste systems. As shown in Chapters 2 and 7, religion is often used to explain why certain social inequalities exist. These explanations justify both the privileges of the upper classes or castes and the relative nonprivileges of the lower ones. A **caste system** is a social arrangement in which access to power and socioeconomic benefits are fixed, typically from birth, according to certain ascribed characteristics of the individual. In the Hindu caste system, an individual of the highest caste has certain permanent prerogatives, benefits, and responsibilities merely by virtue of having been born into that caste. Medieval feudalism was a similar system, in which individuals born into a given socioeconomic stratum could move up or down relative to others of that stratum but could not aspire to the power and privileges of higher social strata.

Modern societies typically emphasize individual achievements as a basis of socioeconomic status more than traditional societies; however, elements of caste are still important. Race is one ascribed characteristic that still figures into socioeconomic status in most modernized countries (the situation of racial castes in the United States is developed in Dollard, 1949). A black man might aspire to become a doctor (i.e., an achieved status) and rise in status relative to other blacks, but his ascribed status (i.e., race) interferes significantly in his chances of achieving a high status relative to white doctors.

Gender is another major basis of caste in modern societies. Women are generally excluded from access to power and socioeconomic status available to men. The expectation that housework is "women's work" exemplifies women's caste status. No matter how great a woman's achievement in art, business, or scholarship, the society still expects that she is the one responsible, by virtue of her gender, for the menial tasks of housecleaning. Such gender-role ex-

[5]There is little empirical evidence about exactly how men and women experience religious rituals and events, but some clues appear in Myerhoff, 1978:207–241. Some parallels may also be drawn from studies of how different racial castes, socioeconomic strata, and ethnic groups differently experience similar religious rituals and belief systems.

pectations have socioeconomic implications. Employers often assume that because men who live with women are not expected to do similar amounts of work at home, women will not have as much time and energy to devote to the job; this expectation then becomes the justification for denying women opportunities for economic achievement (Glazer-Malbin, 1976). Women's status in most religious groups is also circumscribed by caste. Gender is far more important than theological or spiritual qualifications in determining whether an individual can perform certain rituals such as carrying the Torah or consecrating the communion elements.

Caste status confers necessary, but not sufficient, advantages to men's chances for recognition, power, and prestige. Not all men obtain these privileges in the social system, but maleness is virtually a prerequisite for them. Men's superior caste status does not necessarily make life comfortable and pleasurable for all individual men. Some men also suffer exploitation and discrimination—but not because of their gender. Religiously legitimated caste distinctions thus do not empower all men; they do, however, disempower all women by virtue of their gender identity (see Ruether, 1975, for an analysis of sexism as a prototype of domination in Western society).

### Religious Legitimation of Caste

**Symbolism and Myth.** Religion contributes to, or legitimates, a caste system in a number of ways. Often religious symbolism embodies a caste-stratification system. It may depict the higher deities as male and the lower deities (or even negative spiritual forces) as female. The Hindu goddess Kali represents "female illusion," symbolizing in the story of Shiva's entanglement with her the eternal struggle that men must wage against the evils of orientation toward material existence (Daly, 1973; Hoch-Smith and Spring, 1978:4).

Similar distinctions are embodied in the language of religious ritual. The ritual words of the Catholic Mass had, for centuries, stated that Christ's blood was shed "for all men." After several years of internal pressures, in 1985 the Vatican approved changing the words of the Mass to the gender-neutral term "for all" (see also Wallace, 1988). The language of Orthodox Jewish prayer makes gender-caste distinctions even more explicit. Daily morning prayer includes "Praised are you, O Lord our God, King of the Universe, who has not created me a woman." Conservative Jews have changed this wording to "Praised are you, O Lord our God, King of the Universe, for making me an Israelite"; whereas Reform Jews do not use this blessing at all (Priesand, 1975:57–60).

Some biblical language lends itself to gender-caste distinctions between men and women, although many such distinctions are derived from translations, which embody cultural assumptions of the language of translation overlaid on the distinctions made by the original writers of the many parts of the Bible. Nevertheless, besides the hierarchical, patriarchal language used to legitimate caste distinctions, both the Old and New Testaments include radically anti-hierarchical images of human relationships with the deity and each other (Ruether, 1989). Religious groups that have become relatively hierarchical or patriarchal tend to "remember" those biblical images that legitimate their status and power arrangements.

Creation stories in male-dominated cultures often assign to women the responsibility for the presence of evil or the troubles of the present world. In these myths, women's presumed characteristics of sexual allure, curiosity, gullibility, and insatiable desires are often blamed for both the problems of humankind and for women's inferior role. The Hebrew myth of Lilith describes that the Lord formed both the first man and the first woman from the ground. The first woman, Lilith, was equal to Adam in all ways, and she refused what he wanted her to do (including his sexual demands). In response to Adam's complaints, the Lord then created Eve from Adam's rib, thus making her inferior and dependent. In one version of the story, Lilith persuaded Eve to eat from the Tree of Knowledge; thus, both the "good" woman through her gullibility and the "bad" woman through her wiliness and willfulness brought about the expulsion of humankind from the Garden of Eden (see Daly, 1978:86; Goldenberg, 1974).

In general, Christianity, Judaism, and (to a lesser extent) Islam and official Buddhism allowed that women were, in principle, equal to men before God, even though they were socially unequal. Thus, unlike other religions of their time, these religions held that women might aspire to salvation or spiritual perfection as much as men. This idea was not only revolutionary but also quite tenuous, however; eight centuries after the founding of Christianity, church authorities continued to express doubts about whether the souls of women were equal to those of men (Weber, 1963:105; see also McLaughlin, 1974).

**Moral Norms.** Religion typically creates or legitimates the **moral norms** that define what is appropriate to male and female gender roles. Thus, each culture has different ideas about appropriate clothing for males and females, and religious norms of modesty often give these concepts a moral connotation. Earlier Christian missionaries in Polynesia and Africa introduced different norms for clothing (especially for women) to the native peoples not merely because they thought Western customs of dress superior but especially because, according to their moral norms, the natives' state of dress or undress was sinful.

Historically, religious moral norms have been very important in defining the appropriate gender behavior of men and women. Many moral norms of religious groups are gender-specific (i.e., separate expectations for women and men). Religion shaped norms pertaining to sexual behavior (e.g., regarding premarital and extramarital sexual activity). It influenced norms of dress, physical activities, entertainments, and drink. Religion also legitimated gender distinctions in work roles, home responsibilities, child-care responsibilities, education, marriage responsibilities, political responsibilities, and legal status.

**Organization.** Religious organization is also a framework that supports gender-caste systems. Access to positions of authority and to central ritual and symbolic roles (e.g., the priesthood) is limited or denied to women in most historic religions. Control of ritual and symbolic roles is an important source of social as well as "spiritual" power. Women's lack of power is often symbolized and communicated through their inferior ritual status (see Wallace, 1975, 1988).

**Ritual Expression.** Ritual and symbolic roles (e.g., the role of celebrant of the Mass or reader of the Torah) have become the focus of some women's dissatisfaction with their religion's treatment of women because these roles are important symbols of power and status. Most official religions have allotted women a ritual position in the home. Women light Sabbath candles, lead family rosary, arrange family altars, participate in family Bible readings and mealtime grace, and lead young children in their prayers. Women are typically responsible for arranging the observance of religious holidays in the home, including keeping a kosher kitchen at Passover, preparing feast and fast meals, and bringing home blessed palms and holy water for home use. These ritual activities are seen as consistent with her gender roles as mother and homemaker.

In public rituals, women's traditional roles have been clearly subordinate. Men have traditionally held all important leadership and symbolic roles; women were allowed to be present (under certain circumstances) but silent. Women and children were historically segregated in the congregation, sitting in separate pews or separate parts of the room from men. Sometimes they were further hidden by screens or veils. One explanation of these arrangements is that they were intended to prevent women's presence from distracting men. Another explanation is that because women are not regarded as real *persons* in the central action, they are merely allowed to observe as unobtrusively as possible. Whatever the historical background, the segregation and veiling of women clearly communicated their inferior status (Borker, 1978). In many religious groups, women were not allowed in certain sacred places (e.g., on the altar platform or behind the screen where central ritual actions occurred). Most religious groups traditionally held that women required special purification after menstruation or childbirth before they could return to participate in group worship or key rituals (e.g., receiving communion).

Religious groups are inherently conservative because they base their beliefs and practices upon the tradition or scripture produced in an earlier era. Thus, Christian groups often justify their exclusion of women from central ritual roles by reference to Saint Paul's views of the proper place of women:

> I desire then that in every place the men should pray, lifting holy hands without anger or quarreling; also that women should adorn themselves modestly and sensibly in seemly apparel, not with braided hair or gold or pearls or costly attire but by good deeds, as befits women who profess religion. Let a woman learn in silence with all submissiveness. I permit no woman to teach or to have authority over men; she is to keep silent. For Adam was formed first, then Eve; and Adam was not deceived, but the woman was deceived and became a transgressor. Yet woman will be saved through bearing children, if she continues in faith and love and holiness, with modesty (1 Tim. 2:8–15, RSV).

Similar texts exist in other historic religious traditions. Although the Buddha's approach to women's roles was revolutionary relative to Hindu tradition (e.g., he allowed women to become monks and considered them capable of aspiring to spiritual perfection), he prescribed a greatly inferior role for them in the *saṃgha* (religious order), where they were forbidden authority and leadership roles. When his favorite disciple Ānanda asked him why women should not be given the same rank and rights as men in public life, the Buddha replied, "Women, Ānanda, are hot-tempered; women, Ānanda, are jealous; women, Ānanda, are envious; women, Ānanda, are stupid" (cited in Schweitzer, 1936:95).[6]

**Changing Religious Roles for Women.** Acceptance of changes in religious roles for women frequently hinges on whether a group accepts religious traditions as literal truths and divinely ordained. Groups that emphasize orthodoxy to literally interpreted religious traditions are resistant to any kind of change. Thus, Orthodox Judaism, Eastern Orthodoxy, some strains of Roman Catholicism, orthodox Islam, and fundamentalist Protestantism generally oppose changing the ritual and symbolic roles of women. Other religious groups interpret their traditions or scriptures differently. Thus, some groups believe that Saint Paul's restrictions, as previously quoted, are merely the product of his culture's gender-role distinctions rather than part of the central message of the Christian tradition. Generally, those religious groups that are least open to change in women's roles are also more conservative toward other social changes as well. A group's approach to defining women's roles merely reflects its general attitude about the relationship of its tradition to the conditions of modern society. (We will further describe these diverse stances in Chapter 5.) For example, the turn-of-the-century controversy in many U.S. synagogues over gender-mixed seating (i.e., a whole family sitting together on a pew, in contrast to traditional segregation of women and men) was an important symbolic struggle over accommodation to wider U.S. cultural patterns (Sarna, 1986).

In many religious groups, women are pressing for changes in rituals and symbols in order to reflect their experiences as women and to change the images of women held by the whole group. For example, some Jewish women substitute new, women-affirming blessings for the traditional ones. Others have developed an annual Seder ritual that expresses women's voices in the story of the Exodus (Brozan, 1990). Similarly, in 1990 the Mormon church adapted some of its central ritual, dropping the oath of obedience to their husbands that women previously had taken in place of the men's oath to obey God and the church (Steinfels, 1990a).

Thus, the religious heritage of a group does not determine its arrangements. Many contemporary religious groups have integrated women more into the central ritual actions; most Protestant and Reform Jewish congregations allow laywomen the same privileges as laymen. Roman Catholic and Conservative Jewish congregations are more divided in their treatment of laywomen. There is some resistance in Roman Catholic churches to allowing laywomen to serve as

---

[6]No full biographies of Gautama, the historical Buddha, were written until several generations after his lifetime. Early Buddhist scripture suggests that his message was radically egalitarian, but later legends and tales, adopted into the canons of Buddhist schools, portray misogynist attitudes (Barnes, 1987).

lectors, altar servers, and Eucharistic ministers (who distribute communion, especially to homebound and hospitalized parishioners). Similar controversy exists in Conservative Jewish congregations over whether to admit women to central lay ritual roles, such as being called to the Torah (Umansky, 1985).

The issue of the ordination of women is the most controversial because of its great symbolic importance and because the role of the clergy is more powerful than lay roles. The significance of the ordination of women is that it presents an alternative image of women and an alternative definition of gender roles. Thus, the ordination of women has an impact on laypeople as well as clergy.

Women began to press for ordination in the mainline Protestant denominations beginning in the 1880s (about the same time that the Women's Movement began, demanding women's right to vote, own property, etc.). Although some women were ordained in a few denominations in the early part of the twentieth century, only after the 1950s were a substantial number of women granted ordination. The Methodist and Presbyterian churches allowed women clergy beginning in the 1950s, and in the 1970s the Lutheran and Episcopalian churches voted (after considerable controversy and some defections) to ordain women (Hargrove et al., 1985; see also James, 1980; Ruether and Keller, 1981; Ruether and McLaughlin, 1979). Despite increasing numbers of women in seminaries, women are still greatly underrepresented in the profession; the proportion of women in the ministry (4.2 percent in 1980) was less than half that of women in medicine or law (cited in Carroll et al., 1983). The proportion of women clergy in England is even smaller (see Nason-Clark, 1987a).

At the same time, studies in the United States and England show lay church members in mainline Protestant and Catholic churches to be generally positive about women in the ministry (Hoge, 1987; Lehman, 1985, 1987; Nason-Clark, 1987a). Laypersons' opposition to the ordination of women in the Roman Catholic church has dropped dramatically since Vatican II to only 35 percent of U.S. Catholics (*N.Y. Times*/CBS Poll, reported in Berger, 1985). The Catholic church retains its ban on women clergy, however. In 1976 the Vatican issued a "Declaration on the Question of the Admission of Women to the Ministerial Priesthood," basing continued refusal to ordain women on theological grounds that the sacramental expression of Christ's role in the Eucharist required the "natural resemblance" of Christ to the (male) minister (Sacred Congregation, 1977). Although Catholic theologians in the United States and elsewhere roundly ridiculed the theological basis of the document, it has been supported at the international level of church organization (Iadarola, 1985). National groups of bishops (many of whom individually condemn sexism) have not collectively challenged the policy (Steinfels, 1990b). Pope John Paul II reiterated the restriction during his visit to the United States in 1979, and the 1985 extraordinary Synod of Bishops in Rome carefully avoided the issue.

Nevertheless, the experience of women who have been ordained so far (e.g., in the Presbyterian and United Methodist churches and in Reform Judaism) is far from equality. Women are underemployed, paid lower salaries, and are less likely to be considered for the better positions. A church may employ a woman as "assistant minister" with special responsibility for youth programs. Although she may occasionally lead worship services, she is not seen as having an important leadership or decision-making role; being in charge of youth activities is consistent with traditional roles of women (Carroll et al., 1983; Umansky, 1985; Verdesi, 1976).

Women who have had the opportunity to serve in pastoral capacities have often redefined those roles; their style is generally more relational, less authoritarian, and more democratic, involving the laity in leadership (Nason-Clark, 1987a; see also Adriance, 1990, on women religious and pastoral roles in Brazil). Although the Catholic church does not allow women to be ordained, in many regions nuns and laywomen have taken over many functions of the clergy (due to insufficient numbers of ordained clergy). Similar redefinition of leadership roles appears among these women pastors, who develop collaborative relationships with parishioners rather than hierarchical ones (Wallace, 1991).

Resistance to women's roles as clergy or ritual leaders is not due merely to the weight of traditional beliefs and practices. Some groups are afraid of harm to the local congregation or denomination that might be caused by internal controversy over the issue (Lehman, 1985). Other organizational factors figure into the resistance to hiring ordained women clergy as head or sole minister. For example, many churches traditionally assumed that by hiring a married minister they would also receive the (unpaid) labor of his wife. If they hired a married woman minister, in all likelihood her husband would have a career of his own to pursue. In some denominations, fellow clergy appear to be more of an organizational barrier to women's ordination and hiring than are laypersons. For example, in the Church of England (Episcopalian), the opposition has come from the House of Clergy, rather than the House of Bishops or House of Laity. One study of Church of England clergy found that these attitudes were more strongly correlated with people's notions about women's proper roles in family and society than with any theological or scriptural foundation (Nason-Clark, 1987b). In those denominations that have ordained women, however, women clergy have been increasingly accepted by their congregations and colleagues (Carroll et al., 1983).

### Power and Sexuality

Evidence from relatively simple societies suggests that religious distinctions between males and females accompany social distinctions between them, especially in the division of labor. An interplay of power is involved in establishing social inequalities. Religion and human sexuality are both important sources of power, partly because of the mystery, awe, and intensity of the experiences they engender. Those who possess religious power in a social group frequently attempt to control the use of sexual power because they view it as a threat to their power base. Similarly, on the individual level, many religions view the use of sexual power as diminishing or distracting from spiritual or religious power. The regimen of the 3HO movement (Happy-Healthy-Holy—a Western version of Sikh/Hindu religion) severely restricts the frequency of sexual intercourse of members, not because members believe there is anything intrinsically wrong with intercourse but because sexual activity is believed to drain the individual's spiritual energy (Gardner, 1978:120–133; Tobey, 1976).

Female sexuality has historically been perceived as a source of dangerous power to be feared, purified, controlled, and occasionally destroyed by men. General ideas of carnal female sexuality have existed for centuries. Some Western cultural forms of these ideas include images of the enchantress, the evil seductress, the movie queen, and the prostitute. Other cultural images of dangerous female power include the witch, the nagging wife, and the overpowering mother (Hoch-Smith and Spring, 1978:3).

There are several plausible hypotheses for such emphasis on the powers of female sexuality. Douglas (1966) suggests that female sexuality is perceived by males as dangerous and impure because it symbolically represents females' power to counteract the dominance of men. Societies in which there is ambiguity about the legitimate sources of male dominance over females are more likely to develop beliefs in sexual pollution by females than are societies in which women's position is unambiguous; that is, if formal and informal social power is unambiguous (whether or not a society has established a subordinate position for women), men are less likely to fear sexual pollution. Fear of female sexuality is particularly characteristic of societies in which male dominance over females is a primary feature of social organization. The Pygmies, for example, have a division of labor and power structure in which gender is relatively unimportant and have little concern over sexual pollution (Friedl, 1975; Sacks, 1974).

Douglas (1966) cites the contrasting example of the Lele tribe of Zaire. The men of this tribe use women as something of a currency of power, by which they claim and settle disputes. They fight over women and use their dominance over women as a symbol of their status relative to other men. Women of the tribe challenge total male dominance by playing off one man against another and by manipulating men. The men of the Lele have considerable fear of pollution concerning sex and menstrual blood. Women are considered dangerous to male activities, and contact or intercourse with a menstruating woman is believed to be contaminating. Fear of the power of female sexuality is related to the disruptive role that women sometimes play, upsetting the order of a male-dominated system of rewards and privileges, in the society. Female sexuality is therefore symbolized as chaotic, disordering, and evil.

Another feared power related to female sexuality is regeneration—the power of life and death. Many cultures observe with awe and some fear the power of female sexuality to create and sustain new life and, by implication, the dangers of female renunciation of this role. For example, the celibate Hindu goddesses are those with feared malevolent powers, whereas goddesses with male consorts are not so dangerous because their power *(sákti)* can be regenerative and benevolent (Marglin, 1985). The association of female sexuality with fertility, regeneration, and the power of birth makes it both dangerous and potentially fruitful. Thus, religious rituals and social sanctions try to control this source of power.

### Sexuality and Religious Images of Women

The theme of sexual pollution is prevalent in more developed, historic religions as well. The dominant religious images of women in Western societies are built on a dualistic image of female sexuality: The evil side is the temptress/seducer/polluter, and the approved side is the virgin/chaste bride/mother. Religiously approved roles for women are rather narrowly limited to expressions of these images and are defined and legitimated by reference to Scripture, tradition, and role models (e.g., biblical women or saints). Approved roles emulate the motherhood of a biblical Sarah or Mary, the wifely faithfulness of a Rebekah or Saint Elizabeth of Hungary, or the virginal chastity of Mary or Saint Agatha (sources of female images in the Judeo-Christian heritage are explored in Bynum, 1986; Gold, 1985; Ruether, 1974; Hyman, 1973). In nineteenth-century Catholic piety, Mary was elevated to considerable importance—for example, as a source of miracles and prophetic apparitions. Her power, however, derived from attributes of motherhood and virginity, not fertility or sexuality (which figured into some earlier Mary cults). The promulgation of the controversial dogma of the Immaculate Conception (1854) promoted the new piety for a religious image of Mary as a pure and passive vessel (Pope, 1985).

The narrowly sexual base of these dominant symbols of femaleness contrasts dramatically with representations of men in ritual and imagery. Whether a man is a virgin matters little in his religious status or participation in ritual. Women are represented as one-dimensional characters in ritual and imagery; their images are based almost exclusively on sexual function (Hoch-Smith and Spring, 1978:2–7). The Roman Catholic church honors a number of both female and male saints by dedicating the Masses of certain days to their memory and commemoration of their particular virtues. Men are remembered as martyrs, popes, bishops, doctors of the church (i.e., recognized scholars and teachers), kings, abbots, and confessors of the faith (i.e., typically missionaries). Women are remembered as virgins, virgin martyrs, martyrs, holy women, and queens. Saint Agatha is one woman so commemorated because she was tortured and killed for resisting the sexual advances of the Roman governor. The prayer said on a virgin martyr's feast day reads: "O God, one of the marvelous examples of your power was granting the victory of martyrdom even to delicate womanhood" (Maryknoll Fathers, 1957:807; recent changes have altered the significance of these commemorations but not the relative treatment of men and women). Even women renowned for their organizational skills, intellectual ability, literary accomplishments, and spirituality (e.g., Saint Theresa of Avila, a famous mystic who also brought about the reform of her order and administered its convents) are venerated as "virgins."

### Women's Religious Identity

**Sense of Self-Worth.** Such imagery serves to reinforce the distinctions of the gender-caste system. On the one hand, it places some value on "good" women's roles. Such symbolism can enhance the individual's sense of self-worth by showing her existence as part or imitative of some larger reality. She may think,

"Although my position is lowly, it is part of something bigger"—a larger cosmic picture. Religion often explicitly links spiritual rewards with the fulfilling of caste obligations. A Hindu woman learns that she may be rewarded in the next life with a higher status (perhaps as a man) if she obediently fulfills the role requirements of a woman in this life.

On the other hand, although the religious legitimation of women's roles gives meaning to their subordinate position, it does not necessarily make them happy or give them a positive identity. In the Sinhalese religion in Sri Lanka, for example, both the Buddhist doctrine and the folk religious imagery describe women as vehicles for impermanence and sorrow. The law of karma holds that all actions and thoughts have results in the future, including future rebirths of the individual. The desire for life and the illusions of the world perpetuate rebirths and thus suffering. Suffering refers not only to pain but also to the impermanence, emptiness, and insubstantiality of this existence. Buddha taught that escape from this suffering is possible only by detachment from desire, especially for life in this world.

The Buddhist imagery of women portrays them as particularly caught in this web of rebirth. Because women represent birth, they are themselves a symbol of the force of karma that maintains suffering in the world. They are also seen as the source of men's karma because they are among the things that men most desire. Yet women's desirability (i.e., sex appeal) is as impermanent and illusory as the rest of this world. Thus, women are especially entangled in the web of karma and have greater difficulty detaching themselves (AmaraSingham, 1978). In Buddhism, as in Western religions, women are used to symbolize to men and women alike the undesirable qualities of human existence. Women represent suffering and evil.

We know very little about the actual impact of these perspectives on women's identities. Such imagery and symbolism surely influence how a woman feels about herself. Does she, by participating in such a religion and culture, come to perceive herself as inferior and impure, more prone to sin, and a cause of males' sin and suffering? If participation in such rituals and symbols does not lower a woman's self-image, it must be because she is able to separate herself from them or experience this part of religion in some unofficial way. Do women selectively relate their identities with different aspects of rituals or with different rituals altogether? Do they use gender-linked symbols differently from men?

One study of the writings of religious people during the later Middle Ages—a period rife with misogynistic theological, philosophical, and scientific ideas—found that female and male spiritual leaders used gender images very differently. Male spiritual writers were more likely to view the genders as dichotomous, whereas women writers were more likely to use androgenous images, if they addressed gender at all. Men's spiritual ideals focused on conversion and renunciation of the world of wealth and power (for which they invoked gender inversions to symbolize giving up those statuses), whereas women's religiosity emphasized continuity and a relationship to Christ's humanity (and thus physicality). Gender distinctions created genuine restrictions on how women could express their opinions and experiences or achieve heights of spiritual "achievement." For example, both men and women sought renunciation of "the world," but women did not control enough power or wealth to give them up. Thus, women's renunciation was often limited to giving up food or other aspects of daily life over which they did have control. Nevertheless, the religiosity of these outstanding female medieval theologians and mystics was not merely the passive product of the misogynistic ideas of their time. Rather, they appropriated cultural symbols differently, creating a positive spiritual identity for themselves and their followers (Bynum, 1986).

At the same time, religious traditions have provided women with some positive roles and images (albeit severely limited ones), which bring reward and recognition in their fulfillment. For a woman to aspire to achievements in men's arenas would be deviant and punished by the social group. She may, however, aspire to achievements in women's spheres with no competition from men and be rewarded. These values, too, are internalized. The woman who is successful in fulfilling the mother role comes to perceive herself positively in those terms. She feels good about herself because she has achieved a desired status or has performed valued roles.

In Victorian society, the cultural and religious legitimations of motherhood and women's role in the home achieved a new synthesis. The Victorian ideal of womanhood, as reflected in women's magazines and public statements in the nineteenth century, was one of "piety, purity, submissiveness, and domesticity" (Welter, 1966).[7] The emerging capitalist-industrialist society of the nineteenth century made both the home and religion increasingly marginal to the arenas of power and production. Ideological responses of many religious organizations and revivalists to this privatization struck a responsive chord among middle-class women who were enmeshed in the privatization of the home. An "evangelical domesticity" resulted that emphasized the home and women as the primary vehicles of redemption. The rhetoric of preaching and the themes of gospel hymns of this period portrayed the home as the bastion of tranquility in a turbulent world. Woman's image was that of the keeper of this refuge who, through her piety and purity, was the foremost vehicle of redemption, in opposition to the aggressiveness and competition of the public sphere (Sizer, 1979; see also McDonnell, 1986; Ryan, 1983; Taves, 1986; Welter, 1976). Thus, the religious ideology of evangelical domesticity celebrated the privatization of the home and women's roles, further reinforcing the caste distinctions between men and women. At the same time, it enhanced this narrow role in women's eyes by claiming redemptive significance for it.

**Sanctions for Deviance.** The religious definition of gender roles is supported by religious sanctions (i.e., punishments) for deviance from them. Men as well as women are subject to negative sanctions for failure to fulfill gender-

[7]The instance of Victorian gender roles illustrates the fact that men's roles too are often limiting and burdensome. In this case, men were given the full burden of providing for the family by work away from the "comforts of home." The moral expectations of the husband/father as sole provider for a family are relatively recent in Western history.

role expectations, though conformity to the broader, more powerful male social role is less onerous than complying with the restrictive, less powerful female role. Religion, as we shall see in Chapter 7, is a powerful force for social control. One of the most effective sources of such social control is the individual's own conscience. The person who has been socialized into a certain way of thinking about gender roles feels guilty if he or she fails to fulfill these expectations. One young woman who had learned that "women naturally and instinctively feel great affection for their children" felt guilty that her initial reaction to her newborn child was not a great outpouring of maternal love.

Those who control religious power (especially in official religion) frequently use specifically religious sanctions to enforce gender roles. Recent history provides the examples of a Mormon woman excommunicated for her active support of the Equal Rights Amendment to the Constitution; and Roman Catholic priests and nuns silenced by Vatican officials for challenges to the church's treatment of women (among other issues). In Iran, the political power of fundamentalist Islam has resulted in strong restrictions of women's public roles and appearances, as symbolized by the mandate of returning to the *chador* (i.e., veil).

Historically, the intertwined power of religious organizations with other social institutions gave religion more influence in the control of deviance than in modern society. The persecution of witches in the Middle Ages (and until as recently as the end of the seventeenth century) was focused primarily on women, who were presumed weaker and therefore more vulnerable to the influences of the devil. Authorities typically singled out women who deviated from religiously established norms for females—wise women, healers, midwives. Suspected women (e.g., widows and "spinsters") were disproportionately likely to be relatively independent of men. Witchhunts such as the massive medieval persecutions were primarily concerned with purification. Women in particular were considered threats to social purity because of cultural assumptions about their sexuality and because anomalous (i.e., independent) women were perceived as dangerous and disruptive. A substantial number of women were put to death or tortured as witches; scholarly estimates suggest that several hundred thousand witches were killed during the Inquisition.[8] There has been no parallel persecution of men *as men*, although men were persecuted as members of other minorities (e.g., Jews).

### Alternative Definitions of Women's Status and Roles

**Women's Religious Groups.** One of the most effective social controls of women has been the limitation and prevention of separate, autonomous

---

[8]Figures vary greatly from 30,000 to 9 million (sources cited in Daly, 1978:183). See especially Daly's essay on European witchhunts, as well as Szasz (1970); and Anderson and Gordon (1978:2). Daly notes the widespread religious legitimation of violence against women. The Indian *suttee*—cremation of the living wife along with the body of the dead husband—and African clitoral circumcisions are among the ritual "gynocides" examined in her provocative book. She notes that male-dominated scholarship has not shown as much concern over such historical events as the European witchcraze and institution of the *suttee* as over comparable genocides such as the Nazi Holocaust, observing that scholars have glossed over gynocide such as the *suttee* by calling it a "custom." Daly doubts that they would ever refer to comparable atrocities in the Holocaust as "Nazi customs."

women's religious groups. Religion has historically had the potential for promoting social change among colonial peoples and subordinated racial and ethnic groups because, in organizing around a religious focus, such groups have often mobilized to press their social and economic concerns. Unlike racial and ethnic minorities, however, women have not generally formed separate religious groups in Western societies. Male-dominated religious organizations have been understandably threatened by the possible power of autonomous women's groups.

Medieval sisterhoods (including religious orders, lay orders, and numerous unofficial bands of religious women) offered socially approved roles for women who could not or would not fit the primary gender roles of woman as wife and mother. Socially anomalous women (again, typically widows and unmarried women) found in these sisterhoods both security and opportunities for achievement not available to other women in society. Medieval abbesses, for example, were often relatively powerful leaders of large organizations. Church officials, however, kept these groups under relatively tight control (e.g., by regulations restricting sisters' freedom to come and go in "the world"). Like other dissenting, change-oriented groups (e.g., early Franciscans), women's groups that asserted alternative roles were often coopted and controlled.

Feminism had a specific role in the development of sisterhoods in the Anglican church during the nineteenth century (M. Hill, 1973a). The Victorian image of women, as we have seen, emphasized domesticity. Furthermore, the middle- and upper-class Victorian "lady" was educated to be an ornament in the home; her enforced leisure was evidence of her husband's social status. Women such as Florence Nightingale who asserted any other role for themselves were severely criticized. The Anglican church had dissolved monasteries and convents in its initial break with the Roman church in the sixteenth century, and in the nineteenth century strong voices were raised against re-establishing brotherhoods or sisterhoods in the church. Nevertheless nascent feminism, combined with a considerable demographic surplus of women, exerted pressures to create acceptable social roles for women who would work in "unladylike" jobs such as nursing, running settlement houses, teaching the poor, and running hospices for derelicts. The creation of acceptable deviant roles for women was perceived as an attack on the Victorian home and women's proper duties. One bishop stated:

> The rules which I have myself laid down as most necessary in my dealing with such communities [the sisterhood] have been the following:—To point out that the first of all duties are those which we owe to our family. Family ties are imposed direct by God. If family duties are overlooked, God's blessing can never be expected on any efforts which we make for His Church. Every community, therefore, of Sisters or Deaconesses ought to consist of persons who have fully satisfied all family obligations (cited in M. Hill, 1973a:277).

Even among those who approved of the idea of women's religious groups, there was disagreement about the degree of autonomy to allow them. Religious sisterhoods were much more problematic than brotherhoods to church authorities because sisterhoods allowed the possibility of women directing their own work and running their own organizations. The less radical alternative of the office of deaconess (in which women workers would be under the direct supervision of male clergy) was proposed because "women need more support than men" (cited in M. Hill, 1973a:276). Women were gradually allowed into "unladylike" areas of work, partly through the transitional institution of the sisterhood. The potential for feminist protest was, however, somewhat coopted, especially as women's groups were brought under greater control by authorities and their work was gradually redefined as "women's work." Nursing, for example, was consistent with the serving and nurturant image of women as dependent on doctors (male) for decision making. As women's professions became more respectable, women's religious groups became less important as vehicles for feminist aspirations.

Specifically, women's religious groups still remain potentially expressive of women's social and religious interests. Within Roman Catholicism, much of the pressure for the ordination of women comes from the ranks of women's religious orders. Some women's groups have creatively used their greater education and structural independence. But the main arena for the expression of feminist concerns has become public-sphere organizations. The privatization of religion means that religious achievements are similar to domestic accomplishments: Neither really "counts" as a symbol of achievement in the larger society. Conflicts within religious collectivities over appropriate roles for women primarily result from religion's tendency to continue legitimating traditional roles for women in *both* public and private spheres.

**New Religious Movements.** New religious movements are often a vehicle for the assertion of alternative religious roles for women. Religious movements of nonprivileged classes have typically allotted equality to women, at least in their formative years (Weber, 1963:104–105). Many of the medieval millenarian movements initially allowed women virtual equality. The Protestant Reformation's principle of the "priesthood of all believers" was initially interpreted as opening important roles to both laywomen and laymen. Many religious movements of the eighteenth century, and later the Great Awakening, granted equality to women. The Shakers believed that their founder, Mother Ann Lee, was the Second Coming of Christ, thus completing the manifestations of Christ (male and female). The Society of Friends (Quakers) was emphatically egalitarian, encouraging women to equal education, leadership, and public-speaking roles. Quaker women, such as Susan B. Anthony, Alice Paul, and Lucretia Mott, were leaders in the women's suffrage movement; they were also prominent in movements for abolition of slavery, prison reform, fair treatment of Native Americans, and other human rights causes (Bacon, 1986). The late Victorian period was one of nascent feminism, and alternative women's roles were an important part of the beliefs and practices of Pentecostal and Holiness religion, Salvation Army, Seventh Day Adventists, Spiritualism, Christian Science, Theosophy, and New Thought (Bedharowski, 1983; Braude, 1985, 1989; Robertson, 1970:188, 189).

New religious movements are more amenable to alternative gender roles because they are based on an alternative source of authority. Traditional ways of doing things are protected by the established religious collectivity, but new religious movements are often based on charismatic authority (see Chapter 7), which is not bound by tradition. The charismatic leader says, in effect, "You have heard it said that . . ., but I say to you . . . ." This new source of authority allows a break from tradition. Weber (1963:104) points out that as the emphasis on charisma fades and the movement becomes established, these movements tend to react against keeping women in roles of authority, typically redefining women's claims to charismatic or other authority as inappropriate or dishonorable behavior. This process is illustrated in the contemporary pentecostal movement among Roman Catholics. These groups believe that the Holy Spirit speaks and works directly through human vehicles, men or women. Initially, women were generally equal to men in the movement, but as the movement became larger and more organized, much pressure was exerted for women to become subordinate to men. The "gifts of the Spirit" that men claimed to have received began to be different and more powerful than women's. Some (male) leaders of the movement borrowed from their conservative Protestant counterparts a theology of "headship" to legitimate the developing subordinate role of women (Fichter, 1975; McGuire, 1982).

The role of women in developing religious movements is thus a result of alternative sources of authority and power. Many of these movements are understandably appealing to women who are excluded from status and privilege in the dominant religions. Weber (1963:104) accounts for some of the appeal of Christianity and Buddhism by noting that relative to the religious competitors of their times and regions, their subordination of women was less extreme. Similarly, the religious movements of the late Victorian era appealed to women because of their alternate definitions of women's roles.

Several emerging religious movements also focus on gender roles but generally reassert traditional rather than new ones. Thus, the Jesus People, neo-pentecostal movements, evangelicalism, Hare Krishna (Krsna Consciousness), and the Unification Church (of Reverend Moon) define women's roles very conservatively. One study found that many women who "converted" to Orthodox Judaism wanted a traditional religion that offered unambiguous and unchanging gender-role expectations and legitimated their desires for a family (Davidman, 1990b). Thus, some newer religious movements may attract women precisely because they reassert traditional, nonmodern gender roles, whereas other religious movements attract other women because they offer totally new ways of thinking about gender and spirituality.

Another emerging religious movement, the goddess movement, explicitly redefines women's spiritual, social, emotional, and physical roles. The goddess is an image, or symbol, of the divine that is contrary to the image used by patriarchal religions (such as much of the Judeo-Christian heritage). Although much of the goddess movement is self-consciously outside traditional religious groups, some adherents consider themselves Christians or Jews who are seeking new bases for spirituality in their traditions. Some theologians and groups use the conception of the goddess to balance the gender of the deity as both male and female, often combining the idea with a call to a nonhierarchical religion. Others emphasize the goddess as a basis for women's power. One theologian writes:

> The basic notion behind ritual magic and spellcasting is energy as power. Here the Goddess is a center or focus of power and energy: she is the personification of the energy that flows between beings in the natural and human worlds (Christ, 1987:127).

Some groups take the image of goddess literally, trying to enact older pagan practices dedicated to goddess worship. Others use earlier pagan goddess-imagery more loosely, eclectically weaving old symbols and practices together with new ones to create a new, nonpatriarchal women's spirituality (Neitz, 1990; see also Jacobs, 1989b, 1990).

The availability of nonreligious modes of asserting new gender roles may have eclipsed the religious modes of dissent, however. Even for many religious feminists, religious organizations are not "where the action is." The reassertion of traditional gender roles in some contemporary religious groups, such as fundamentalist movements in Christianity, Judaism, and Islam, may be a reaction against these nonreligious feminist movements and their attempt to remove gender issues from the purview of religious authority.

Can official religion create or encompass new or changing roles for women? Sociological analysis suggests the answer is a qualified yes. On the one hand, religious images and symbols are elastic; they can be changed and reinterpreted. On the other hand, traditional images and symbols of women are generally negative and resistant to change. Because much of the power of symbols and ritual is in their reference to tradition, they are essentially conservative. Although religious movements are able to express women's dissent and create new social roles, religious movements have historically returned to traditional, hierarchical, or bureaucratic forms of authority as they become settled—and in so doing have reverted to less innovative and more submissive roles for women. Women's religious groups have the potential of focusing and asserting their social concerns; yet they have been historically controlled and coopted by (male) authorities of the dominant religion.

**Nonofficial Religion.** While the dominant religious groups have allocated women few if any important ritual roles, women have often found important symbolic roles outside the approved religious structure. Women have been prominent as healers, mediums, and midwives. These roles often paralleled approved religious or medical roles and functions. Just as men have been prominent in official religion as it became a differentiated institution, women have been important in "common" or nonofficial religion. Nonofficial religion has provided women with a *chance to express their own specific concerns for meaning and belonging.* Midwifery, for example, has traditionally been a highly spiritual role (see Dougherty, 1978). In nonofficial religion, women have opportunities for leadership and power, as exemplified by the status of women mediums, faith healers, and astrologers (Bednarowski, 1983; Braude, 1985, 1989; Haywood, 1983). Some nonofficial religion is organized as specifically feminist dissent (e.g., some contemporary witchcraft covens). Much of the nonofficial religion discussed in this chapter may be understood as a way in which individuals assert alternative definitions of reality. Women throughout the ages have asserted alternative roles for themselves through religious expression.

### Conclusion

Religious groups' treatment of women's roles and women's sexuality is essentially an issue of *power*. If women obtained greater power than they have traditionally had, they could redefine themselves religiously and socially. On the other hand, religious legitimations and organizations presently work against their obtainment of that power.

This Extended Application, while focusing on women's roles, illustrates more broadly how religion comes to embody and further legitimate caste stratification and power. Our analysis suggests that nonofficial religion sometimes represents one form of counterassertion (albeit often incoherent and fragmented) of power and self-worth by those excluded from power in official religion. If this is true, we need to pay more serious attention to nonofficial religion and its complex relationship with official religion.

## Summary

This chapter has contrasted official religion and its characteristic forms of individual religiosity with nonofficial religion and religiosity. Official religion is a set of beliefs and practices that are prescribed, regulated, and socialized by organized, specifically religious groups. Nonofficial religion, by contrast, is a set of religious and quasi-religious beliefs and practices that are not accepted, recognized, or controlled by official religious groups. Official religion is relatively coherent and organized, but nonofficial religion includes an assortment of unorganized, inconsistent, and heterogeneous beliefs and practices.

The official model of religion is clearly identifiable. Its elements—doctrine, cultic expression, ethics, and organization—are mediated by religious institutions. These institutions consolidate and control the religion's definitions of what it stands for. The relative specificity of these official models of religion enables researchers to develop measurable criteria of religiosity. Such measures are, however, bound to the orthodox official models of religion, making them of limited use in understanding religious change and nonofficial religion. Nonofficial religion is more widespread than generally recognized, and it overlaps official religion in both content and adherents.

The Extended Application demonstrates how religion influences women's gender roles and identities. Religious organizational structure, myths, and symbols embody and legitimate subordinate caste status for females—a point of conflict in many religious groups today. As the Extended Application suggests, some aspects of nonofficial religion involve ways by which persons who are unrecognized or even disempowered by official religion have asserted themselves. Thus, nonofficial religion and religiosity are both serious and relevant to our understanding of religion as a whole.

## Recommended Readings

### Articles
The following articles tap some of the methodological issues discussed in this chapter:

Charles Glock. "On the Study of Religious Commitment." Research Supplement to *Religious Education* 57(4), 1962; reprinted in Glock and Stark, *Religion and Society in Tension* (Rand McNally: Chicago, 1965).

Mary Jo Neitz. "In Goddess We Trust." In *In Gods We Trust: New Patterns of Religious Pluralism in America*, T. Robbins and D. Anthony, eds. New Brunswick, N.J.: Transaction, 1990, pp. 353–372.

Wade Clark Roof. "Concepts and Indicators of Religious Commitment: A Critical Review." In *The Religious Dimension: New Directions in Quantitative Research*, R. Wuthnow, ed. New York: Academic, 1979, pp. 17–45.

Joseph Ryan. "Ethnoscience and Problems of Method in the Social Scientific Study of Religion." *Sociological Analysis* 39(3), 1978:241–249.

### Books—Nonfiction

Hans A. Baer. *The Black Spiritual Movement: A Religious Response to Racism*. Knoxville: University of Tennessee Press, 1984. An ethnography of a widespread syncretic religion in the context of U.S. urbanization and racism.

Michael H. Ducey. *Sunday Morning: Aspects of Urban Ritual*. New York: Free Press, 1977. An analysis of the ritual of Sunday morning worship services of four churches (Evangelical United Brethren, United Church of Christ, Roman Catholic, and Missouri Synod Lutheran); especially useful for appreciating the implication of ritual action, patterns of authority, power relations, and tension with the dominant society.

Phillip E. Hammond, ed. *The Sacred in a Secular Age*. Berkeley: University of California Press, 1985. This collection, organized under the auspices of the Society for the Scientific Study of Religion, consists of short essays and useful bibliographies on most of the major topics describing contemporary religion.

Wade Clark Roof and William McKinney. *American Mainline Religion: Its Changing Shape and Future*. New Brunswick, N.J.: Rutgers University Press, 1987. Using survey and historical data, this book describes the socioeconomic location, demography, contemporary moral issues, and social changes of mainline Protestant denominations in the United States.

### Books—Fiction

Margaret Atwood. *Handmaid's Tale*. New York: Fawcett Crest, 1985. A dystopian vision of a society governed by a fundamentalist theocracy that has drastically reshaped gender and family roles.

# CHAPTER FIVE

# The Dynamics of Religious Collectivities

Religious groups have a variety of organizational forms. Some have loose boundaries and relaxed authority structures; others have clear-cut boundaries and rigid authority structures. Some groups are large and relatively anonymous; others are small and face-to-face. Many collectivities expect their members to organize their day-to-day life around their religious commitment, whereas many others expect only partial and segmental commitment. The task of classifying organizational patterns is not an end in itself, but understanding the varieties of religious collectivities, as they change in time, helps us analyze the internal dynamics of a religious group and its relationship with the larger society.

## Types of Religious Organization

Weber (1963:65) first introduced the sociological concept of **sect,** referring to an association that accepts only religiously qualified persons. Weber was especially interested in how different kinds of religious collectivities were related to social change. The religious community of Jesus and his followers, for example, was qualitatively different from the religious community within the established Jewish religion of his time. Weber noted that while charismatic leadership frequently stimulated social change, the process of trying to maintain and defend innovation typically resulted in routinization. **Routinization of charisma** is the process by which the dynamism of charismatic leadership is translated into the stability of traditional or bureaucratic organization. The very organization of a community of followers is thus part of this process of routinization (Weber, 1963:60, 61; see also Weber's essay "Sect, Church and Democracy," 1968:1204).

Weber's distinction between church and sect was amplified by his contemporary Ernst Troeltsch, a theologian and church historian. Troeltsch was interested in tracing the social impact of certain ideas or "teachings" of Christian churches in Europe. He thought that three types of religious organization were implicit in Christian teachings, and a given group developed along one of these three lines depending upon its relationship to its social environment (Steeman, 1975:202–203). Troeltsch's three types were church, sect, and mysticism. His application, however, focused mainly on church and sect as a dichotomy. The residual concept of mysticism reflected important insights but is not properly a separate type of organization because mysticism exists within all other types of collectivity.

### The Development of the Church–Sect Dichotomy

Troeltsch characterized the **church** as a type of organization that is essentially conservative of the social order and is accommodated to the secular world. Its membership is not exclusive but incorporates the masses. The **sect,** by contrast, is exclusive, aspiring to personal perfection and direct fellowship among members. The church is characteristically an integral part of the social order, whereas the sect stands apart from society in indifference or hostility. Thus, the church tends to be associated with the interests of dominant classes, whereas the sect is connected with subordinate classes. The church has an objective institutional character. One is born into it, and it mediates the divine to its members. By contrast, the sect is a voluntary community of fellowship and service; relationships with fellow members and the divine are more direct. According to Troeltsch, both types are logical results of Christian beliefs.

The church type, exemplified by the Roman Catholic church of the thirteenth century, is of diminishing significance. The medieval synthesis of church and society was undermined, partly because of the challenges of the sects. Later church-type organizations like Calvinism in Geneva or the free churches (e.g., Congregationalism) in England and the United States achieved considerable synthesis with their societies, but the synthesis was based on a general Protestant consensus rather than upon a monopoly of legitimacy (Steeman, 1975:194).

Troeltsch (1960:381) forecast that modern society would be increasingly characterized by his third type of religious association: "idealistic mysticism" or "spiritual religion" (see Garrett, 1975). This is essentially a radical religious individualism. It lays no emphasis on the relations between believers; any association among them is based on a "parallelism of spontaneous religious personalities." Such groups are indifferent to sacraments, dogmas, ethical norms, and organization. Mysticism tends to resist control by authorities and is thus a threat to organized religious forms. Nevertheless, because it tends to be uninterested in changing the church or the world, it does not have the social impact of sect-type groups. While Troeltsch developed the category of "idealistic mysticism" to describe a particular historical phenomenon—eighteenth- and nineteenth-century German Lutheran mysticism—he suggested that it also described the religion of contemporary educated classes. He viewed this trend with pessimism because he considered some organizational and historical continuity necessary to the continued existence of religion (Campbell, 1978; Troeltsch, 1960:798, 799).

Troeltsch's categorization was intended primarily for comparing and contrasting concrete historical developments rather than for creating a classification scheme. Later formulations, built uncritically on Troeltsch's, inherited the historical limitations of his focus. Thus, in Troeltsch's scheme, the church-type and sect-type are polarized extremes, reflecting the historical situation of late medieval Christendom, in which class structure was generally polarized and religious institutions were relatively undifferentiated from economic and political ones. This lack of differentiation made it possible for religious dissent to be simultaneously politico-economic dissent. Any attempt to apply this dichotomy to modern societies stumbles because the modern situation is vastly more complex (Robertson, 1970:117).

### The Development of Other Types

This difficulty of applying the church–sect dichotomy to the U.S. situation prompted Niebuhr (1929:25) to add the type **denomination**, which was characterized by accommodation to society comparable to that of the church but lacking the ability or intention to dominate society. Niebuhr's main interest was the process by which sects were transformed into other organizational forms. To him, the initial enthusiasm of the sect appeared limited to the first generation (i.e., to those who consciously chose to belong). After the initial fervor, the sect must either compromise with the "world" or be organizationally weakened. A sect that compromises moves toward a denominational stance—organized religion accommodated to society. Niebuhr's typology was less neutral than Troeltsch's. As a theologian, Niebuhr used the schema to criticize some religious groups for their accommodation to the stratification system—a symptom of secularization and divisiveness, in his view (Martin's 1962 clarification of the concept of denomination eliminates much of the value-laden quality of this schema).

Further clarifications were added by Becker (1932) who developed the category **cult**. The cult is characteristically a loose association of persons with a private, eclectic religiosity. The concept of "cult" is very useful sociologically because it describes the form of social organization of much popular (nonofficial) religion. Historically, many cults arose from the popular beliefs about particular local holy persons and places. For example, St. Nectarios of Greece, a bishop on the island of Aighina, was widely recognized as a holy man even before his death in 1920. After his death, his veneration gradually became nationwide and numerous miracles were attributed to him. By the end of the twentieth century, his icons could be found in houses, hospitals, cars, and buses throughout the country, and his shrine on Aighina had become the pilgrimage site for hundreds of thousands of devotees. Widespread public and private devotion to this saint *preceded* his canonization by the Orthodox Patriarchate in 1961 (Kokosalakis, 1986).

Although many cult devotions historically took place in a group (e.g., a whole village making an annual pilgrimage), they tended to be highly individualistic. Participants sought personal intentions, and the desired religious experience was not expected to bind members into any long-lasting sense of community. Structurally, cults and cultic devotions, such as pilgrimages and healing rituals, could be threatening to church authorities' control. Analyzing several historical pilgrimages, Turner and Turner (1978:32) observe:

> . . . there is something inveterately populist, anarchical, even anticlerical, about pilgrimages in their very essence. They have at times been linked with popular nationalism, with peasant and anticolonial revolt, and with popular millenarianism. They tend to arise spontaneously, on the report that some miracle or apparition has occurred at a particular place, not always a place previously consecrated. Pilgrimages are an expression of the communitas dimension of any society, the spontaneity of interrelatedness, the spirit which bloweth where it listeth. From the point of view of those who control and maintain the social structure, all manifestations of communitas, sacred or profane, are potentially subversive.

The response of the authorities was, if not to suppress the cult, then to routinize its practices, for example, by putting church officials in charge of shrines and by inserting approved rituals of penance and Eucharist as the culmination of a pilgrimage.

Historical cults also had loosely defined boundaries: Doctrine was absent or minimal, and membership was loose, undefined, and voluntary. These groups tended to be very pluralistic, so one might belong to several different cults, depending on one's needs at the time. For example, a person might adhere to one saint's cult for healing arthritis, use another set of devotions for the blessing of crops, make a pilgrimage to a holy well to seal a vow to stop drinking, participate in a different saint's cult for fertility, and use icons and amulets of yet another cult to protect loved ones from harm.

Because of this pluralistic tolerance among historical cults, the groups coexisted—indeed flourished—alongside the churches. Part of the historical strength of Roman Catholicism and Eastern Orthodoxy may be due to their flexibility in absorbing and, often, coopting these cults. By absorbing local popular cults as optional devotions, these churches were able to build organizations that transcended familistic, localistic, or tribal barriers. For example, early Christian bishops encouraged the cults of the saints but restructured the practices, then prevalent among wealthy Christians, of privately controlling the remains (relics) of martyrs for sanctifying family burial plots and familistic rituals with them, such as feasts celebrating the dead. By placing the relics in church shrines and altars and transforming these saints into mediators for the whole congregation, the bishops shifted the balance of social power in the community, transcending the narrow kin-based patronage system of Christian communities in late antiquity (Brown, 1981).

Furthermore, the cults of the saints and shrines at holy sites, such as wells, caves, mountains, and stones, enabled Christianity to absorb and transform pre-Christian, pagan elements of indigenous cultures. It was only with the gradual consolidation of *official religion* (see Chapter 4), with its characteristic boundaries around approved, "pure," orthodox religious belief and practice, that cults and cultic adherence acquired a more negative status in Western societies. Sociological analysis needs to pay particular attention to periods of high levels of cult-type religious expression as evidence of certain social changes. Might there not be parallels between today's cultic activity and that of earlier times, for example, in how cults make possible a new mode of individuation, changes in gender and family roles, and a breakdown in older social barriers embedded in the orthodoxy from which cults deviate?

Since the 1960s, the use of the concept "cult" has been confused by the media's pejorative misapplication of the term to all unfamiliar religious movements. In sociological terms, most of these movements are more accurately characterized as sects, but the concept of "cult" has become a negative label for religious associations considered deviant by the larger society (and by the sectlike anti-"cult" movement).[1] The sociological concept of "cult" is, by contrast, neutral, merely describing characteristics of organization and individual orientation of members.

### Usefulness of the Typology

It is debatable whether church–sect typologizing can be applied outside the Christian context. Some sociologists argue that the basic theme of the category (i.e., religious group versus the "world") is specific to the development of Christian churches. The theology of the Incarnation (i.e., God-as-human) illustrates this tension. Thus, Christianity is more likely to develop a wide range of religious collectivities (e.g., church, sect, cult, denomination) than other religions such as Hinduism, which does not have the same tension with the "world" (Berger, 1967:123; Martin, 1965:9, 10, 22). If, on the other hand, a typology is applicable *in principle* to a certain range of social phenomena, the fact of which types are *not* found in certain contexts is significant (Robertson, 1970:121). For example, if "denomination" is defined in a way that could in principle be applied to all societies, the fact that denominations do not exist in, say, Iranian Islam is worth discussing.

Similarly, it is sociologically interesting if a type appears significant in one period of history and not in another. Although the Buddha preached nonengagement in sociopolitical activities, Theravada Buddhism (one major strain of Buddhism) became closely linked with most Southeast Asian monarchies (i.e., Siam, Burma, Laos, Ceylon, Cambodia) from about 1000 to the mid-1800s. Its churchly hegemony was interrupted not by sectarian strife but by European colonialism. The impact of colonialism was to put Buddhism—still the majority religion—into a position of marginality relative to the dominant colonial society. Buddhism became more sectlike relative to society and thus a vehicle for nationalist dissent. Recently, however, Southeast Asian societies have experienced a differentiation of religious institutions from economic and political spheres because of both capitalistic and socialistic influences so that religious collectivities do not have their former potency as vehicles for dissent (Houtart, 1977). These examples suggest that typologies of religious organizations can be

---

[1]Because of this pejorative distortion of the concept, I seriously considered substituting another term in this edition of the book (see Ellwood, 1986; Richardson, 1990b). The term *cult* is retained, however, because it captures the historical continuity with earlier Christian and non-Christian cults and highlights the social processes leading to an era, such as our own, in which highly individualistic popular and other nonofficial religion is both prevalent and suspect.

extended beyond the Christian context if cautions are taken to base categories on sociological rather than on doctrinal characteristics.

The church–sect categorization of religious groups has been heavily criticized, and some sociologists suggest doing away with it (Goode, 1967). Some critics note the historical and cultural limitations of the model, even with its later refinements. Others criticize the parsimony and lack of precision in the construction of types. The typology is not clear-cut, so it has limited empirical usefulness. Criticism is especially directed at the static uses of the concepts. Rather than viewing religious collectivities as dynamic and changing, some sociologists have applied the concepts as a rigid taxonomy (see especially critiques by Johnson, 1963, and Robertson, 1970:116 ff.). Nevertheless, the church–sect categorization has formed the basis for some of the most important work done in the sociology of religion. Sociologists have investigated sects and other organizational forms as vehicles for social dissent and change. They have studied why people join sects, how socioeconomic status of members is related to the type of collectivity, and the dynamics of sects and other organizational forms in relationship to their social environment. Thus, this categorization, although analytically imperfect, has been a productive and stimulating basis of inquiry.

### Refining the Typology

The church–sect line of typologizing appears to have considerable analytic value, so we will now consider some necessary clarification.

First, we can distinguish between forms of *organization of a religious collectivity* and modes of *orientation of individual members*. The concepts "church," "sect," "denomination," and "cult" have been used variously to describe both of these aspects. This usage leads to confusion because members of one type of religious group (e.g., a denomination) may very well have several different modes of religious orientation; some may be sectarian, others denominational, in orientation. For this reason, we will separate discussion of organizational forms and member orientations.

Second, categories should be based on *sociologically* important characteristics. Two criteria used by sociologists appear especially useful: the relationship between the religious group and the dominant society (Johnson, 1963); and the extent to which the religious group considers itself to be uniquely legitimate (Robertson, 1970:120–128). The relationship between religious collectivity and its social environment has been a key theme since Troeltsch and Weber. Furthermore, it reflects a basic tension between values derived from religious sources of authority and other values in the society. No organization could neutralize this tension, though each type responds differently to it. The tension derives in part from the peculiar nature of religious sources of authority, but the general typology also applies to nonreligious organizations (Figure 5.1). The dominant medical system has church-type characteristics in contrast to sect-type organizations of alternative healing systems (Freund and McGuire, 1991). Similarly, political splinter groups often display sect-type characteristics relative to churchly or denominational dominant organizations (Jones, 1975; O'Toole, 1976).

Basically the church–sect typology is a general developmental model for any

| Self-conceived legitimacy | Group/society tension | |
|---|---|---|
| | Positive | Negative |
| Unique | CHURCH | SECT |
| Pluralistic | DENOMINATION | CULT |

**Figure 5.1**  Organizational "Moments" of Religious Collectivities

organization defining itself as uniquely legitimate and existing in a state of positive (i.e., church) or negative (i.e., sect) tension with society (Mayrl, 1976:28). The additional categories of denomination and cult present a general model of similar positive (i.e., denomination) and negative (i.e., cult) tension, but in this case either the organization does not claim unique legitimacy or the society prevents it. This model is depicted in Figure 5.1.[2]

**Church.** The church considers itself uniquely legitimate and in a relatively positive relationship with society. The church does not recognize the legitimacy of claims of any other religious group. Its message is *"Extra ecclesiam nulla salus"* (Outside the church there is no salvation). Its response to competing claims may be to ignore, suppress, or coopt the competing group. The church is basically accepting of society but—to the extent that religion is differentiated from other institutions—it exists in tension with other institutions. Thus, for example, the church might use its status to attempt to change the political or economic arrangements of society. Overall, however, church-type organizations tend to support the societal status quo.

At the same time, the model emphasizes the extent to which society also accepts the church. Membership in a church-type organization is not considered deviant or unusual. The church type is exemplified by the Roman Catholic organization of thirteenth-century Europe and the Theravada Buddhism of feudal Southeast Asia. In both instances, the church-type organization claimed an effective monopoly of legitimacy (although Buddhism was less aggressive

[2]This schema is similar to that of Swatos (1975, 1979). His model refers to monopolistic or pluralistic *societies*, whereas this model refers to pluralistic or monopolistic *claims to legitimacy*. Part of the tension with the dominant society depends on the extent to which the collectivity attempts and succeeds in getting its claims to legitimacy recognized.

toward minority religions), and there was positive tension between the church and society. The church type legitimated political power and the economic system, accepted the goals and values of society, and, in turn, was approved and protected by society.

**Sect.** The sect is a religious organization that considers itself uniquely legitimate and is in a relatively negative relationship with the dominant society. The sect does not accept the legitimacy claims of the church or of competing sects; like the church, it proclaims itself the only way. The sect, too, may adopt any number of strategies to deal with competing groups: withdrawing from their influence, aggressively attacking, or trying to convert and absorb them. By contrast, however, the sect is also in conflict with society. This relatively negative tension between the sect and its social environment is both a threat to the continued existence of the sect and a potential basis for social change. Essentially, the sect is a form of social dissent, exemplified by the Anabaptists of Reformation times and Shakers, Mormons, Rappites, Quakers, and Christadelphians in eighteenth- and nineteenth-century America. Contemporary sectlike groups in American society include the Amish, Mormons, the Jewish Hassidim, Jehovah's Witnesses, Seventh Day Adventists, and Pentecostal and Holiness groups. Some new religious groups such as the Children of God (later called the Family of Love), Unification Church, Synanon, and People's Temple are also sectlike (see Barker, 1982; Bromley and Shupe, 1979a; Chidester, 1988; Richardson et al., 1979; Robbins and Bromley, 1991).

**Denomination.** The denomination is in a positive relationship with society and accepts the legitimacy claims of other religious collectivities. It is perhaps even more generally accepting of society than the church because, with its pluralistic power base, it is less able to control society. Also in tension with the "world," the denomination acts on and reacts to society but usually from the sense that the "world" is "okay" (see Berger, 1961). Examples of denominations include most major U.S. religious groups—Presbyterians, Congregationalists, American Baptists, Methodists, Lutherans, Episcopalians, Roman and Eastern Catholics, Reform and Conservative Jews. Similarly, in Japan, religious groups such as the Pure Land and Zen divisions of Buddhism exemplify denominational pluralism (cf. Morioka and Newell, 1968).

**Cult.** The cult accepts the legitimacy claims of other groups and is in a negative relationship with society. Like the denomination, the cult does not claim to have *the* truth and is tolerant of other groups' claims. At the same time, the cult is critical of society. Like the sect, the cult is a form of social dissent; however, its dissent is likely to be less extreme because of its pluralistic stance. Historical examples were medieval healing cults and nineteenth-century Spiritualism. In contemporary society, there are numerous cults, including Arica, Vedanta, astrology, flying saucer groups, Silva Mind Control, Rosicrucianism, and American Zen. Other movements such as Divine Light

Mission and Transcendental Meditation were cults in some stages of their development or at some but not all organizational levels (see Downton, 1979; Wallis, 1985; Westley, 1983; Woodrum, 1982).

**A Dialectical Process.** Finally, our model illustrates the constant tension and dynamism in the process by which a group tries to effect its religious convictions. There is an ongoing dialectical process between the religious collectivity and its social environment. The religious group responds to changes in its social situation; it can also effect changes in that situation. The types delineated in Figure 5.1 are not fixed categories but are *moments* in a dialectical process.[3]

Because the model is based on fluid processes, to categorize a collectivity, specifying (1) societal context, (2) level of analysis (e.g., national, regional, or local), and (3) time period is necessary. Thus, we can classify the Roman Catholic organization as a church relative to French society of the thirteenth century. Relative to U.S. society of 1840, however, the Roman Catholic organization was more of a sect, because it neither accepted nor was fully accepted by society. Much of nineteenth-century American Catholic devotional literature, preaching, and practices of popular piety reflected sectarian strategies in response to the perceived hostility of the Protestant majority (Taves, 1986). Later in the United States, especially after Vatican II, the Roman Catholic organization appears to have become increasingly a denomination because it is more accepting of and accepted by society and is more tolerant of other groups' claims to legitimacy (see Seidler and Meyer, 1989).

Specifying the level of analysis is necessary because the relationship to society of a national group may be different from that of a local group affiliate. In some isolated rural areas, small local churches, although affiliated with a national sect, are churchlike organizations relative to their subcultural setting. These groups are often the only church serving their community. Many members do not even know about religious alternatives, and the groups are able to enforce their monolithic status. They are fully at home with the culture of that locale, often the only one that its members know. Although such groups may be nationally part of a sect, at the local level they may operate very much as a church. Another example is the People's Temple of Jim Jones, affiliated at the national level with a denomination (Disciples of Christ); but in California and later Guyana, it increasingly became a sect.

### Variables Related to Organization Type
Some typologies of religious collectivities have included criteria that we will present as descriptive variables rather than as definitive. One such variable is

---

[3]Recent changes in Western societies have reduced the impact of religion on society (see Chapter 8), but the model offered in Figure 5.1 covers societies and historical periods in which religion's impact is more evident. This model of religious processes can be compared with the Marxian theory of *praxis*—the ideological context of the relationship between theory and practice in social movements. Mayrl (1976:28) notes the irony that Marxism as a social movement has also experienced a parallel (i.e., sect-to-church) shift from critical praxis to adaptation to the social situation.

the *degree of organizational differentiation, specialization, and complexity.* Churches and denominations have generally tended to be larger, more complex organizations than sects and cults. The development of specialized ministry, bureaucracy, and formal organizational structure seems to accompany churchlike and denominationlike forms. It is unwise to include this feature as definitive, however, because it would detract from recognizing the real dissent of some highly organized groups. Both Jehovah's Witnesses and Scientology are highly differentiated and complexly organized, yet both have retained their negative tension with society. To allow for this variable, the terms **established sect** (Yinger, 1970:266–273) and **established cult** (or "centralized cult"; Nelson, 1969:233) may be used for highly organized groups that retain their sectarian or cultic stance. There are no terms in use for the parallel types—relatively simple, loosely organized churches and denominations such as Unitarian Universalists and some Baptist groups.

Another variable, treated as definitive by some theorists, is the degree of *elitism* of a collectivity (see Robertson, 1970; Wilson, 1970; Yinger, 1970). This variable is a common characteristic of sectarianism as a mode of religious orientation, but it is probably a by-product of the two definitive variables in the model; that is, groups viewing themselves as uniquely legitimate are more likely to consider themselves superior to "nonlegitimate" groups. Troeltsch (1960) observed that the church tended to have unrestrictive membership requirements and that members were often "born into" it rather than joining it, as in a sect. This depiction, however, underplays the extent to which the church, too, could be exclusionary when that suited its purposes. In the height of its power, the Roman Catholic church in Europe excommunicated dissidents, labeled heretics and witches, and controlled who could become members. The early part of the Inquisition in Spain (mid-fifteenth to mid-sixteenth centuries) was directed largely at ferreting out *Conversos* (i.e., converts, principally from Judaism) who were not "good Christians." These "converts" were suspect because they had joined the church to escape death at the hands of Christian mobs (in the years preceding the Inquisition, 50,000 Jews had been killed, and an estimated 2 million had "converted"; Longhurst, 1962:39). Whether a group is inclusive or exclusive depends largely on its relationship with its social environment at any given time.

Another variable is whether the religious group functions to integrate its members into society or whether it results in disintegrative functions (Robbins et al., 1975). As the last part of this chapter will show, the same religious movement may serve integrative and disintegrative functions simultaneously. Thus, some 1960s and 1970s Jesus movement groups socialized their counterculture youth members into values of the dominant culture—hard work, deferred gratification, marriage, and obedience to authority. At the same time, however, the sect-type structure of the groups removed members from the larger society, biological family, and former friends. Integrative and disintegrative functions of religious groups are important, but they are not sufficiently clear-cut to use in defining typologies.

## Modes of Religious Orientation

It is helpful to distinguish sectarian, cultic, denominational, and churchly *orientations* from sect, cult, denomination, and church as *organizational types* because several different orientations can occur within the same collectivity. The model of religious orientation types shown in Figure 5.2 is derived from the previous typology of religious collectivities. It is based on the characteristic commitment patterns of members belonging to the types of religious collectivities discussed. Although each type in Figure 5.2 corresponds to a type of group, members of a group often display several different orientations.

### Types of Religious Orientation

The two key characteristics for defining "typical" orientations of individual members are (1) the extent to which the member's role as a religious person is segmented into a separate role or is expected to be diffused throughout every aspect of the person's life and (2) the extent to which the individual judges self and others according to standards of "mass" or "virtuoso" religiosity. These qualities are present in varying degrees, so we can speak of relatively "mild" or "strong" sectarian, cultic, denominational, or churchly orientations.

Figure 5.2 ties together a number of theoretical and empirical observations about sectarianism. Some theorists (Parsons et al., 1961:251) have suggested that one of the main characteristics of sectarianism is much less differentiation of members' religious role from other social roles (e.g., work or family). The sectarian orientation means that the religious role is the master role, pervading

Religious emphasis

Religious "role"

|  | Mass religiosity | Virtuoso religiosity |
|---|---|---|
| Diffused | CHURCHLY | SECTARIAN |
| Segmented | DENOMINATIONAL | CULTIC |

**Figure 5.2**   Modes of Individual Religious Orientations

and subordinating all other aspects of the person's life. By contrast, a denominational orientation is more segmented and the religious role is separate from other aspects of life; this does not mean that the religious role is not important, but it has less influence in the total picture of daily life.

The second variable refers to a related normative aspect: the extent to which one judges oneself and others by standards of "virtuoso" religiosity. **Religious virtuosi** are those who strive for religious perfection (however defined) and are not satisfied by the normal levels of religiosity of the masses (Weber, 1963:163–165). The person may define religious perfection in terms of some moral, ascetic, mystical, or even body-emotional ideal. Individual orientations are therefore guided by whether the religious role applies to all spheres of life and by whether it is judged by the norm of perfection.

**Sectarian.** The sectarian orientation is characterized by normative virtuosity and diffusion of the religious role (cf. Knudsen et al., 1978). The sectarian orientation toward perfection is part of the reason for the sect's negative tension with the "world." Sectarian dissent is a form of judgment of the imperfection of the rest of society. The sectarian orientation insists on the pervasiveness of the religious role, which is, ideally, the organizing principle of work, family, leisure activities, political stance, and personal life. This commitment to perfection and the diffuseness of the religious role explain why sectarianism is frequently characterized by total commitment (Snow and Machalek, 1983; Wilson, 1967:1–45, 1970; cf. nonreligious institutions described in Coser, 1974). The sectarian orientation is also more prone to religious extremism precisely because of its characteristic emphasis on applying a virtuoso religious norm in all spheres of social life.

**Denominational.** The key features of a denominational orientation are an acceptance of ordinary levels of religiosity and a religious role segmented from other aspects of life. This orientation is satisfactory for most members of a denomination because their beliefs and practices are less at odds with society than are those of the sect. Because denominations consider their worldview as pluralistically legitimate, commitment tends to be more individualistic. Members feel free to select not only a "church of their choice" but also which elements of that religious view they wish to accept. The social acceptability of denominational membership, together with this individualism, make the characteristic orientation of denominationalism somewhat resemble commitment to other voluntary organizations such as Kiwanis, Elks, or League of Women Voters. This does not mean that the religious content of denominationalism is not important, but commitment is much less intense or urgent than in sectarianism. One study found that denomination members were less intensely involved in their religious group than sect members. Members of denominations were much less likely than sect members to consider church membership very important, to restrict their friendships to fellow members of the congregation, or to restrict their membership to religious community organizations (Stark and Glock, 1968:101, 166, 172).

**Cultic.** The cultic orientation is also prevalent in contemporary religion, either by itself or in combination with denominational modes. It is characterized by separation of religious roles from other aspects of life, as in denominationalism. By contrast, however, a cultic orientation also involves seeking a higher level of spirituality, comparable to sectarian "perfection" except that the definition of perfection is less specific and more eclectic. Members' commitment is that of the seeker, permanently searching for particles of truth from a wide variety of sources. Cults do not assert unique legitimacy, and members with a cultic orientation often draw on several different groups, either simultaneously or serially. A person deeply involved in Spiritualism may be simultaneously "into" Silva Mind Control and also be an active Methodist.

This individualism is a distinctive feature of cultic adherence because it contributes to the precariousness of cults as organizations. Cultic individualism means that no clear locus of authority exists beyond the individual member, and there is thus no way to handle "heresy" and therefore no clear boundaries between members and nonmembers. In the cultic mode, as in the sectarian, the individual finds some mutual support from fellow believers, but this group contact is less strong because of the characteristic pluralism and individualism (Campbell, 1977; Martin, 1965; Wallis, 1974, 1977).

**Churchly.** The major characteristics of the churchly orientation are acceptance of a "mass" standard of religiosity and a religious role diffused throughout everyday life. Religious commitment is very general, sustained by the whole fabric of society, and does not simply comprise the religious group itself. Religious activity may be considerable or negligible, but rarely does the individual consciously choose to be religious. Although church-type religious organizations do not exist in U.S. or Canadian society, individual churchly orientation is possible in certain subcultural settings (e.g., a Utah Mormon enclave).

The differentiation between types of religious collectivity and characteristic modes of individual orientation helps explain the diversity of orientations within a given group. Churches and denominations in particular, but also sects and cults, often have members with very heterogeneous orientations. This diversity is important in understanding the dynamics of religious groups. The interaction of these members often leads to important transformations in the collectivity itself. Such diversity, however, complicates the task of analyzing religious collectivities.

In one study, considerable diversity was found within the same denomination, with some members having highly sectarian patterns of involvement. Sectarian members of one denomination also resembled sectarian members of other denominations in several important ways—especially socioeconomic status. Table 5.1 illustrates some of these findings. These figures suggest that sectarian orientations exist even in denominations that are not sects or sectlike (e.g., Congregationalism). Regardless of denomination, furthermore, sectarian orientation varies according to socioeconomic status. The percentage of members with strong sectarian involvement increases as social status decreases

Table 5.1    Sectlike Religiosity and Individual Status among Congregationalists, Presbyterians, Disciples of Christ, and Baptists

| Religiosity High on Sectlike Involvement | Individual Status | | | | Denomination |
|---|---|---|---|---|---|
| | Upper | Middle | Working | Low | |
| % | 22 | 27 | 32 | 37 | Congregationalists |
| Number | 1,386 | 1,145 | 572 | 102 | |
| % | 33 | 41 | 51 | 52 | Presbyterians |
| Number | 711 | 674 | 471 | 129 | |
| % | 36 | 45 | 43 | 56 | Disciples of |
| Number | 387 | 396 | 284 | 52 | Christ |
| % | 42 | 48 | 50 | 56 | Baptists |
| Number | 171 | 233 | 218 | 75 | |

Source: Adapted from N. J. Demerath, III. *Social Class in American Protestantism*, Rand-McNally, Chicago, 1965: p. 119; used by permission.

(Demerath, 1965). The sectarian pattern of involvement may, however, be a product of an intervening variable, localism (discussed in Chapter 3). Persons of higher socioeconomic status may be generally less sectarian because their social mobility and educational levels promote a cosmopolitan orientation. Thus, the relationship between sectarian orientations and socioeconomic status may be, at most, indirect. As noted in Chapter 3, sectarianism is a cognitive minority's strategy for maintaining the plausibility of its belief system. The sectarian orientation would therefore serve believers of any socioeconomic status.

The religious group must cope with the conflict produced by these differing orientations; but if it can balance the opposing impulses of sectarianism and denominationalism, the organization may emerge stronger. A parallel is the strength gained by medieval Roman Catholicism when it incorporated the sectarian impulse in the form of monasticism. Conflicting modes of orientation often exist in developing religious movements, too. Thus, in the early years of the Catholic Pentecostal movement (now called the Catholic Charismatic Renewal), sectarian and cultic orientations were both prevalent. Sectarian adherents tended to be highly involved in the prayer group and likely to view pentecostalism as the only way to be "really" a Christian. They were more likely than cultic adherents to consider their religious role as the master role in their lives. Cultic adherents, by contrast, were more likely to have segmental commitment to their prayer group and were less likely to view the movement as having the exclusive truth (Harrison, 1975; McGuire, 1982). These two orientations have produced considerable tension within the movement, with the cultic adherents decrying the others' elitism and authoritarianism and the sectarian adherents denouncing the lack of commitment and the independence of those with a more cultic orientation (cf. Neitz, 1986; Warner, 1988). The resolution of conflicts such as this is a significant part of the transformation of religious collectivities.

### Conflicting Orientations: Dissent Within

The coexistence of conflicting religious orientations within the same collectivity is described by Wach (1944:173–186) in his concept of **ecclesiolae in ecclesia** (i.e., little churches within the church). These groups consist of members of the church (or denomination) who are dissatisfied with the spiritual or moral tenor prevailing in the church and advocate a different kind of religiosity, often aiming at eventual conversion of the entire church. Essentially, these groups embody sectarian or cultic dissent *within* the larger collectivity (i.e., church or denomination). Their members press for a different tension between the collectivity and the "world" and are dissatisfied with the imperfection of the church. Wach distinguished three general kinds of *ecclesiolae*.

**Collegium Pietatis.** The simplest form is only a step beyond individual protest: the **collegium pietatis** (i.e., an association of religious colleagues), which is essentially a cultic orientation within the larger collectivity. According to Wach's description, these collegia do not see their beliefs and practices as uniquely legitimate. They are loosely organized, stressing individual attitudes and seeking greater personal perfection. Historic examples of collegia include German pietism in the early part of the Reformation, early years of the Society of Friends (i.e., Quakers), and the Oxford Group Movement in nineteenth-century Anglicanism. Similarly, Karaism (eighth and ninth century), Kabbalism (especially sixteenth and seventeenth century), and neo-Kabbalism (twentieth century) among Jews spawned groups like the collegia. Contemporary examples of collegia include loosely organized prayer or fellowship groups within larger religious organizations. Societies related to some special method of spirituality or specific religious practice (e.g., rosary society, a healing cult, or adherents of a shrine) could also be collegia.

**Fraternitas.** Although common religious ideals and experiences are sufficient to draw people together as in collegia, many groups seek a closer association—a sense of community. The establishment of groups with some measure of communal life marks the **fraternitas** (i.e., brotherhood). Their communality, however, is integrated more by "spirit" than formal organization; thus, they are often in conflict with the larger collectivity. Such groups often have egalitarian concepts of fellowship. The fraternitas also represents a somewhat cultic orientation within the larger collectivity. Examples include the Brethren of the Common Life (fourteenth century), early Moravians (eighteenth century), Methodist fellowship groups (especially nineteenth century), and the earliest years of the Bruderhof and kibbutzim (twentieth century). The "underground church" movement in American Roman Catholicism (1960s and 1970s) and the havurah and schtibel movements in Judaism (1970s and 1980s) are contemporary *fraternitates* (see Bubis et al., 1983; McGuire, 1972, 1974; Pinsky, 1986).

**The Order.** Wach's third type of internal dissent is the **order**, characterized by greater stratification and specialization of functions and a more rigid conception of community. The order is essentially the organization of sectarian dissent *within* the church (or denomination). It expects the individual's total commitment. Members are bound in their adherence to the religious community and in their renunciation of competing ties. Monasticism is the most obvious and universal example of the order. Through the institution of monasticism, the church absorbs and segregates sectarian protest. At the same time, the order may bring about reform in the larger collectivity, such as the reforms in medieval Roman Catholicism brought about by Cistercians (twelfth century). Monasticism is a feature not only of Christianity but also of Islam, Buddhism, Jainism, and Taoism; but in different cultures, it has different meanings because the characteristic tension between church and social environment is different. There is often much tension between the order and the larger collectivity of which it forms a part. Even though the order may compromise its dissent, it represents a considerable source of independent authority within the collectivity.

Contemporary examples of orders include modern descendants of medieval Roman Catholic monastic orders and nineteenth-century Anglican orders. Other order-type collectivities include "covenant communities" among Catholic Pentecostals and the Ecumenical Institute, Iona, and Taize, which are ecumenical Protestant communal groups in the United States, Britain, and France, respectively. The order is an expression of the sectarian elite within a larger collectivity and is the ultimate social organization of virtuoso religiosity. The collegium and fraternitas similarly express virtuoso religiosity but in much less organized and homogeneous form.

Members of all these "ecclesiolae in ecclesia" tend to be religious virtuosi, as are members of sects and cults. Part of the nature of their negative tension with "the world" is their dissatisfaction with ordinary religiosity. Even where religious virtuosi have not banded together in "little churches within the church," there is a marked sectarian impulse among a religious elite (Berger, 1958; Fichter, 1951, 1954). The main difference between orientations of the order and the sect is the order's willingness to accept a dual standard of religiosity— one standard for the elite and one for the masses. Sectarians, by contrast, do not tolerate this distinction and see their standard as the only way (M. Hill, 1973a). Protestantism has historically tended to spawn sects rather than to incorporate dissent into orders because it is more likely to reject the dual standard (Martin, 1965:191–193).

The pressure to change brought by religious virtuosi is analytically different from the impetus to change in the challenge of a charismatic leader and followers. The charismatic leader proclaims a new message and new basis of authority. Religious virtuosi, by contrast, generally call for a return to the "true" tradition; authority is legitimated by reference to tradition. While this latter pressure for change is less threatening to the larger collectivity, it still has considerable potential for stimulating dramatic changes through something of a "revolution by tradition" (M. Hill, 1973a:3). Thus, dynamic processes within religious organizations themselves sometimes bring about major changes in the collectivity; and the existence of conflicting orientations within the same groups helps account for some of these transformations.

# Dynamics of Religious Collectivities

An interesting subject in itself is how a religious group comes into being. What brings people together? What attracts them to the group? What repels them from other alternatives? What holds them together? What shapes their organization and collective direction? Even more important, however, is the relationship between the formation of the group and larger social processes. What social factors contribute to the formation of a new religious group? What factors promote or inhibit its success, and how does the developing religious group interact with its larger social environment? Developing religious groups have often had impact both on society and on the collectivities against which they were dissenting, so the dynamics of religious collectivities are linked with the process of social change. At the same time, social change has historically been an impetus for religious change, so religious dynamics indirectly reflect larger social processes.

## The Formation of Sects and Cults

Social factors, as well as theological ones, appear to be very important in the formation of new religious groups. The formation of religious movements and new religious collectivities has generally been explained by three social factors: socioeconomic or other deprivation, social dislocation, and socioeconomic change.

**Deprivation.** Sect membership is disproportionately drawn from the lower classes (Aberle, 1962; Lanternari, 1963). Sects are a vehicle for dissent from society as well as from the church, so persons who are relatively comfortable with society are unlikely to participate. On the other hand, the most destitute stratum, partly because of its characteristic fatalism and sheer lack of resources, is rarely involved. Furthermore, economic deprivation is not sufficient to explain such movements because middle- and upper-class persons are also drawn to sects. Thus, in the latter part of the twentieth century, sects such as Hare Krishna (Krsna Consciousness) and the Unification Church have had wide appeal among comfortable middle-class youth. If deprivation is a factor in the formation of sects, the concept needs clarification to fit these diverse empirical situations.

Toward such clarification, Glock (1973:210) proposed the notion of **relative deprivation**, which refers to "any and all of the ways that an individual or group may be or feel disadvantaged in comparison either to other individuals or groups or to an internalized set of standards." People's subjective assessment of their circumstances, rather than objective circumstances themselves, is the key criterion. Glock also notes that economic deprivation is only one of several ways that a person can feel deprived; a person can also feel social (i.e., status), organismic (i.e., physical), ethical (i.e., philosophical), or psychic deprivation. Ethical and psychic deprivation are somewhat ambiguous. Ethical deprivation refers to the experience of value conflicts, especially those of intellectuals trying to reconcile their philosophical standards with the realities of the larger society. Psychic deprivation refers to lack of a meaningful value system by which to interpret and evaluate one's life. Glock suggests that the type of deprivation felt influences whether a movement will survive and what kind of organization it develops. Another factor is the external environment, that is, whether societal sources of deprivation persist as, for example, an economic depression (Glock, 1973:210–213).

The existence of relative deprivation does not *cause* a movement of dissent, but it does create a pool of dissatisfied persons who, if mobilized, would express dissent through a political or religious movement. The deprivation model is helpful in conceptualizing the relationship between sources of dissatisfaction and a movement's social base. It does not, however, adequately explain why dissatisfied persons should choose a particular movement. Furthermore, in applying the model empirically, there is a tendency to reason from after-the-fact evidence. Some interpretations suggested that because a movement provided X, that provision constitutes evidence that recruits were deprived of X. Nevertheless, the deprivation model is useful in understanding not only sect formation but also subsequent transformations in the organization of the collectivity.[4]

**Social Dislocation.** Another interpretation suggests that sects and cults arise to express and respond to the problems of social dislocation of some people. The Holiness movement spread rapidly in the 1930s among rural people who had recently migrated to urban environments. One study of this process interpreted the movement as a result of "culture shock" produced by rapid transition from the stable rural environment to the city (Holt, 1940). Sects and cults allow members to protest against their new environment while simultaneously overcoming some of the difficulties of dislocation by finding a new community among group members. Movements responding to social dislocation would thus be organized as small, close-knit communities. This interpretation holds that sects and cults offer members a surrogate community or family in the face of the impersonality, isolation, or confusion experienced in society. Like deprivation theories, the social dislocation model is useful in describing why an available pool of potential recruits may find a religious movement attractive. It also illustrates how a movement can express dissent and simultaneously resolve some of its members' problems. On the other hand, this model does not account for many religious movements. Relatively few members of neo-Pentecostal movements among Roman Catholics, Presbyterians, Episcopalians, and Luther-

---

[4]A student of Glock subsequently used an extreme form of this reasoning to attempt a definitive explanation of religion. Accordingly, religion is defined as "a system of general compensators based on supernatural assumptions" (Stark and Bainbridge, 1979:121; see also Stark and Bainbridge, 1980; Bainbridge and Stark, 1984). Compensators refer to intangible promises that *substitute* for certain rewards that people desire, such as prosperity, success, and physical and mental health. From this definition, the authors derive a typology of church, denomination, sect, and cult, distinguished by the type of compensator each offers. This approach has been severely criticized for its *reductionism;* i.e., it tries to reduce religious belief and practice to overly simplistic this-worldly considerations: cost versus benefit, whether real reward or substitute compensation (Wallis and Bruce, 1984, 1986). Like relative deprivation theories, this approach tends to argue backward: this religion offers X; therefore, X must be a compensation for desired reward Y. It imputes to religious adherents motives and meanings which they would probably neither recognize nor affirm as theirs (McNamara, 1985). Critics further argue that the theory is not adequately supported by empirical evidence (Wallis and Bruce, 1984; Bibby and Weaver, 1985)—especially not by the kind of evidence produced by sound ethnographic research that listens to believers' accounts of their motives, meanings, and experiences.

ans have experienced serious social dislocations such as migration, unemployment, or family disruption. Similarly, members of some groups may rely little on fellow members as family or community surrogates. Thus, although the dislocation model is useful, it is not sufficient to explain these sects and cults.

**Socioeconomic Change.** A third approach to understanding the impact of the larger society on the formation of sects and cults is the emphasis on socioeconomic change. One study of historical millenarian movements suggested that the key factor was the ambiguity surrounding a disaster (natural or social) that produced the impetus for such movements (Barkun, 1974). Further evidence for this interpretation comes from analysis of a period of prolific movement development, the early nineteenth century in upper New York State. The revivals, experimentation, and development of new movements in this time may have been correlated with dramatic economic fluctuations, which had produced alternating rising expectations and disappointments, as well as other disruptions like epidemics (Barkun, 1986). Another interpretation suggests that religious and political movements arise in response to major changes in the world economic order. The type of movement that develops depends, according to this theory, on whether its adherents are central or peripheral to emerging sources of economic power (Wuthnow, 1978, 1980).

All available evidence suggests that a strong relationship exists between social conditions and the development of sects and cults. None of the existing models alone is adequate to explain this relationship, though they are not necessarily wrong. What is needed is a sufficiently complex model that incorporates varied causes. Focusing on socioeconomic concomitants of religious movements, however, may lead to ignoring or downplaying the reasons given by members for the group's formation. Religious aspirations themselves are important sources of religious movements. Religious virtuosi seeking to perfect their spiritual lives, cultic seekers looking for new techniques for spiritual experience, persons dissatisfied with the level of religiosity available to them in other collectivities—for these people, *religious* aspirations are foremost among their reasons for forming or joining a developing religious movement. Indeed, parent organizations such as a church often give impetus to such movements by raising members' aspirations. Vatican II and changes within the Roman Catholic church in years immediately following the Vatican Council may have raised religious expectations of church members to such a degree that they subsequently sought developing movements either within the church (e.g., Cursillo, Underground Church movement, and Charismatic Renewal) or outside it.

### Movement Organizations

Just as our typology of religious collectivities can be viewed as a general typology of organizations in tension with their social environment, so too can the processes by which religious groups change be understood as general. Some useful insights into these processes come from the theory of movement organizations (see Beckford, 1975a; Lofland and Richardson, 1984). This approach explores how social movements organize themselves toward achieving their goals; how they recruit, consolidate, and deploy members; how they maintain their organization and arrange leadership. This approach has been fruitfully applied to a study of the Young Men's Christian Association (YMCA) (Zald, 1970), a Jesus movement commune (Richardson et al., 1979), Jehovah's Witnesses (Beckford, 1975b), Pentecostalism (Gerlach and Hine, 1970), Catholic Pentecostalism (Harrison, 1975), and Sokka Gakkai (Snow, 1987).

The factors of deprivation, social dislocation, and massive social change may provide the impetus for a religious movement, but aspects of the movement organization itself determine whether the movement will develop and spread. One study of the spread of Pentecostalism identified several movement organizational factors including (1) the relationship among local groups of the movement—flexibly linked networks of local groups promoted the movement's spread; (2) recruitment along the lines of pre-existing class social relationships—family and friends; (3) emphasis on a clearly defined commitment act or experience; (4) an ideology offering a simple, easily communicated interpretive order, a sense of sharing in the control and rewards forthcoming, and a feeling of personal worth and power; and (5) emphasis on the perception of real or imagined opposition (Gerlach and Hine, 1968). The researchers found that this model applied not only to expressly religious groups but also to other dissenting movement organizations such as the Black Power movement (Gerlach and Hine, 1970).

Not all of these contributing factors are present or significant for all movements. Jehovah's Witnesses, a rapidly spreading movement, recruit very little among pre-existing networks of social relationships (Beckford, 1975b). Some cults recruit almost totally anonymous persons. In 1975 two itinerant preachers, who proclaimed salvation through the aid of extraterrestrial beings with flying saucers, recruited over 100 members (who gave up jobs, family ties, and material possessions) in four brief meetings, usually after less than six hours of contact with any member of the group (Balch and Taylor, 1977). Nevertheless, the approach is helpful for analyzing some of the ways in which the internal organization of a movement influences its outcome. It also suggests some of the ways (e.g., increasing the perception of opposition) that the group itself influences the tension between it and its social environment.

### Organizational Transformations
Considering the dynamic forces built into our model of religious organization types (i.e., church, sect, denomination, cult), it is not surprising that groups change from one type to another. Forces for change exist within the organization itself (as demonstrated by the "ecclesiolae in ecclesia") and in the relationship of the group to its social environment. The processes by which a religious group changes from one type of organization to another are sociologically important because they reflect both the impact of the group on society and the influence of society on the group.

### Institutionalization: O'Dea. 
A movement cannot have any lasting impact on society without some measure of **institutionalization**—that is, the creation of

some objective structure by which it tries to realize its values and goals. O'Dea (1961) delineates five inherent dilemmas that institutionalization often poses for any social movement, religious and nonreligious alike.[5] On the one hand, institutionalization enables the emerging movement to express and spread its perspective; on the other hand, it contains built-in tendencies toward stagnation and disorganization, separating members from the initial shared experiences of beliefs that held them together. Religious reform or revival movements are prototypes of this process, but other movements such as civil rights, environmentalism, communes, and political reform movements also wrestle with the dilemmas of institutionalization.

One such dilemma is the organizational contradiction of **mixed motivation**. According to O'Dea, the earliest stages of a movement are characterized by a singlemindedness of followers; either the cause or the charismatic leader unites members' motivations. As the movement stabilizes, other motives come to the fore: desire for prestige, expression of leadership and teaching abilities, drive for power, search for security, respectability, economic advantages, and so on. The institutionalized movement is stronger precisely because it is able to mobilize the self-interested as well as the disinterested, but members' pursuit of vested interests often conflicts with some of the movement's original ideals. A movement that relies totally on members with "pure," disinterested motives has little chance for survival, especially after the departure of the charismatic leader who initially united the members. On the other hand, mixed motives frequently lead to compromise and corruption of the movement's ideals.

A second dilemma involves the movement's **symbol system**—its language, ceremonies, and physical symbols of central meanings. In the earliest stages of a movement, members develop symbols by which they attempt to share significant experiences. These special words, ceremonies, and objects express (however incompletely) what it means to be a part of the movement and its distinctive way of life. Without this objectification (i.e., transforming subjective experiences into concrete, shared images), there would be little basis of commonality among movement members. Shared symbols hold the members together and become part of the message they carry to prospective recruits. At the same time, however, there is an inherent tendency in the process of institutionalization for these symbols to become remote from the experiences and meanings that shaped them. They are cut off from the subjective experience of members who use the symbolic objects, group language, or ceremonies; they become ends in themselves or "things" to be manipulated for achieving ends. For example, the "kiss of peace" was an early Christian symbol of communal love, forgiveness, and unity. In many settings now, however, the "kiss" is merely a ritualistic handshake "performed" at a designated part of the worship service. Without symbolization, the central meanings of the movement cannot be transmitted or shared, but symbolization (i.e., objectification) contains the inherent risk that symbols will become alienated from the subjective life of the members.

Movements also face the third dilemma of **organizational elaboration of**

[5]The following synopsis of five dilemmas from O'Dea (1961) is used with permission.

**structures versus movement effectiveness**. As Weber (1947:358ff.) pointed out, the long-term impact of a movement hinges on transformation of bases of authority and leadership from a charismatic mode to either traditional or legal-rational structures. When a movement becomes established, there is a strong tendency for the organization to calcify around the memory of the early dynamism; its own tradition becomes the rationalization for why things should be done a certain way.

Early stages of movement organization involve simple structures such as the charismatic leader and followers or leader, core followers, and other followers. The transition to legal-rational structures is typically accompanied by the elaboration and standardization of procedures, the emergence of specialized statuses and roles, and the formalizing of communication among members. The early years of the Divine Light Mission in the United States were characterized by rapidly growing, loosely affiliated local ashrams (i.e., groups of devotees, usually living communally), united mainly by devotion to the ambiguous charismatic figure of Guru Maharaj Ji. As the DLM became increasingly structured and centralized, leadership and power focused in the movement's Denver headquarters. The guru's desire to consolidate his power and authority over the movement in the United States resulted in greater formalization: rules and regulations for ashram living, standards for recruited "candidates," and pressure toward certifying movement teachers (Pilarzyk, 1978:33–37). Although such division of labor and formalization of structures make the growing movement more efficient, the process may also make it less effective in the long run. The bureaucratic impersonalism and complexity may obstruct some of the movement's original goals. Members come to feel more peripheral to the movement, and their motivation becomes problematic.

A fourth internal dilemma is the **need for concrete definition versus legalism**. On the one hand, the movement's message must be translated into concrete terms in order to apply to people's everyday lives. On the other hand, the concretization that occurs as the movement becomes institutionalized promotes a legalistic approach to those very terms. The ethical insights of a religious movement might be translated into concrete moral rules (e.g., a norm against consuming alcohol); then, if a legalistic approach to this rule arises, members are reproached for breaking the "letter of the law" (e.g., using vanilla extract, which contains alcohol, in baking). Legalism reduces the insight of the ethical message to petty conformity.

The fifth dilemma of institutionalization is that of **power**. The developing movement needs to structure itself so that its message and meaning remain firm. It needs to protect itself, from threats both within and without its boundaries. A certain amount of internal power gives the nascent movement strength against threats that would dissolve it. On the other hand, that same power may be asserted against deviant and dissenting members. Members' compliance may be assured by external pressures or coercion rather than by inner conviction and commitment. Similarly, desire for power and influence may motivate movement leaders to assert themselves in other institutional spheres, diverting the movement from its original message and goals. The direction taken by the leadership of the People's Temple illustrates one (albeit extreme) resolution of this dilemma. A number of its practices (e.g., the move from California to a remote part of Guyana and the requirement that members give up personal property) can be understood as efforts to overcome the inherent precariousness of charismatic leadership. They were strategies to consolidate power, similar in kind to strategies adopted by leaders of other nascent charismatic movements (see Johnson, 1979).

O'Dea (1961) describes these five contradictions as dilemmas to emphasize the extent to which neither direction (institutionalized or noninstitutionalized) is completely satisfactory. The very processes that would enable a social movement to realize its ideals contain inherent contradictions, which simultaneously prevent the movement from accomplishing these goals. The concept of dilemmas, however, is somewhat misleading because rarely does a social movement make conscious choices to institutionalize. Nevertheless, O'Dea's analysis is useful for showing that frequently criticized features of institutionalized religion (or any other institutionalized social movement) are actually results of the ways in which that movement has organized itself for effectiveness and social impact.

All religious organizations, from the small, informally organized sect to the largest bureaucratically organized church, can be analyzed for how they acquire and use the resources of ideas, people, and materials. Organizations can also be compared for how they socialize, recruit, and control their members. And they can be examined for their structural qualities: specialization, centralization, formalization, and authority relationships (Beckford, 1975a:34–92). Transformations in religious groups frequently involve changes in some of these organizational aspects. Thus sects developing into denominations frequently become more bureaucratically organized. These organizational qualities are not definitive (e.g., not all sects that become bureaucratically organized are transformed into denominations) but rather reflect the inherent tension between the religious group and its social environment. Thus, examination of how change occurs in different types of religious organization focuses on the two key characteristics defining our model of organizational types. Transformations in religious groups can be analyzed in terms of two questions: (1) How does the group's sense of its monopolistic or pluralistic legitimacy change, and (2) what are some of the forces affecting the group's tension (negative or positive) with its social environment?

**Sect Transformations.** The dynamism of sectarian dissent makes the sect a potential force for social change. Yet successful sects seem to accommodate to society with great regularity, eventually becoming indistinguishable from denominations (Niebuhr, 1929). The transformation of sect-type groups into denomination-type groups is more prevalent in U.S. society than other possible transformations such as sect-to-cult or sect-to-church. These latter directions are also theoretically possible, however. Sects can become churches, both by accommodating to the larger society and by consolidating their position to achieve a monopoly of legitimacy. Christianity began as a minority sect within Judaism, and in its spread to other cultural settings, it gradually consolidated its position, achieving an effective monopoly of legitimacy under Constantine.

Similarly, sects can change into cults. Because this transformation involves giving up claim to unique legitimacy, the process is often the result of authority dissolution within the group. If a charismatic leader has successfully organized the group around his or her personal authority to present "the truth," the death or downfall of that leader might delegitimate that "truth." If the group comes to view those teachings as pluralistically legitimate, it is likely to become a cult (if it survives the leadership crisis at all).

A good example of this dissolution is the development of one flying saucer cult (Festinger et al., 1956). The group initially began as a cult, one of several groups from which members drew their inspiration, but gradually became a sect, especially as its role in the prophesied imminent end of the world was made clear by the increasingly authoritative pronouncements of group leaders. The failure of the prophecy undermined the authority of key leaders and eliminated a major basis for the group's distinctiveness—its position as the "elect" to be saved by flying saucers. Although failure of the prophecy did not destroy the group altogether, it gradually became more cultlike again, receding into the general cultic milieu.

Sects and cults are organizationally more precarious than denominations and churches because dissent and deviance are difficult to maintain in the face of opposition from other religious groups and the larger society. Sects, in particular, are likely to clash with the church and society because of their claim to unique legitimacy. On the other hand, history provides several examples of groups that sustained their sectarian dissent and never developed denominational patterns. The Old Order Amish have for many generations preserved their distinctive beliefs and way of life, largely by withdrawing into their own communities.

Such groups may be considered *established sects* (Yinger, 1970:266–273), groups engaged in sectarian dissent that have organized their internal and external social arrangements toward the perpetuation of their distinctiveness. Some internal arrangements that promote the success of social movements—the quality of leadership, commitment and recruitment processes, social control, and ideology—enable sects to become more stable. Arrangements for the relationship of the group to society also promote the establishment of sects; like the Amish (see Kraybill, 1989), the group may withdraw physically from the rest of society and systematically control which elements of cultural change to allow "into" its everyday experience. Or it may migrate to a more favorable environment, like the Hutterites and Bruderhof (Peters, 1971; Zablocki, 1971). Or it may insulate itself by regulating members' contact with nonmembers and the rest of society, like the Hassidim (Poll, 1969; Shaffir, 1974), censoring or forbidding exposure to mass media, limiting occasions of visiting with nonmembers, or absorbing members so totally that outside influences cannot affect them. Such insulation may intensify the group's sectarianism, as in the case of the Exclusive Brethren (Wilson, 1967:287). Established sects have effectively developed boundaries—physical, ideological, or symbolic—between themselves and the threatening "world." Interestingly, pluralistic societies make it difficult for sects to establish these boundaries because such societies are more likely to be tolerant of the sect and not intensify its sense of opposition. In a totalistic society, by contrast, social opposition can fortify the sect's boundaries, indirectly promoting its survival.

At the same time, however, established sects appear to be somewhat more accommodated in some aspects than new sects. They may be less strident in their claims of uniquely legitimate authority, or they may develop a slightly more positive tension with "the world," perhaps relaxing part of their stand. While maintaining a measure of sectarian withdrawal, some Hutterite colonies have accepted modern technology, resulting in changes in community structure, work, and gender roles (Peter, 1987). Sects' relationship with society depends partly on how society relates to them. The dominant social or religious group may become more tolerant, allowing the sect greater freedom to maintain its distinctive ways. The development of Mormonism into an established sect resulted partly from internal organization and partly from gradual tolerance by society. Interestingly, one of the most significant accommodations by Mormonism was socioeconomic: the abandonment of its early nineteenth-century working-class ideals of cooperation, common ownership, and classlessness. The religious group was transformed into a major financial force, a "theocratic corporation," espousing conservative politics, individualism, and free enterprise (Baer, 1988).

Sects become denominations by giving up their claim to exclusive legitimacy and by reducing their dissent, accommodating to society. Certain kinds of sects tend more to such changes than others. Sects that express their dissent by withdrawing from society are less vulnerable to accommodation pressures than are sects that aim to convert society (Wilson, 1967:26–27). Similarly, sects that dissent against the evils of society are more resistant to transformation than sects emphasizing individual sin or anxiety (Yinger, 1970:272).

The process of accommodation is not necessarily all on the part of the sect; the dominant religion or social group can change as well (B. Johnson, 1971). As the society or established church changes, the sect's tension with that larger group also changes, and the sect may be thrust toward greater or less negative tension with society.

The transformation of sect to denomination is not always in one direction. A group may move toward denominationalism, then have an internal "reform" and move back toward sectarianism. In the 1880s, the Society of Friends (i.e., Quakers) underwent a crisis, in which some members criticized the group's growing compromise with "the world." There were both sectarian and denominational orientations within the group, which had gradually become prosperous and denominationlike. During the crisis, members returned to their historical roots and decided to eliminate a number of their more denominationlike practices (e.g., having ministers and birthright membership) (Isichei, 1967:181). Our model of types of religious collectivities consists of dynamic forces on two continua. Thus, we can imagine groups moving in either direction on either continuum: becoming more or less pluralistic and having more or less negative tension with society.

The prosperity of the Quakers illustrates another factor in sect transformation. Does the very success of a group make it difficult or impossible to retain original ideals and dissent? It is easier for group members to reject the things of "the world" if they are too poor to afford them. As the sect members become established and increasingly respectable in the community, there is pressure to relax restrictions and become more like their neighbors.

These changes are further promoted by the fact that children of the generation that founded the sect have not had the distinctive experiences of their parents. The second generation's commitment to the group may be serious, but certain issues that were very important to their elders may not hold much significance for them. The "sabras" (i.e., subsequent generation) of the Israeli kibbutzim did not have the intense group experience of hardship and fellowship of the initial Zionist pioneers who settled on the land in the newly created Jewish homeland. The pioneers had forged a group vision of an ascetic communal existence, but subsequent generations of members do not have that unique experience as the basis of their commitment (Spiro, 1970:38–59).

A number of studies have suggested that the values and social interaction of some sects, while dissonant from those of the larger society, have the unintended effect of making members more prosperous and acceptable by the standards of society. Certain fundamentalist Christian sects encourage their members to work hard and reliably, to respect authority, and to refrain from drinking, smoking, gambling, and entertainments. Thus, sect values increase the likelihood of members' earning steady wages while not spending that money—except on religious activities. The overall effect is that members are socialized into acceptable roles in society and are able to enjoy material comforts (Johnson, 1961); their sectarian dissent inadvertently brings dissidents "back" into the dominant society. For example, the Black Muslim movement changed from a violently sectarian separatist stance to a more denominational status within Sunni Islam. A major factor in the change was the increasingly middle-class character of the movement, brought about by the Black Muslim work ethic and morality (Mamiya, 1982).

This process does not occur in all groups, however. Sects that live communally withdrawn from society are better able to control the socialization of their members and to disperse material comforts over the entire religious community. Some sects, in fact, socialize their members into roles that are extremely dysfunctional in the larger society, making exit from the sect very difficult (Beckford, 1978b). Nevertheless, material success is a built-in threat to some sects because it is more difficult to dissent from a society in which one has a stake and is enjoying some benefits.

Similarly, organizational success is a threat to sect stability. If a sect is successful in attracting many members and expanding into new geographical areas, it may grow more denominationlike. The group may become bureaucratized and rely on religious specialists—ministers, fundraisers, religious teachers, and so on. Some established sects (e.g., Jehovah's Witnesses) demonstrate the possibility of retaining the dissenting stance under a complex organization. Nevertheless, organizational success brings structures that make it difficult to retain the initial sectarian enthusiasm and small-group intensity.

**Cult Transformations.** Cults are highly precarious organizational forms. Their characteristic pluralism together with their members' general individualistic mode of adherence make them more unstable than other types of religious collectivity. The typical cult transformation is from cult to sect, but some cults can stabilize their groups as established cults. Spiritualism, now nearly 135 years old as a movement, has retained its cultic form. It is loosely organized around independent Spiritualist "mediums," some of whom are affiliated in a network of contacts but with little organizational control or authority (Nelson, 1969).

By contrast, a large number of groups that began with a cultic phase have transformed into sects. The Jesus People (Richardson, 1979), Christian Science (Wallis, 1973), the Bruderhof (Zablocki, 1971), Scientology (Wallis, 1977), and the Unification Church (Lofland, 1966) all began as cult-type groups. The fragility of cults as social organizations results from their indistinct and pluralistic doctrine, their problem of authority, and the individualistic, segmental mode of commitment characteristic of cultic adherents. In the transformation from cult to sect, the key feature is the arrogation of authority; certain members successfully claim strong authority, thereby enabling them to clarify the boundaries of the group belief system and membership. The successful claim to strong authority gives leaders a basis for exercising social control in the group and for excluding persons who do not accept the newly consolidated belief system. Movement toward sectarianism may be a deliberate strategy of cult leaders or simply the by-product of other organizational decisions (Wallis, 1974, 1977).

Other factors promoting this cult-to-sect transformation, besides internal group factors, include factors relating to individual members and to the social environment. Interaction between the cult and other elements of society sometimes promotes the transformation. Largely because of the material and moral encouragement of branches of certain established sects and denominations, the Jesus movement's initial cultic independence was transformed. It became increasingly sectlike, consolidating its belief system along the lines of these outside groups (Richardson, 1979). One element of this belief system was a greater emphasis on authority in the religious community and in families.

Other cult transformations (to denomination or church) are theoretically possible but not common. It is unlikely that a group with cultlike organizational characteristics would have any propensity toward becoming a church. On the other hand, the cult's characteristic pluralism would seem to have strong affinity with the pluralistic tolerance of denominations; yet there are few if any documented instances of a cult transforming directly into a denomination.

Two groups that have made considerable changes toward a denominationlike accommodation with worldly success are 3HO and Vajradhatu, but their pronounced roots in Eastern religion (3HO is derived from a yogic/Sikh tradition, Vajradhatu from Tibetan Buddhism) leave these cult-type movements far from denominational acceptance by the larger society (Khalsa, 1986). Similarly, Hare Krishna (Krishna Consciousness in America or ISKON) has both sectarian and cultic characteristics, and, in its first decades, its leadership moved between emphasizing sectlike or cultlike stances. After coming under attack from the anti-"cult" movement, however, leaders decided to try to become a "denomination of the Hindu church" (Rochford, 1985). While the group's religious freedom is more likely to be respected if it is identified as a Hindu ethnic church, U.S. religious pluralism does not as yet grant equal legitimacy to religions outside the Judaeo-Christian traditions. Thus, no matter how accommodating these groups are, societal attitudes would impede acceptance of groups based on Eastern traditions as having full denominational status in the United States.

A more common pattern is for cults to first undergo internal organization changes, often becoming more sectlike, and later edge toward being more denominationlike. Christian Science began as a cult but experienced the early consolidation of authority of Mary Baker Eddy. Its organizational form was somewhat sectlike for most of its history, though its middle-class members were often more cultic in their style of adherence. With increasing social acceptability, the group now appears to be moving toward the denomination end of the typology.

**Denomination Transformations.** The denomination is the predominant organizational form of religious groups in the United States and Canada today. Widespread religious pluralism is, however, a relatively recent phenomenon, so there is little historical evidence concerning denominational changes. History provides several examples of groups arising as denominations and becoming established as denominations. The Reform movement among Jews was from its outset more tolerant and positively oriented toward society than its Orthodox predecessor (see Steinberg, 1965). The movement required organizational stability and popular acceptance before it could become an established denomination. Similarly, Congregationalism and Methodism were born as denominations rather than sects (Martin, 1965:4). Some of the mainline black denominations (e.g., the African Methodist Episcopal church) also began as denominations. Theirs was not a theological protest against the parent group but mainly a reorganization to allow blacks to achieve leadership and other important roles in their church (see Gravely, 1989).

A group that begins as a denomination but moves away from its stance of pluralism is likely to develop a more sectlike organizational form. The Salvation Army, for example, became a sect after its more pluralistic beginnings (Robertson, 1967). Similarly, recent clashes among Southern Baptist leaders have resulted in greater sectarianization of that denomination (Ammerman, 1990). Just as sectarianization is a strategy for consolidating control in cults, so too are denominations susceptible to sectarianization. But societal pluralism makes it highly unlikely that a denomination could achieve a sufficient monopoly of legitimacy to become a church.

Changes in society appear to be influencing some denominations toward a more sectlike status. To the extent that society does not support or accept key beliefs of groups, they may develop a sectlike organization to protect those beliefs and practices. Societal changes are creating more negative tension between many established religious groups and society, thus encouraging sectlike responses (see Roof, 1978:215–227). For example, increasing sexual "permissiveness" in society presses denominations that do not accept this attitude to retreat toward a more defensive, negative posture.

**Church Transformations.** Similarly, changes in social setting create changes in modes of organization in church-type collectivities. If a group is unable to maintain its monolithic status, it cannot continue as a church-type collectivity. Churches that retain their claims to unique legitimacy are likely to take on sectlike characteristics. The Roman Catholic church of early twentieth-century immigrants in the United States maintained a highly sectarian stance to protect its members from the beliefs and practices of the predominantly Protestant culture. More recently, its greater tolerance of other religions has moved American Roman Catholicism toward a denominational form.

In Canada the Catholic church has more legal recognition by the state but tends to be more geographically concentrated and less culturally assimilated into mainstream values (Westhues, 1978). Thus, even in a society very similar to that of the United States, the Catholic church has a different organizational form. In England the transformation of the Church of England (i.e., Episcopalian) from its churchlike form to a denominational form was primarily the result of increasing societywide pluralism (Swatos, 1979). Church-to-sect and church-to-denomination are the two main historical transformations of church-type collectivities.

### Implications of Organizational Transformations
The types of religious collectivity characteristic of various periods or societies vary. Changes in religious collectivities often reflect responses to society. Religion may be undergoing some significant changes in its social form and location (as Chapter 8 will describe in more detail). An examination of transformations in religious collectivities can help us grasp some emerging patterns.

**Sectarianism in Contemporary Society.** Whether as a deliberate strategy or simple adaptation, increasing sectarianism may well be necessary for certain religious groups in contemporary society. The primary, close-knit, face-to-face relationship is an important part of the sectarian orientation because it supports the believer in the face of real or perceived opposition. The group helps maintain the believer's assurance of being part of the "one way" despite a hostile or unbelieving "world." The close-knit group provides support for a **cognitive minority**—a group of people whose worldview differs from that of the dominant society. Primary group relationships provide a structure within which the believers' distinctive worldview is plausible, whereas outside the group that worldview is disconfirmed (Berger, 1967:163).

To the extent that *any* religious group attempts to maintain its distinctiveness and applicability to all aspects of life, it runs counter to developments in society—especially the compartmentalizing of religion into the private sphere.

Even established religious groups take on some qualities of the traditional sect as they organize to protect their beliefs, practices, and values in the face of society's lack of support. The group must provide its own social support for its beliefs and way of life. Members become a cognitive minority and therefore structure their interactions with each other and the "outside" to protect their worldview (Berger, 1967:163ff.). This change toward sectarianism is, however, primarily a *reaction* to changes in society. One comparative study of strategies of a mainstream Orthodox and a Lubavitch Jewish congregation found that both groups had to adjust their teachings in response to issues posed by modernity, specifically individualism, pluralism, and feminism (Davidman, 1990a). Any group, regardless of how extreme or mainstream its beliefs, that seeks a different role for religion in society is thrust into a minority position. Even doctrinally "orthodox" groups find their posture shaped in response to societal forces.

A related reaction to the "modernizing," pluralistic impetus of "the world" takes the form of religious extremism. In the 1970s and 1980s, there was a marked increase in media attention to religious sectarian pressures in the political sphere. In the United States, the "New Christian Right" has attempted to mobilize sectarian dissent into political action. In Israel similar impulses have been expressed in the form of religious ultranationalism. The Shi'ite branch of Islam in Iran has been notably successful in gaining political clout to turn back the forces of "modernity." These developments are discussed further in Chapters 7 and 8.

The sectarian orientation thrives on a sense of opposition. Ironically, backlash movements (such as the anti-"cult" movement) often inadvertently contribute to the cohesion of the very groups they attack. Similarly, although the larger society may not appreciate the polarization, the homosexual rights groups and their "moral majority" opponents are often actually symbiotically linked. Sectarian groups often provoke their opposition, partly because of their sense of their own rightness and partly because they typically enjoy greater cohesion and sense of purpose when they feel their values and goals are under attack.

**Privatized Religiosity in Contemporary Society.** Other types of religious collectivity are well suited to privatized religiosity. Denominations that focus on religiosity in the private sphere (i.e., family, community, parish)—and cults, with their characteristic individualism and eclectic approach to beliefs and practices—fit very well into their privatized slot. Neither claims exclusive legitimacy, and neither expects its beliefs and practices to govern behavior in all aspects of social life. A member of an Eastern meditation group may find in it greater interior peace and joy, relief from stress and tensions in everyday life, and more satisfying ways of relating to others. At the same time, the person may be an effective corporate manager whose religious beliefs and practices do not challenge or guide behavior in the business world. The cult can perform an adaptive role, providing meaning and a sense of belonging without conflicting with members' roles in a highly differentiated society (Robbins et al., 1975).

In this context, we can examine the significance of emerging religious movements in contemporary Western society. The last two decades have witnessed the emergence of several new religious movements and the rebirth and transformation of some old ones. Although most of these movements are clearly still religious minorities, they have attracted significant numbers of people. In 1976 the Gallup Poll found that about 12 percent of those polled were engaged in one or more of certain recent movements (i.e., mysticism, Oriental religions, yoga, Transcendental Meditation, and the charismatic renewal). Cultic movements, TM and yoga, accounted for 4 percent and 3 percent of the response, respectively. Mysticism (which could be practiced in combination with most other religious perspectives, including traditional religions) accounted for 2 percent, as did the charismatic renewal, which includes members of many traditional religions such as Episcopalianism, Lutheranism, and Roman Catholicism. A further 1 percent were involved in an Eastern religion. Projecting from this sample, these figures suggest that approximately 19 million Americans were involved in these various religious movements (Gallup Poll, 1976).

The sheer number and diversity of new religious movements make it difficult to generalize about these developments. Some groups (e.g., Zen Buddhism) are based on elements borrowed directly from traditional Oriental religions. Others (e.g., Sokka Gakkai and the Unification Church) are syncretic religions that emerged in the Orient and took on further dimensions in their American translation. Some emerging religions such as the Jesus People, the charismatic renewal, and neo-Evangelicalism are adaptations of older Christian movements. Other movements such as Arica and Kerista took their inspiration, as well as some of their language, from the Human Potential movement—an amalgam of religious and psychotherapeutic beliefs and practices.

The following Extended Application suggests a sociological interpretation of these emerging religious movements.

## Extended Application: Emerging Religious Movements

The image of the new religious movements that have caught the attention of the press, public, and social scientists is of a diverse, colorful, strange, or exotic assortment of religious groups. The picture includes saffron-robed, head-shaven youths on street corners singing "Hare Krishna." It includes middle-aged women "channeling" messages from 35,000-year-old spiritual guides. Further images include flashy televangelists, door-to-door witnessing, Eastern gurus, suburbanite witchcraft covens, sales of crystals and incense, and tape cassettes with new hymns or religious messages by group leaders. In the late 1960s and early 1970s, some observers identified new religious forms as part of the youth culture or the counterculture, but by the late 1970s into the 1990s, it became clear that emerging religious movements were appealing to middle-aged (and older), economically comfortable, educated persons, as well. Getting beyond the sensationalistic treatment that such movements have received in newspaper or television coverage is difficult. While some of the emerging religious fervor is the product of media hype or crass commercialism, much of it represents genuine religious experience and commitment.

Analyzing sects, cults, denominations, and churches is more than a mere exercise in typologizing. The foremost purpose of categorizing religious collectivities is to conceptualize a situation and then to ask: Of what larger phenomenon is this an example? The types of collectivities characteristic of emerging religious movements are related to their cultural setting. Indeed, the types of social organization and characteristic religious orientations developed by these emerging religious movements are particularly well-suited to the social structural place of religion in modern society. The types of collectivities characteristic of these new religious approaches are linked with peculiarly modern forms of individual-to-society connections (as discussed further in Chapter 8; see also Beckford, 1984; Beckford and Richardson, 1983; McGuire, 1985).

## The Variety of Emerging Religious Movements

Writing around the turn of the century, Durkheim (1965:475) made some predictions about the future of religion that ring surprisingly true today:

> If we find a little difficulty today in imagining what these feasts and ceremonies of the future could consist in, it is because we are going through a stage of transition and moral mediocrity. The great things of the past which filled our fathers with enthusiasm do not excite the same ardour in us, either because they have come into common usage to such an extent that we are unconscious of them or else because they no longer answer to our actual aspirations; but as yet there is nothing to replace them. . . . In a word, the old gods are growing old or already dead, and others are not yet born. . . . it is life itself, and not a dead past which can produce a living cult.

The new religious movements appear anomalous because they have emerged precisely at a time when religion, as this society has traditionally known it, seems weakest. While theologians were exploring the meaning of the "death of God" and church administrators were trying to cope with membership slippages and general drops in attendance—and while social scientists were debating the thesis of "secularization" (see Chapter 8)—new religious movements emerged, and many of them flourished. Some of these movements are particularly interesting because they appealed to societal sectors that seemed least likely to be attracted to conventional religious movements—young, educated, cosmopolitan, comfortably middle-class persons.

The apparent newness and uniqueness of these developments are, however, misleading. New religious movements have been emerging throughout America's history, in part the result of U.S. religious and cultural pluralism. The diversity of sects that characterized America's early religious pluralism brought one Puritan minister to call the sect-filled commonwealth of Rhode Island "the sewer of New England." Later periods of great religious fervor were the Great Awakening (1730s) and the Second Great Awakening (1790s onward). The latter continued for some years and stimulated pioneer revivalism in the Western expansion well into the nineteenth century. The early nineteenth century saw many religious and communal experiments such as Ralph Waldo Emerson's Transcendental Fellowship (which spawned the Brook Farm commune) and

John Humphrey Noyes's Perfectionists (out of which the Oneida commune was born). Millenarian groups such as Seventh Day Adventists and, later, Jehovah's Witnesses were founded and became especially active as the turn of the century approached. A related movement, Fundamentalism, arose within Protestantism and caused considerable controversy. The Holiness movement, out of which Pentecostalism emerged, also brought about conflict in some churches. A very different, late Victorian development was the rise of Spiritualism and related cults such as "animal magnetism" (i.e., hypnosis) and various healing groups, including New Thought and Christian Science. A number of nonofficial religions flourished, though the actual size of their followings is unknown; many of these groups drew on occult traditions of Europe such as Gnosticism, kabbalism, astrology, Rosicrucianism, and so on (Ahlstrom, 1972, 1978; Albanese, 1991).

Recent new religious movements are, from this perspective, simply the latest in a long history of eruptions of religious fervor. Even the more exotic new religions emerged from a pre-existing social base. The two main sources of new religious movements are the Judeo-Christian heritage and the more amorphous nonofficial religious area or "cultic milieu" (Ellwood, 1978). The Judeo-Christian strain is a fertile source of new religions because of its built-in tendency for cycles of renewal, reform, and schism. In recent times, there is an inherent potential for religious retrenchment to organize itself in sectarian forms, especially as the dominant culture in Western societies becomes less consistent with conservative Jewish/Christian ideologies.

Another segment of new religious movements emerged from the already present but unorganized stratum of nonofficial religion. Most ideas and many ritual practices of numerous new religions were already present in the cultic milieu, but the new movements shaped the ideas and adherents into an organized form. Some new religious movements claim to be totally new; others emphasize that they are older even than historical religions such as Christianity. For example, many new religions borrow from Eastern religions such as Zen and Tibetan Buddhism, various Hindu forms of yoga and meditation, or Sikh, Sufi, and Taoist traditions. Other emerging movements borrow heavily from Celtic lore, Native American religions, and shamanism. Other sources in inspiration include various psychotherapeutic group movements (often called the Human Potential movement), the "Green" movement (especially its spiritual, "deep ecology" version), and various identity movements (particularly, feminist groups, but also movements asserting social identity alternatives for racial and ethnic minorities, for persons with handicaps or chronic illnesses, for homosexuals, etc.). Nevertheless, as a social phenomenon, even the more exotic new religions have their roots in identifiable clusters of prior beliefs and practices (Albanese, 1990, 1991; Melton et al., 1990).

Nonofficial religion tends to spawn cults (as defined earlier in this chapter), though some of these are eventually transformed into sects through changes in the authority pattern. One main reason for this tendency is that nonofficial religion itself is seldom a coherent, authoritative, organized belief system; adherents are often loosely affiliated, drawn together only by vaguely similar perspectives or parallel beliefs. Cults emerge from the amorphous background of nonofficial religion; they organize around a teacher or a technique and are inherently fragile because cultic adherence is highly individualistic and eclectic. Thus, they typically dissolve back into the diffuse milieu from which they came, and members find other focuses for their beliefs and practices within the larger nonofficial religious setting. From the perspective of official religion then, with its emphasis on organizational permanence and numbers of adherents, cultic religious movements are insignificant. This view, however, overlooks the considerable staying power of general cultic ideas and of the numerous adherents who subscribe simultaneously to official and nonofficial religion (Ellwood, 1978).

### Sectarian and Cultic Religion in Modern Society

An examination of two contemporary religious groups exemplifies some of the characteristics of modern sectarian and cultic religiosity in the United States.[6] The Women of Truth prayer group is an interdenominational fellowship of about sixty women who practice charismatic (or pentecostal) Christianity. The Meditation Circle is a group of approximately twenty men and thirty-five women who meet regularly for meditation and a variety of other spiritual practices. When we began the study, both groups had been in existence for several years, and both were steadily attracting new members.

Members of the Women of Truth were more homogeneous in age (about 35–55), education level (most had some college education, but only about a third had graduated and none had advanced education), marital status (all were married or widowed), and social class situation (virtually all were homemakers whose husbands earned middle-class incomes in business or lower-professional occupations). The prayer group met as a whole once a week, usually in a church to which some members belonged. Most activities were scheduled during daytime hours when members had the fewest obligations to their families. Most members also attended meetings of two or more subgroups that also met weekly for Bible study, visiting the sick, praying for each other, or for mutual support (e.g., a new mothers' group and a widows' group).

The Meditation Circle had a greater age range (about 30–60, with the average age a little lower than Women of Truth); members were generally better educated (nearly all had or were studying for college degrees, and about half had advanced degrees, as well). Fewer than half of the members were presently married; the rest were mainly never married or divorced. All but two of the members were employed out of the home. The income range was also greater, with three members who worked in the fine and performing arts earning near-poverty incomes and some established professionals (architect, lawyer, psychotherapist) earning enough to be in the upper-fifth income bracket in the region. The entire Circle gathered every Wednesday evening in a rented carriage house, and about half the group also met on Sunday mornings. Some members participated several times a week in other group-sponsored meditations, healing circles, classes, workshops, and social gatherings.

---

[6]This information is based on my own research, fully reported in *Ritual Healing in Suburban America*, 1988; these names of groups are fictitious, and these descriptions are composites drawn from several similar groups in order to protect the anonymity of respondents.

The sectarian orientation of the Women of Truth is particularly evident in their preoccupation with maintaining the boundaries of their truth. They emphasize a strong sense of right versus wrong, teaching constantly about sin, repentance, confession, forgiveness, and salvation. They reaffirm strict traditional moral norms regarding gender roles, marriage and divorce, sexual behavior and procreation, work, and childrearing. They interpret their daily lives as arenas of a war against Satan, a battle of powers. For example, one lengthy informal discussion shortly before Christmas was devoted to preventing members from buying for their children "dangerous" toys and games, like Dungeons and Dragons and fantasy books. Prayer meetings typically include prayers to protect members from the influence of Satan.

Like other pentecostals, the Women of Truth emphasize knowing God through experience, rather than just knowing about God. Personal experience (e.g., praying in tongues) and group experiences (e.g., hearing a prophecy— believed to be God speaking directly to the group through one of its members) are of primary importance. The belief system holds that the Holy Spirit is powerful and that this power is available to the individual through the charisms. Accordingly, the Holy Spirit empowers the individual for better prayer, personal transformation, fighting Evil, and personal healing. Similarly, the group is empowered in its battle with Evil and in its prayer, praise, and healing. Thus, religious experience is seen as both validating the group's belief system and empowering its members in their personal lives and relationships with others.

Another sectarian characteristic is the group's strong authoritarianism. Accordingly, there is only one truth; it is not up to each member to discover but is spoken authoritatively by pentecostal leaders, and the authority structure of the group is God-given. Specifically, the authority structure places women in submission to men, so the officers of the prayer group must have their husbands' permission to serve and their business meetings must be held under the supervision of a (male) pastor.

Within the context of the group, members found considerable support for integrating their pentecostal beliefs into their daily lives. One woman witnessed to how her husband's job entailed frequent moves from one city to another, and how Christ led her to a group like this in each new city to keep her close to the Christian way of life and to give her new friends. Another woman received prayers and moral support as she described the anguish of trying to stay married to an abusive man who had not yet "found the Lord." Yet another member described the "healing" of her relationship with her teenage daughter, who had stopped running away from home and had not dropped out of school. Prayer meetings were full of everyday concerns: children and husbands, relatives and friends, neighbors and pets, illnesses and personal problems, housework, churchwork, yardwork, family vacations, routine leisure activities. Supported by the close network of the prayer group, these women tried to apply their faith to all aspects of their lives. At the same time, the unity of the small, close-knit group of fellow believers served to protect members from the evils of "the world," shoring up members' defenses against those aspects of modern society their religion decries.

The Meditation Circle was, by contrast, highly eclectic in its beliefs and practices. Individual members were free to choose among many modes of spiritual development, including those not taught within this circle. The group had begun with a combination of Jain Yoga and Meditation, together with other meditations borrowed from Psychosynthesis and Arica, and other practices borrowed from Tibetan Buddhism and Reiki. Individual members had tried virtually every "New Age" or "Human Potential" movement introduced in that region: rebirthing, crystal healing, Shiatsu, neurolinguistic programming, Cornucopia, Reichian therapy, T'ai Chi, Transcendental Meditation, and so on.

No single path was considered right for all persons at all times. Rather, the group emphasized choosing change and developing one's self. One woman explained:

> Self-awareness and taking responsibility go together hand-in-hand. If I walk around unaware, then I feel like bad things just "happen" to me. I'm a victim! The more I understand myself and my situation, then the more control I have in my life. I realize that I have choices—like I chose not to accept a promotion if it meant sixteen hour days, . . . or like choosing to be an adult and not a little girl to my mother for the first time in thirty-seven years. . . . And I can choose my responses to situations and to other people. Like I don't have to get caught up in stress or in anger or whatever.
>     . . . But there's no single right choice that's for everybody all the time. A lot of paths are good paths. Like, I'm a vegetarian these last few years, but that doesn't mean you aren't making the right choice for you if you aren't vegetarian. And maybe ten years from now, [being] vegetarian won't be what I need then. My responsibility is to choose which paths are best for me for now . . . . We keep growing by using that responsibility for ourselves.

Thus, there was little or no focus on fixed (much less traditional) moral norms, guilt, or sin. Rather, the good life was one of continual growth in understanding of self and others, greater awareness of one's body and emotions, greater balance and harmony in one's relationships to inner self, to others, and with the cosmos. While the "sacred" was conceived to be greater than the individual, it was tapped by going within oneself. Thus, the realm of the sacred was relatively imminent, subjective, and not fixed in one place, belief system, or religious object.

The group's ideal was a self that could be flexible and adaptable without being determined or overwhelmed by others. Group rituals served to free the self from learned constraints and negative effects of guilt and to open new potentials for choice and change. They proffered energizing experiences without expecting permanent commitment for the direction of that energy.

Group members were very loosely committed to the group and its practices. The practices changed continually, generating little long-term commitment. Members cared about each other as individuals and often went out of their way to support or help each other, but the group itself was not an end, much less the only possible source of spiritual belonging or meaning. Furthermore, many members belonged simultaneously to other groups (e.g., mainline denominational churches), which fulfilled some of their needs; through their jobs, professional ties, and other networks, they participated in wide, relatively cosmopolitan supports for other aspects of their identities.

## Implications of New Religious Movements

The emergence of numerous, diverse religious movements at this point in history may have come as a surprise to many observers, especially those predicting the demise of religion. New religious movements have emerged, as we have seen, from a broad background and long history of similar movements. Nevertheless, in interpreting this latest eruption of religious movements, we need to ask why they are emerging now and why these particular types of movements are finding such wide appeal.

One approach suggests that new religions develop in response to crises in world economic order and that the type of religious movement varies according to the location of adherents relative to the bases of economic power (Wuthnow, 1978b). Another possible interpretation views new movements (in modern Western societies, at least) as responses to the disintegrating old bases of moral or political order. This interpretation explains dualistic movements such as the charismatic renewal and modern Christian evangelicalism as protests against the moral relativism and permissiveness of U.S. culture; other new movements (e.g., Human Potential and imported Oriental religions) turn the moral relativism into a virtue (Robbins et al., 1978; see also Tipton, 1982b).

Another approach suggests that these movements represent two kinds of responses to the demands of modern societies on individuals. Cultlike groups offer the ability to control and change one's social "face"—to manipulate one's social identities—while reaffirming a hidden-core self (Bird, 1978, 1979; Dreitzel, 1981). Sectlike groups address the same problem of self by consolidating all social identities into a single, central, religiously defined self and by strengthening the institutions (e.g., religious group and family) that control and support that self.

One interpretation of the upswing of interest in cult-type religions such as the Meditation Circle is that they are possibly fulfilling Durkheim's prophecies about the shape of religion in modern pluralistic society: the "cult of man." Durkheim (1969:26) stated:

> As societies become more voluminous and spread over vaster territories, their traditions and practices . . . are compelled to maintain a state of plasticity and instability which no longer offers adequate resistance to individual variations. These latter, being less well contained, develop more freely and multiply in number; that is, everyone increasingly follows his own path.

Elsewhere Durkheim (1951:336) predicted that in complex societies religion would focus on the enhancement of the human personality because this would be all that members of such diverse societies would have in common. Traditional religions locate the sacred outside the individual, reflecting (according to Durkheim) the self-consciousness of the whole social group. Although the "cult of man" would express and dramatize social relations, sacred power in such religion would be located within each individual (Westley, 1978:139). The idealization of individual humanity lends itself to cult organization, with its characteristic pluralism and individualism.

Although all of the new religious movements assert alternate meaning systems and practices, their opposition to the dominant culture and religiosity is somewhat limited. Cult-type groups, with their characteristic pluralism, are particularly accepting of many dominant culture values. They propose an alternative worldview and practices that emphasize neglected aspects of reality, but this alternative conveniently enhances adherents' chances for success or, at least, happiness in the context of the dominant society (Bird, 1978:181). Sectlike groups, while encouraging members to see a gap between their way and the "ways of the world," are often similarly accommodated to the values of the dominant culture. Members of pentecostal prayer groups frequently witnessed that God had blessed them with success on the job or with material goods, as evidence of God's power. The tension of contemporary religious movements with the rest of church and society appears to be segmented or privatized, and the apparent impotence of such protest may be related to wider issues, as discussed in Chapter 8 (see Bellah, 1976; Wallis, 1978; Wilson, 1976).

This tension of new sects and cults with the society raises the issue of their long-term impact. Religious movements are potential sources of social transformation (explicated further in Chapter 7). Are the new religious movements transformative? Or do they serve merely to integrate their adherents into the dominant society (see Hargrove, 1978; Robbins et al., 1975)? Many of the movements themselves explain their privatistic situation as a positive feature. They assert that real social change will emerge through their movement by the transformation of individuals. The organizational features of sectarian and cultic religion, such as exemplified by the Women of Truth and the Meditation Circle, are particularly well-suited to the social structure of modern societies in which religion is privatized (see Chapter 8). The sectarian forms preserve religious purity by shoring up the supports of belief (usually in a transcendent sacred reality) by a close-knit enclave of fellow believers, protected from the threats of "the world." The cultic forms turn the individualism and freedom of choice, characteristic of the modern world, into a religious virtue and an advantage in the spiritual realm, consciously developing changeable selves, in tune with a more subjective, imminent sacred reality.

Nevertheless, social movements such as these have in the past often influenced the society indirectly. Their ideas filter into the mainstream culture, and they serve as experimental models for the rest of society. The idea of holistic health and treatment of illness has filtered (in a watered-down form) from religious movements to mainstream medical schools. Whether such influences are ultimately transformative cannot yet be known. Are the new religious movements bringing about a "new age" or ushering in the millennium? Or are they ways of soothing such unpleasant aspects of modern life as loss of freedom and control in the public sphere and doubts about self and interpersonal relations in the private sphere? These implications suggest why an understanding of new religious movements and their impact is important for gaining an understanding of the society as a whole.

# Summary

Religious collectivities and their variety have been the subject of much analysis and typologizing. The dynamic model presented in this chapter emphasizes the ever-changing form of religious collectivities. The church–sect typology is a general developmental model for any organization defining itself as uniquely legitimate and existing in a state of positive (i.e., church) or negative (i.e., sect) tension with society. By contrast, cults and denominations do not claim unique legitimacy.

This model distinguishes between types of religious collectivity and types of religious orientation of members. The two main criteria for defining typical orientations are (1) the extent to which the person's religious role is diffused throughout all aspects of life and (2) the extent to which the person adopts standards of virtuoso religiosity for judging self and others. Diverse orientations can exist within the same collectivity. The tension between two or more different religious orientations sometimes produces change within the organization or stimulates formation of a new group.

Religious collectivities typically change as they interact with their social environments; organize themselves toward achieving their goals; recruit, consolidate, and deploy members; and arrange their organization and leadership. As a new movement becomes institutionalized, it faces several dilemmas. The very processes that would enable the movement to realize its goals simultaneously change the movement and reduce its chances of attaining those goals. Emerging religious movements illustrate this model of religious collectivities and orientations.

# Recommended Readings

### Edited Collections

Eileen Barker, ed. *Of Gods and Men: New Religious Movements in the West.* Macon, Ga.: Mercer University Press, 1983.

Thomas Robbins and Dick Anthony, eds. *In Gods We Trust: New Patterns of Religious Pluralism in America.* New Brunswick, N.J.: Transaction, 1990.

### Case Studies

Mary Jo Neitz. *Charisma and Community: A Study of Religious Commitment Within the Charismatic Renewal.* New Brunswick, N.J.: Transaction, 1986. An ethnography of Catholic charismatic social reality, conversion, and community experience.

Margaret Poloma. *The Assemblies of God at the Crossroads: Charisma and Institutional Dilemmas.* Knoxville: University of Tennessee Press, 1989. Using surveys, case studies, and extensive interviewing, Poloma examines the beliefs and practices of a large pentecostal denomination, whose very institutional growth and boundaries pose problems for its future.

R. Stephen Warner. *New Wine in Old Wineskins: Evangelicals and Liberals in a Small-Town Church.* Berkeley: University of California Press, 1988. A highly readable ethnography of a single Presbyterian congregation, as it underwent significant internal changes, growth, and divisions; this book received the 1989 Distinguished Book Award of the Society for the Scientific Study of Religion.

## Fiction

Umberto Eco. *The Name of the Rose*. New York: Harcourt Brace Jovanovich, 1980. Set in a fourteenth-century Italian monastery, this novel is full of mystery, historical and philosophical webs, occult and esoteric allusions, theology and religious history, semiotic riches, and, above all, the love of words—spoken words, written words, read words, experienced words.

Frank Herbert. *Dune* (1965); *Dune Messiah* (1969); *Children of Dune* (1976) all published by Berkley Books, N.Y. *God Emperor of Dune* (1981), *Heretics of Dune* (1984), and *Chapterhouse Dune* (1985) are published by G. P. Putnam & Sons, N.Y. This science-fiction series includes many excellent illustrations of religious movement formation, mobilization, charisma, routinization, sectarianization, and so on.

# CHAPTER SIX

# Religion, Social Cohesion, and Conflict

Religion contributes to both social conflict and cohesion. Some people find it surprising or objectionable that religion should be a force for conflict as well as cohesion. In many respects, however, the aspect of conflict is merely the obverse of social cohesion; some conflict is an integral part of what holds groups together. Religion's importance as an expression of a group's unity also makes it significant as an expression of that group's conflict with another group.

We must keep a neutral conception of conflict and cohesion. A tendency exists in our society to think of conflict, in the abstract, as "bad"; cohesion, in the abstract, seems "good." Yet when we examine some concrete instances of cohesion and conflict, we see that we evaluate the *content,* not the process, as good or bad. Was Moses's confrontation of the pharoah necessarily bad because it was conflict?

In this chapter, we first examine the ways in which religion reflects or contributes to the cohesion of a social group. Particularly problematic is the issue of whether religion contributes to the integration of complex societies such as the United States, and we explore this issue by examining the civil religion thesis. Then we analyze the relationship between religion and social conflict, focusing on those aspects of religion and society in general that contribute to or reflect social cleavages. Finally, we apply these understandings to the case of "the troubles" in Northern Ireland, where religion is an important factor in civil strife.

## Religion and Social Cohesion

*Blest be the tie that binds*
*Our hearts in Christian love.*
*The fellowship of kindred minds*
*Is like to that above.*
(Hymn, John Fawcett, 1782)

The theme of social cohesion is central to sociology. What makes society possible? What integrates separate members into a larger whole, the identifiable entity we call "society"? Society is more than an aggregate of people who happen to share a certain time and space. Although made up of individuals, society is not reducible to individual beliefs, values, and behavior. Social norms and traditions existed before the individual and have a force that is external to the individual. In a society with a norm against cannibalism, for example, persons known to violate the norm will be punished regardless of whether or not they agree with the norm. Another evidence of the external quality of society is the process of socialization, in which the child is confronted with the given expectations, language, and knowledge of that society.

What, then, is the nature of the unity of society that gives it this powerful quality? How is the individual linked to the larger society? How does the society gain the commitment and cooperation of its members? According to integration theories, societal cohesion and stability are assured by the functioning of institutions (e.g., religion, education, and the family) that represent the larger social reality to the individual and enable the individual to accept personally that definition of reality.

Religion is one important contributing factor in societal integration. Religious symbols can represent the unity of the social group, and religious rituals can enact that unity, allowing the individual to participate symbolically in the larger unity they represent. The Christian ritual of communion is not only a commemoration of a historical event in the life of Jesus but is also a representation of participation in the unity ("communion") of believers. Also, to the extent that a religion imparts important values and norms to its members, it contributes to their consensus on moral issues. Especially significant is religion's ability to motivate believers' commitment and even sacrifice to the group's purposes. By referring to a sphere that transcends everyday life, religion encourages individuals to seek the good of the group rather than their own interests. This consensus and commitment of members is largely a positive effect, but religion also wields powerful negative sanctions for noncooperation. Again, by reference to a transcendent realm, religious sanctions are more potent than earthly punishments.

Integration theories of society stress the equilibrium and harmony of the group. They show the ways that religion helps to maintain equilibrium whenever events threaten it. For example, funeral rituals help a group regain its social balance and morale after the death of a member (Malinowski, 1948:18–24). Religious healing is often a way of reintegrating the deviant member into the group (McGuire, 1988). Changes of social status (e.g., marriage, adulthood, or adoption) are integrated into existing status arrangements through religious symbols and rituals (Van Gennep, 1960). The balance of political and economic power is often ritually expressed and confirmed (Kertzer, 1988). Sometimes religious symbolism and ritual action overcome sources of group cleavages and conflict. In a contemporary Mexican village, the celebration of the fiesta of the Virgin of Guadalupe served to express ethnic cleavages (Mayo and Mestizo) within the community but then to overcome them by affirming the Virgin's protective relationship to all Mexicans (O'Connor, 1989).

Most examples of religion contributing unambiguously to social cohesion are from relatively simple, homogeneous societies. If religion is coextensive with society, its contribution to social cohesion is generally clear. Many societies, however, are not coextensive with a single religion. What is the basis of cohesion in a society such as the United States, in which many competing religions exist and a relatively large number of persons participate in no religious group? What is the function of religion for cohesion where conflicting societies presumably share the same religion (e.g., two warring Christian nations)? How can integration theories explain situations in which a religion arises in a society to conflict with the established ways of that society? Although religion does contribute to social cohesion in complex, heterogeneous societies, its role is not very clear-cut. In such societies, the basis of societywide integration is problematic.

### Religion as an Expression of Social Cohesion: Durkheim

One approach to social cohesion and religion holds that religion is the *expression* of social forces and social ideals. This perspective emphasizes that *wherever there is social cohesion, it is expressed religiously*. The classical statement of this theoretical approach is Durkheim's, and his insights are sufficiently important to explore in some detail. Throughout Durkheim's works—on the division of labor and on deviance, education, and religion—the theme of social cohesion is central (Bellah, 1973). The relationship of the individual to the larger society was relatively unproblematic for Durkheim because he identified the larger social group as the source of individuation. His theory of religion illustrates this resolution.

Durkheim based his discussion of religion on anthropological evidence about the beliefs and practices of Australian aborigines because he believed that their religion would illustrate the most elementary form. Although some of the material he used has been discounted by later anthropological investigations, Durkheim's general theory of religion is still useful for understanding some aspects of religion, particularly in relatively homogeneous societies. Durkheim's definition of religion rests on a distinction between the "sacred" and the "profane." Religious beliefs express the nature of sacred things and human relationships with them; religious rites establish the proper conduct in the presence of sacred things.

According to Durkheim (1965:22), religion is in its very essence *social*. Religious rites are collective behavior, relating the individual to the larger social group. And religious beliefs are **collective representations**—group-held meanings expressing something important about the group itself. Durkheim may have overstated the social aspect of religion, as Malinowski (1948:65) later pointed out, because subjective and even purely individual religious experiences figure importantly in virtually all religions. The experiences sought by the Christian mystic or a participant in a Sioux initiation rite, for example, are highly individualistic, though their meaning is derived from group-held beliefs and imagery. Furthermore, Malinowski noted, not all times of collective effervescence are religious, nor are all religious gatherings necessarily unifying (see Geertz, 1957). Bringing people together for a periodic religious ritual creates the potential for friction, especially in times of stress or hunger. Nevertheless, Malinowski agreed with Durkheim that religion provides a basis of moral cohesion to the social group.

Durkheim observed that a sense of *force* was central to primitive religions. The totemic principle that represented abstract force to the Australian tribes was comparable with other primitive religions' awe of a force (e.g., "mana," "orenda," and "wakan" in Polynesian and North American tribes). Durkheim emphasized that religious force was not an illusion. Although the symbols for expressing this force are imperfect, the force that people experience is real—*it is society*. Durkheim described religion as a system of shared meanings by which individuals represent to themselves their society and their relations to that society. Thus, religious meanings are metaphorical representations of the social group, and participation in religious ritual is experience of the transcendent force of society itself (1965:257).

Thus, Durkheim resolved the individual-to-society relationship by presenting individuals as transcending themselves in communion with the greater reality—society itself. But Durkheim added that this force is not entirely outside the individual; it must also become an integral part of the individual's being because society cannot exist except through individual consciousness. By this twofold relationship, religion assures the commitment of society's members and empowers them to act accordingly. Thus, religious beliefs are idealizations by which society represents itself to its members. Religious rites renew these representations by rekindling the group's consciousness of its unity. At the same time, they strengthen individuals' commitment to the group's expectations and goals (Durkheim, 1965:especially Book 3, Chapter 4).

How do collective representations and collective rituals link the individual to the larger social group? Durkheim's emphasis on language and ritual suggests two ways of understanding this individual-to-society link. First, language and other symbol systems (e.g., religious symbols) depend on shared meanings, and meaning requires a shared reality. By exploring how language and other symbol systems articulate a group's reality, sociologists may come to a better understanding of how people subjectively share that reality (see Fabian, 1974:249–272). Second, studies of ritual can point to how the individual is related to the larger society. Participation in certain religious rituals appears to reduce the sense of boundaries between participants, producing an experience of unity (see Douglas, 1966, 1970; Kertzer, 1988; Turner, 1969, 1974a, 1974b). Both language and ritual articulate the unity of the group and serve to separate that group from others.

Durkheim's emphasis on the significance of ritual for both the individual and the collectivity points to some interesting problems of describing religion in modern society (see Pickering, 1984:442–456). Moving from primitive religions to world and civil religions, Durkheim (1965:474, 475) asserted that all societies need regular events to reaffirm their shared meanings and central ideas. He saw no basic difference between specifically religious commemorations, such as Passover or Christmas, and civil rituals, such as Independence Day or Thanksgiving. Thus, Durkheim implies that in modern nations (e.g., his own country, France) religious representations and rituals may comprise the civic religion of the national collectivity. This hypothesis has been applied to U.S. society too, as the following discussion shows.

### The Civil Religion Thesis

The civil religion thesis is important because it proposes that a religious form exists for the unity of even highly differentiated, heterogeneous societies. **Civil religion** is "any set of beliefs and rituals, related to the past, present, and/or future of a people ('nation') which are understood in some transcendental fashion" (Hammond, 1976:171). Civil religion is the expression of the cohesion of the nation. It transcends denominational, ethnic, and religious boundaries. The civil religion has its own collective representations, by which the nation represents an ideal of itself to its members. It has its own rituals, by which members commemorate significant national events and renew their commitment to their society. Despite many debates over definitions of civil religion, the thesis appears useful for describing a number of social phenomena.

Many civil ceremonies have a marked religious quality. Memorial Day, Fourth of July, presidential inaugurations, all celebrate national values and national unity (Bellah, 1967; Cherry, 1969; Lerner, 1937; Warner, 1953). There are national shrines such as the memorials in Washington, D.C., the Capitol itself, the birthplaces of key presidents, war memorials, and other "special" places. It is not their age or even historical significance but their ability to symbolize the transcendence of the nation as a "people" that inspires awe and reverence. A visitor to Independence Hall said, "Just standing here sends chills down my spine." National shrines are "sacred," in Durkheim's sense of the word. Similarly, there are sacred objects of the civil religion—especially the flag. Interestingly, the Bible is probably also a sacred object in civil religion, not because of its content but because it signifies an appeal to God as the ultimate arbiter of truth and justice. The extent to which these ceremonies, shrines, and objects are set apart as sacred can be seen in the intensity of outrage at inappropriate behavior or "desecration." Some people were arrested during the 1960s for wearing or displaying a copy of the U.S. flag improperly (e.g., on the seat of their pants).

Civil religion also has its myths and saints. Lincoln is a historical figure who particularly symbolizes the civil religion. His actions and speeches contributed to the articulation of that religion in a time of crisis, and his life from his humble birth to his martyrdom typifies values of the civil religion. Other "saints" include key presidents (Washington, Jefferson, Wilson, Franklin D. Roosevelt, Kennedy), folk heroes (Davy Crockett, Charles A. Lindbergh), and military heroes (MacArthur, Eisenhower, Theodore Roosevelt). Similarly, there are stories that exemplify valued traits (e.g., the Horatio Alger rags-to-riches genre) and images (the frontier). Socially important myths include the American Dream—the land of plenty—unlimited social mobility, economic consumption, and achievement.

Although these shrines, saints, and ceremonies are not religious in the same sense as, for example, Greek Orthodox shrines, saints, and ceremonies, they are still set apart as special and not to be profaned. They are an important element of nonofficial religion (as described in Chapter 4) and exist alongside—separate, yet related to—official church religion.

If civil religion is the expression of the integration of a nation, we might expect its especially powerful articulation at the resolution of a conflict. Just as tribal rituals "heal" internal strife and celebrate the unity of the tribe, so too does the rhetoric of inaugural speeches and court pronouncements represent the resolution of conflict and appeal to the overarching integration of the group (Hammond, 1974). The symbolism of civil religion is also very evident when the nation is believed to be threatened by an enemy. During wartime, members' commitments and sacrifices are given special significance. Even one's vegetable garden becomes a symbol of patriotic effort—a "victory garden."

**American Civil Religion.** According to Bellah (1967), American civil religion is related to biblical religion, yet is distinctively American. Biblical symbolism (e.g., chosen people, promised land, new Jerusalem, death and rebirth) are prominent themes. On the other hand, the civil religion is genuinely American and parallels the biblical religions, not replacing them. Civil religion and Christianity, accordingly, are clearly divided in function: Civil religion is appropriate to actions in the official public sphere, and Christianity (and other religions) are granted full liberty in the sphere of personal piety and voluntary social action. This division of spheres of relevance is particularly important for countries such as the United States, where religious pluralism is both a valued feature of sociopolitical life and a barrier to achieving a unified perspective for decision making. By having a civil religion for the public sphere and a diversity of particular religions in the private sphere, the social structure has cohesion with the sense of individual freedom of choice. The success of this division is, however, problematic in U.S. society.

What emerges from sociological descriptions of American civil religion is a picture of diverse—even conflicting—values associated with what is central to the people. Bellah's version implies a single, clear-cut, yet ever-developing ideological stance. But is the American civil religion a unified entity? And if not, does it have the capacity to be unifying?

One explanation of diversity in American civil religion is the distinction between "priestly" and "prophetic" versions of civil religion (Marty, 1974). The **priestly version** of American civil religion celebrates the greatness of the nation, its achievements and superiority. The **prophetic version** calls the nation's attention to its offenses against the idealizations for which it stands. Both versions are very much a part of American thought and rhetoric, but they are clearly in conflict. During the 1960s and 1970s, these versions were used to justify opposing stances on U.S. engagement in Vietnam. One set of bumper stickers during that period proclaimed, "America—Love it or leave it!" Another set of bumper stickers retorted, "America—Change it or lose it!" Are these expressions of the *same* civil religion?

While Bellah's characterization of American civil religion strongly emphasizes a prophetic strand, nationalistic sentiments and beliefs of the priestly version also have a religious quality. Bellah emphasized that civil religion involved subordination of the nation to principles that transcended it. The priestly version of civil religion, by contrast, frequently devolves to worship of the nation itself or to identification of God's will with the aims of "our kind of people." This version of American civil religion has historically been used to legitimate intolerance, as illustrated by the U.S. history of agitation and discrimination against Asian-Americans since 1850, culminating in the internment of American Japanese during World War II (H. Hill, 1973).

Even the most vulgar forms of American civil religion have considerable appeal and motivating power, exemplified in the prosecutor's speech to the jury at the trial of some communist labor organizers in 1929:

> Do you believe in the flag of your country, floating in the breeze, kissing the sunlight, singing the song of freedom? Do you believe in North Carolina? Do you believe in good roads, the good roads of North Carolina on which the heaven-bannered hosts could walk as far as San Francisco? . . . Gastonia—into which the union organizers came, fiends incarnate, stripped of their hoofs and horns, bearing guns instead of pitchforks. . . . They came into peaceful, contented Gastonia, with its flowers, birds, and churches . . . sweeping like a cyclone and tornado to sink damnable fangs into the heart and lifeblood of my community. . . . They [the people of Gastonia] stood it till the great God looked down from the very battlements of heaven and broke the chains and traces of their patience and caused them to call the officers to the lot and stop the infernal scenes that came sweeping down from the wild plains of Soviet Russia into the peaceful community of Gastonia, bringing bloodshed and death, creeping like the hellish serpent into the Garden of Eden (quoted in Pope, 1942:303–304).

There is a religious quality to these images of the nation, representing people's sense of their unity and capable of motivating them to action for the group. The legitimating capacity of these ideas and sentiments, whether in their most noble or most vulgar expression, suggests that a broad concept of "civil religion" is useful.

**Religion and Nationalism.** So far we have focused primarily on American civil religion. Is the concept of civil religion useful for describing similar phenomena in other societies?

Essentially, civil religion may be a very general form of nationalism. A useful distinction at this point is between state building and nation building. **State building** refers to developing an authoritative, utilitarian organization for expediently conducting a country's internal and external business. **Nation building** refers to developing a country's sense of solidarity and identity as a people. An example of this analytical distinction is the Spanish peoples. Most Spaniards both participate in the Spanish state and identify with the Spanish nation, but the Catalonian minority do not feel part of the Spanish national identity or solidarity, even though they are subject to the authority of the Spanish state (see essays in Eisenstadt and Rokkan, 1973). Although nationalism has been a continuing political force in history, civil religion may be the response of modern and modernizing peoples to their peculiar problems of nation building.

Civil religion is clearly an element in nation building. It can shape a national vision, sacralizing the ideals and "destiny" of a people. It can give national solidarity and identity a religious quality, enabling peoples of diverse tribal, regional, ethnic, and religious groups to come together in a central, unifying cultural experience. It does not always successfully accomplish this task of unification, but successful nation building often entails civil religious solidarity. This nation-building potential can be seen in the attempt to develop an Arab civil religion in the Middle East. Lacking an Islamic tradition of effective political participation, some Arabs are consciously trying to develop Islam into a national (i.e., civil) religion. They present Muhammad as an Arab hero whose convictions inspired him to initiate a new civilization, create an Arab culture, and unify the Arab peoples. The deity of Islam is portrayed as mandating an Arab destiny (Hermassi, 1978; see also Ayubi, 1983). This kind of Arab civil religion stands in a similar relationship to its Islamic roots as American civil religion does to Protestantism. Like American civil religion, Arab religious nationalism has its potential for prophetic criticism. Both Arab and American civil religions can contribute to the process of nation building in the face of internal divisions.

Several factors affect the potency of a religious tie with nationhood. One obvious factor is the degree of religious *homogeneity* in the nation. If a people are already of a single particular religion, that religion can (but does not necessarily) serve as the link with nationhood. Where historically that single religion has been the sole available means of expressing nationality against foreign domination, that religion is especially likely to develop as a potent form of civil religion, as well (D. Martin, 1978:107; see also Ramet, 1989). This religion–nationalism relationship is exemplified by the situation of Palestine, Cyprus, Poland, and Ireland (described in some detail in the Extended Application later in this chapter). It also illustrates how religion can be internally unifying, yet a salient source of conflict with outsiders.

In more heterogeneous societies, civil religions may be articulated precisely to overcome the divisiveness of exclusive loyalty to particular religious communities. For example, in Malaysia, the king must be both a Malay and a Muslim, according to the constitution. To promote a unified identity among the entire nation, including large minorities of non-Malays (e.g., Chinese and Indians) and non-Muslims, the government promotes civil rituals and tries to sacralize the ideals of interethnic cooperation and "just" representation of all groups in the heterogeneous population (Markoff and Regan, 1981; see also Mess and Pearce, 1986, and Regan, 1989, regarding the rise of Islamic fundamentalism in reaction to Malaysian civil religion and modernization).

Because religion has considerable potential for legitimating (or delegitimating) the ruling authority of the state, leaders and would-be leaders often try to shape and interpret the religion-to-nationhood linkage in a way that would favor their own power and legitimacy. Sensing the growing influence of opposition Shiite fundamentalist leaders in the 1960s and 1970s, the Mohammad Shah of Iran tried to create a civil religion under the control of the state in order to undermine the traditional religious institutions. He attempted to persuade the people that he was the rightful guardian of the faith. He began to replace religious courts with state-sponsored "Houses of Justice," religious educational institutions with state-controlled ones, and mosques under the control of his religious opponents with state-sponsored ones. The Shah's religious establishment had little effect on the loyalties of most Iranians; indeed, many were incensed, believing that the Shah's civil religion was a perversion of their Islamic tradition (Voll, 1982:292–296). The failure of the civil religion contributed to the decline of the legitimacy of the monarchy and its eventual overthrow by revolutionary religious fundamentalists.

Often, representatives of particular religions are highly critical of civil religions, especially when the particular religion seeks *hegemony* (i.e., dominant influence) in the state. Some critics decry the dilution or adaptation of the true (particular) faith and its symbols. Others criticize any accommodation with religious perspectives other than their own. In the United States, much of the conflict over prayer in public schools is due to the tension over the state's pluralistic accommodation of non-Christian or nonreligious minorities. Among the most outspoken proponents of prayer in the schools are those who seek a conservative Christian hegemony in the nation. Such tensions over the "proper" religious national vision illustrate that there may actually be conflicting or competing civil religions.

The development of civil religion in Israel also reflects the influence of particular religious perspectives in trying to shape the image of nationhood. Israel exemplifies a situation in which there are multiple competing national visions, with their respective civil religions. Between 1919 and 1945, the Israeli civil religion could be characterized as "Zionist-socialism," uniting around a blend of socialist-worker beliefs and symbols, together with an image of the state of Israel as the culmination of Jewish history. "Statism" was the second phase, between 1948 and 1956. It was characterized by an emphasis on the state of Israel itself as a centralized political reality and as a symbol of the collective Jewish people. The statist model, however, did not develop adequate symbols and rituals to be effective. Since 1956 a new civil religion has developed; it focuses more on Judaism and Jewish tradition as a basis for collective Israeli identity and on Israel's connection to worldwide Jewry. Thus, Judaism and Jewishness have become increasingly central to Israeliness. The explicit linkage of the particular religion and national identity has been effectively reasserted (Liebman and Don-Yehiya, 1983). Different national images and their relative emphasis on traditional Judaism have profound influence both on the problem of integrating the society, treatment of indigenous (Muslim) Palestinians within the country, and on formulating a viable diplomatic policy relative to Israel's Arab neighbors (Friedman, 1986, 1987; Lustick, 1988; O'Brien, 1987; Ruether and Ruether, 1989).

The comparison of America's and other nations' versions of civil religion is useful for an understanding of the religious link between the citizen and nation. It is also related to church–state and religion–power issues, discussed further in Chapter 8.

**Forms of Civil Religion.** There appear to be at least four forms of civil religion. In the United States, the civil religion is differentiated from both the state and church (i.e., as particular religions). The civil religion, as Bellah (1967) pointed out, is appropriate to the public sphere, and particular religions apply to the private sphere. A second form of linkage is an ostensibly secular version of nationalism exemplified by the U.S.S.R., where civil religion and state are linked but differentiated from particular religions (which are totally privatized). In the two other forms, church (i.e., particular religion) and state are not differentiated. The church is sponsor of the state in one form; and in the other, the state is sponsor of the church (Coleman, 1970; Geertz, 1963). These latter two types are less obvious cases of civil religion because they are so closely linked with particular religions. The first two types are better suited to religiously heterogeneous nations because the civil religion is differentiated from particular religions. This very differentiation, however, may itself cause certain problems of integration.

Robertson (1978:175) suggests that the idea of a civil religion arises in response to a sense of imbalance between the individual and the state or society. He describes this gap as an imbalance between *identity* and *authority*, stating that civil religion expresses "the nature of the personal identity/society authority connection, when individual (at least sub-national) 'identity work' is perceived to make societal functioning ineffective and/or inefficacious." Differentiation between the spheres of life in which this "identity work" occurs (e.g., family, neighborhood, particular religions, ethnic community) and the spheres of public life (e.g., work, law, politics) is very great in modern societies. Civil religion may be so prominent in the United States because that society is the prototype of such a highly differentiated society. Another observer (J. Wilson, 1979:174) suggests that civil religion is a politico-religious revitalization movement in a new system of global order. One observer suggests that certain conservative elements of contemporary Japanese civil religion, such as the attempted restoration of state Shinto symbols and efforts to make the Yasukuni shrine (a Shinto memorial to the nation's war dead) into a national shrine, constitute just such a revitalization movement (Takayama, 1988).

The civil religion thesis is an important sociological concept. It explains certain aspects of American religiosity that are not related to particular religions. It provides a hypothesis for understanding expressions of national unity in a heterogeneous, highly differentiated society. American civil religion, as described by Bellah and Robertson, is related to the apparent weakness of particular religious institutions in the public sphere. This suggests that the development of a separate civil religion may be related to processes of modernization and secularization, discussed further in Chapter 8.

Finally, the civil religion thesis proposes a basis for the relationship of the individual to the larger modern society. It allows, in theory, the societal needs of cohesion and commitment of members and the individual needs of identity and belonging to be met by the same social processes. To what extent American civil religion really accomplishes this linkage remains unclear. We need to know to what extent people identify their interests and sense of belonging with the nation as a whole. Or, by contrast, to what extent do they locate their interests and community in particular segments of the society that are in conflict with other segments? The relationship of the individual to society is a critical issue in understanding modern societies. If a civil religion can effect a special kind of individual-to-society connection, it forms an important clue to that understanding (Bellah, 1978; Robertson, 1978). And if a civil religion is unable to achieve that connection, the reasons for this inability are also useful for our understanding of modern society.

## Religion and Social Conflict

> *Mine eyes have seen the glory of the coming of the Lord.*
> *He hath trampled out the vintage where the Grapes of Wrath are stored.*
> *He hath loos'd the fateful lightning of His terrible swift sword.*
> *His truth is marching on.*
> *Glory, Glory, Hallelujah . . .*
> *His truth is marching on.*
> ("Battle Hymn of the Republic," Julia Ward Howe, 1861)

This hymn, which has as its context both particular (Protestant) religions and American civil religion, illustrates the capacity of religion to inspire and reflect social conflict. In this section, we consider factors that make religion a powerful basis or reflection of social cleavages.

### Conflict as the Obverse of Cohesion

We must keep in mind that cleavage and conflict are, in many respects, merely the "other side of the coin" of cohesion and consensus. We tend to think of conflict as a breach in sociation, but as Simmel (1955:18) reminds us, conflict is one form of sociation. Simmel emphasized that "a certain amount of discord, inner divergence, and outer controversy, is organically tied up with the very elements that ultimately hold the group together."

Thus, our discussion of religious expression of social cohesion is necessarily related to a consideration of conflict as well. For example, the Doukhobors are a religious sect in Canada and the United States whose values and way of life often conflict greatly with those of the dominant society. Members have often clashed with educational, social welfare, and police authorities; for example, in 1953, Canadian Doukhobors protested educational authorities' toughened stand of public schooling for their children by burning schools and making nude pilgrimages (Woodcock and Avakumonic, 1977). Their strong cohesion as a group sets them off from non-Doukhobors. At the same time, their experience of opposition from the rest of society increases their group's cohesion and commitment. So cohesion at one level of association can produce conflict on another level, and conflict from the outside can contribute to internal cohesion.

Religion has been historically related to conflict at several levels. Perhaps the most obvious has been conflict *between* religious groups, especially when religious boundaries are coextensive with political boundaries. Religion played an important part in the reconquest of Spain by "the Christian monarchs" (end of the fifteenth century). This event was accompanied by powerful "us-against-them" sentiments, resulting both in expulsion of the Moors (i.e., Moslems) and suppression of indigenous Jews. Conflict between religious groups within modern nations tends to be more subtle (e.g., much of the anti-Semitism in America); but when religious boundaries are coextensive with other boundaries (e.g., social class, race, or ethnicity), open conflict can erupt.

Another level of conflict arises *within* a religious group. Conflicts during the early part of the Protestant Reformation exemplify this type, as protesting groups were defined as splinter groups within the Roman Catholic church. Similarly, most established denominations and sects began as internal conflicts in parent groups.

Sectarian religious groups also exemplify another level of conflict—between a religious group and the *larger society*. Sometimes the conflict results in reprisals by the larger society, as when courts override family or educational arrangements of the Mormons, Amish, and Doukhobors and jail Quaker conscientious objectors. And sometimes sectarians express conflict only by condemning and withdrawing from the "ways of the world."

### Social Sources of Conflict

**Social Cleavages.** Some religious cleavages have their sources in the organization of society. As noted in Chapter 2, religious belonging is one basis of self-identification. A strong sense of belonging (e.g., to a national, religious, kinship, or ethnic group) entails a sense of barrier between members of that group and those outside it. Because religion is one important basis for group identification in society, it is a potential line of cleavage (Coleman, 1956:44, 45). A closer look at religious conflict, however, suggests that the situation is more complex. Often religious boundaries overlap with other lines of cleavage such as social class, race or ethnicity, political or national allegiance, and so on. What appears to be a religious struggle may be also an ethnic or social class conflict.

When religious divisions are coextensive with other lines of cleavage, it is difficult to distinguish the exact role of religion in conflict. In 1844, Protestants and Roman Catholics in Philadelphia engaged in armed combat. Protestants were angered that the Catholic bishop had persuaded the school board to excuse Catholic children from religious instruction, which was then a standard part of the public school curriculum. Mass meetings were held to attack the change, and a Protestant crowd marched into a Catholic neighborhood. Street fighting and general rioting resulted, and Protestant mobs set fire to several houses and Catholic churches. As violence mounted, the governor sent the militia, against which the Protestant mob fought with its own cannon and muskets. Thousands of Catholics fled the city (Shannon, 1963:44).

On the surface, this historical event appears to be a simple case of religious conflict, but the divisions between Roman Catholic and Protestant were also cleavage lines of ethnicity, economic interest, politics, and neighborhood.

Catholics involved were almost all Irish, and many were recent immigrants. Protestants were largely part of a rising political stream of anti-immigrant fervor, later culminating in the "Know-Nothing" Party (1854). Economic factors included the competition of immigrants with WASP "natives" for jobs. While the extent of anti-Catholic prejudice at that time should not be understated, it is nevertheless difficult to distinguish elements of religious conflict from other sources of divisiveness.

Because religion is often coextensive with other lines of cleavage, it is frequently used as a way of expressing other divisions. A South African can use loyalty to the Dutch Reformed church to express many other loyalties: racial (white, as opposed to black), ethnic (Afrikaner/Dutch, as opposed to English), language (Afrikaans and rabidly anti-English), and political (Afrikaner Nationalist Party) (see Moodie, 1978, and parallels with the Lebanese situation, discussed in Labaki, 1988, and the Northern Irish conflict, described in further detail in this chapter). Overlap of cleavage lines also means, however, that sometimes religious conflict coincides with other, perhaps more fundamental, cleavages. In India, for example, state policy regarding education (and, by extension, economic opportunity) must deal with competing or conflicting demands of groups, defined by coinciding religious-ethnic-language identities (e.g., Aryan Hindus using Sanskrit, North Indian Muslims speaking Urdu, Sinhalese Buddhists using Pali, and Punjabi Sikhs writing Gurumukhi). When these groups conflict over an educational policy, religion is a relevant factor, but it is inextricably intermeshed with ethnic, linguistic, and often economic and regional divisions as well (Oomen, 1989). Because of this kind of linkage with other important sources of group identity, religion has often been a mask or even an overt legitimation for political and economic conflict.

Both the rhetoric and inspiration of religious conflict can cover aspirations for political power or economic gain. Right-wing radio evangelists of the 1930s, Protestant minister Gerald L. K. Smith and Roman Catholic priest Charles E. Coughlin, organized a politico-religious campaign. Virulent anticommunism, anti-Semitism, and nativism were central to their message. Coughlin attracted a radio audience of approximately 10 million weekly listeners, who contributed so much mail and money that he needed 145 clerks. Together Smith and Coughlin organized the Union Party for the 1936 presidential elections. After the defeat of their presidential candidate, Coughlin emphasized anti-Semitism even more, primarily as a symbol of his attack on the country's economic and money system. Although Coughlin received some active church opposition and very little official Roman Catholic support, this right-wing coalition had special appeal to discontented urban Catholics, the elderly, and rural poor (largely Protestant). The political ambitions of the coalition were served by the anti-Semitic and anti-"subversive" messages proclaimed with religious teachings over the radio (see Bennett, 1969; Lipset, 1963; Ribuffo, 1983). In this example, a religious cleavage between Christians and Jews was exploited to further political ends.

**The Marxian Perspective.** Marxian interpretations hold that religious conflict is merely the expression of fundamental economic relationships. Accordingly, the dominant class attempts to impose ideas (including religious ideas) that legitimate its interests. Or it may exploit existing religious divisions among subordinate peoples to prevent them from realizing their true class interests. As we will note in Chapter 7, religion can contribute to legitimation of the existing social system and also express real conflicts in that system (Turner, 1977).

According to Marxian theory, religious dissent is often an expression of economic dissatisfaction, but its religious nature prevents dissenters from fully realizing their economic class interests. Religious conflicts have often divided members of the working class when their true interests, according to Marx, would be best served by uniting against the common enemy, the ruling class. In the late nineteenth century, some 100,000 members of the American Protective Association pledged not to vote for, strike with, or hire Catholics if non-Catholics were available (Ribuffo, 1988). Such movements attracted economically threatened working-class persons who might have otherwise benefited from uniting in political and labor movements with Catholics who shared their class interests.

Part of the reason why Marx's predictions of increasing class conflict have not eventuated in many modern societies is that their members are subject to many **cross pressures** other than social class interest, thus making religion a less volatile source of conflict too. Cross pressures refer to the conflicting loyalties that individuals feel when they identify with several different roles and reference groups (Coleman, 1956:46). An example of cross pressures is the conflicting loyalties felt by a black woman who manages the local branch of a bank, lives in an integrated, middle-class suburb with pleasant Jewish and Irish Catholic neighbors, and belongs to a local Baptist church. When an issue arises in which different attachments conflict (e.g., whether tax monies should be used to fund abortions for poor women), she may acutely feel the cross pressures. Her loyalties to poor blacks or to women may urge her to support funding the abortions; her identification with her religion or the privileged classes may press for the opposite stance.

In many societies, this kind of cross pressure is absent or minimal because all major meaningful kinds of personal identification coincide. If the dominant class is of a totally different race, religion, and cultural background from the subordinate class (as in many colonial societies), the potential for social conflict is great because there are few conflicting loyalties within the person. In the United States, lines of cleavage—religious, ethnic, racial, economic, residential, regional, political—are seldom coextensive. Although real status inequalities exist (e.g., whites and Protestants are considerably overrepresented among the political and economic elites), geographical mobility, relative economic mobility, and mass communications have contributed to blurring the lines of cleavage. The existence of these cross pressures within individuals means that the likelihood of concerted conflict along any *one* line of cleavage (e.g., social class or religion) is reduced (Coleman, 1956:47).

Because religious cleavage sometimes masks other cleavages does not mean, however, that there are not real religious interests involved. Protection of religious interests frequently results in conflict. The struggle for *religious liberty*—

the freedom to hold and practice the religion of one's choice—has been a recurrent cause of political conflict. Violent suppression of the Huguenots (i.e., French Protestants) resulted in continued conflict from their inception in the early 1500s until 1789. The Huguenots persisted in asserting their right to practice their religion, and the conflict included civil wars, mass emigration, massacres, and torture. To many people today, religious intolerance seems impolite and, at worst, unjust. We find it difficult to understand why religion would be the basis of such strong antagonism.

**Conflict over Bases of Authority.** One of the key reasons why religion can result in such extreme conflict is that it not only reflects lines of cleavage in society, but it can also challenge the bases of legitimacy by which authority is exercised. The French rulers believed (probably accurately) that the Protestant worldview constituted a challenge to the basis of their authority because that authority was justified largely by the Roman Catholic worldview. Religious ideas and interests have historically been significant forces in establishing authority.

Authority, in contrast with power relations, requires that subjects consider it legitimate. **Legitimacy** means the social recognition of an authority's claims to be taken seriously, and it implies negative social sanctions for failure to comply with authoritative commands. Conflict is implicit in authority relationships because they involve dominance and subordination. Yet they are based on more than pure power. Other considerations (e.g., scientific or religious knowledge) may also be sources of legitimacy of authority. There is constant potential for conflict between authorities who base their claims on different sources of legitimacy. When religion is related to civil disobedience (e.g., the actions of the Berrigan brothers in the 1960s and 1970s), religion is largely the basis of challenging the civil authority's source of legitimacy. This kind of civil disobedience is saying, in effect, "Your rules and practices may be right according to your authority, but our actions appeal to a higher basis of authority."

The place of religion in societal authority relationships is complex because religion is both a *basis* of legitimacy and, often, the *content* of authoritative pronouncements; thus, the pope pronounces religious messages and bases his claim to speak authoritatively on religious tradition. Religion has historically been an important source of conflict because it offered both a basis for legitimacy of authority and a set of ideas around which conflict centered. The apparent decrease in religion's role as a source of authority in modern society is due partly to its loss of legitimacy—its capacity to compel people to take its claims seriously.

### Sources in the Nature of Religion

The nature of religion and religious groups also contributes to social conflict. In many respects, this capacity for promoting conflict is simply the obverse of religion's ability to engender social cohesion. Religion is one way of expressing the unity of the "in-group"—"our people." This in-versus-out dichotomy, however, applies to conflict both outside and within the group.

**Conflict with Outsiders: Boundaries.** This distinction between in-group and out-group implies distancing oneself from outsiders. Just as religious rituals celebrate the identity and unity of the group, they simultaneously maintain the boundary between that group and outsiders. Not only do religious groups protect their external boundaries, but they are also concerned with internal purification—another potential source of conflict (Douglas, 1966, 1970). The we–they dichotomy is both a structural and cognitive framework that includes how people think about themselves and others. This cognitive boundary is based on the fact that "we" have shared central experiences that "they" have not. In-group language embodies these shared experiences and further distinguishes "us" from "them." "Born-again" Christians consider their religious experience an important distinction between themselves and others, and their ways of witnessing to their special experience of being "born again" symbolize this difference.

Religion figures importantly in the socialization of children in most societies and thus becomes part of the "we–they" cognitive framework from an early age. Developing a sense of religious belonging, children come to think of themselves as part of the in-group and share their group's way of perceiving the rest of society. This cognitive framework may also include the perception of others as enemies or inferior to members of one's in-group. The values and attitudes learned in one religious group often vary considerably from those taught children in another group. Thus, religion's significance in socialization enhances its potential for divisiveness (Coleman, 1956:54, 55; Gorsuch and Aleshire, 1974; Wach, 1944:36).

**Conflict with Outsiders: Particularism.** Certain religious perspectives appear to promote conflict. Particularist worldviews encourage intolerance and prejudice toward the out-group. Religious **particularism** is the viewing of one's own religious group as the only legitimate religion (Glock and Stark, 1966:20). The clash of worldviews in a complex society is resolved in a number of possible stances: "Our way is totally right, theirs is totally wrong"; "Their way is good, our way is better"; "Our ways are both right but appropriate for different people"; "Their way and ours are essentially the same, and apparent differences are only incidental matters." The belief systems of some religions include particularist judgment of nonbelievers. For example, a study of Canadian racist extremists found that social class and education levels were not closely correlated with involvement in extreme racist activism; rather, the most striking correlation was active membership in certain fundamentalist Christian religions (Barrett, 1987). A U.S. study also found a consistent correlation between fundamentalism and the *general* tendency to discriminate against others, as well as actual discrimination toward specific targets, such as women, blacks, and homosexuals (McFarland, 1989).

Indeed, religious particularism seems to require a sense of opposition; one's own religion is seen as triumphant *over* some other. The in-group needs an out-group against which it can compare itself (Glock and Stark, 1966:29; Raab,

1964).[1] Religious worldviews that involve a particularistic stance toward the rest of humankind hold greater likelihood for promoting conflict than less triumphalistic worldviews.

Groups with particularistic worldviews are often able to mobilize the efforts of their members precisely because of this sense of opposition. Particularism enhances militance for one's beliefs. The strong element of religious particularism in Islam promoted its early missionary expansion throughout the Middle East, west to North Africa and Spain, east toward India and Southeast Asia, and north toward Eurasia. This particularism, combined with religiously legitimated militance, was also embodied in the *jihad* (i.e., holy war)—a recurrent feature of Islamic conflict. Contemporary conflicts in Lebanon, Iran, and Indonesia have been viewed as jihads (see Utrecht, 1978). The idea of missionary expansion is foreign to most non-Western religions because relatively few religions combine the particularism of having "the truth" with the mandate to convert the entire out-group. Interesting parallels with Christianity and Islam, however, are the worldviews of contemporary totalistic movements such as nazism and communism. Communism is particularly comparable to traditional Christianity in its particularism and sense of opposition, militance, and missionary activism (see Coser, 1973; Zeldin, 1969).

**Internal Conflict in the Religious Group: Deviance and Control.** This same in-group versus out-group dichotomy applies to the relationship of the group toward its own members who are defined as deviant or heretical. Again, conflict is the obverse of cohesion. The way in which a group treats a deviant member involves some form of conflict by which the group exerts its control. **Deviance** is behavior that is contrary to norms of conduct or expectations of a social group. Because it is the group that sets norms and identifies individual instances of deviant behavior, analysis of deviance describes the group as much as the deviant member (Becker, 1963). If a group sets a norm (e.g., against gambling) and labels a member as deviating from this norm, the group is both punishing the gambler deviant and proclaiming its own identity as a nongambling people.

Especially in small, closely knit societies or religious groups, the group's social control over deviant members can be powerful. Mennonites sometimes "shun" deviant members. Family and neighbors refuse any social interaction, sometimes for years, until the deviant repents of a "sin" or recants a "heresy." Because religion is often a source of the social group's norms and values, it is frequently important in the social response to deviance. The actions of the deviant member are seen as not only hurtful to the group but a violation of things that the group holds sacred.

Durkheim (1965:Book 3) emphasized that "piacular rites" (e.g., expiation for

---

[1]For example, some years ago I was talking with a boy of about twelve who had recently been confirmed in his church. Responding to my query about what that ritual meant to him, he said, "It means being willing to fight and even die for Christ." I asked whom he thought he had to fight, and he replied, "I don't know—I guess the Jews." Nascent anti-Semitism perhaps, but probably only a reflection of the boy's sense that his Christianity had to triumph over some other group. The Jews may have been the only non-Christian group of which he had ever heard.

wrongdoings and cleansing of impurities) are just as expressive of the group's core unity and force as "positive rites" (e.g., communion). By ritually reincorporating deviant members who repent, both individual and group are strengthened in their solidarity around the norm. Threat of deviance from within is potentially more disruptive than opposition from outsiders because insiders "ought to know better"; outsiders can be dismissed as uninformed and ignorant of the "truth." Insiders must be taken more seriously as "one of us." A fellow member who goes against an important group norm becomes an affront to the essential unity of the entire group.

Deviance may contribute to group solidarity even when the deviant member is not repentant. In uniting against the deviant member, the group is strengthened. Even more than external opposition, internal conflict with the deviant member sharpens the group's sense of its boundaries and norms (see Durkheim, 1938:65–75; Mead, 1918). Collective rejection of the person accused of adultery reminds the entire group how much it should abhor adultery. Sometimes norms may be unclear or changing, and collective treatment of the deviant member may actually articulate the norm. In recent years, for example, numerous community or religious groups have attempted to proscribe or punish homosexual practices. Some have gone so far as to stigmatize AIDS patients due to the association of their disease with homosexuality. One study found that intolerance (e.g., prohibition of children with AIDS from schools) toward AIDS patients is correlated with factors of lower levels of education and self-esteem, political conservatism, and religious fundamentalism (Johnson, 1987). These actions are not merely a reflection of antagonism toward homosexuals; more importantly, they are groups' attempts to assert their norms in a situation where societal norms are unclear or changing. By uniting against what they define as deviance, they are confirming their own norms for themselves.

Religious groups with particularistic worldviews appear especially intolerant of deviance. Their certainty of their own total rightness increases their condemnation of the wrongness they perceive of any other beliefs and practices. Particularistic religious groups give deviant members more reason to fear group sanctions. If you have been socialized in a religion that you believe is uniquely true—the only path to salvation—you are not likely to consider leaving the group. You are likely to try very hard to follow its rules for behavior, accept its punishments for infringement, and hope never to be forced to leave. If you accept the group's claims to be the only true religion, expulsion (i.e., excommunication) is the worst possible punishment. In uniting against the deviant, members of particularistic religious groups gain both a sense of solidarity and a sense of their own moral rightness; they are triumphant over external and internal opposition. The core of righteous members of such groups often develops purist and elitist characteristics, which further promote the likelihood of conflict.

**Internal Conflict: Issues of Authority and Heresy.** Internal strife in religious groups can develop over nonreligious interests too, including socioeconomic issues, leadership and power, and other social cleavages within the group. For example, in the late fourteenth and early fifteenth centuries, the Catholic church experienced a major schism, resulting from a purely political split, rather than theological dispute. The same group of cardinals, acting apparently in good faith, elected two rival popes; subsequently, various kings and princes from most regions of Europe sided with one or the other pope to legitimate their own political claims. The subsequent resolution of the schism was due to political, rather than any religious, rapprochement (see Blasi, 1989).

Nevertheless, conflict over religious ideas and practices frequently centers on the issue of **authority**, which implies control both of the religious organization (however small) and over the articulation of central beliefs and practices. The separation of Eastern ("Orthodox") and Roman Catholicism was fundamentally a conflict over authority. Ostensibly, the schism was over theological issues such as whether Christ was "of the same substance" as God the Father or was "similar" to the Father. Behind these theological issues, however, were deeper sources of conflict: Greek philosophy and polity versus Roman philosophy and law. In particular, the split represented the Eastern patriarchs' dissatisfaction with the consolidation of church power and authority in the hands of the single Western patriarch (i.e., the pope in Rome). The specific theological issues, important as they were to conflicting parties, did not cause the split; the issue of authority in the church was the central source of conflict (Niebuhr, 1929:111–117; "Orthodox Eastern Church," 1943, 16:938–939).

Similarly, the main division between the dominant form of Islam (Sunni) and Shiism (the most powerful of the challenging Islamic sects) occurred over the issue of the *imam*, the legitimate successor to the Prophet, Muhammad. According to Shiite tradition, there is a line of succession to the imamate from the Prophet through his cousin Ali to their contemporary leaders; the imam is thus the rightful, divinely ordained leader of Muslims (Voll, 1982:7–31).

Other sources of conflict over authority are religious revelations. Because religious experience and ways of knowing are so intensely private, there is always the potential for believers to receive revelations or come to interpretations that differ from official ones. Revelations, prophecies, insights, and new interpretations of scripture and tradition have all been significant sources of intragroup conflict in most major religions.

When members assert a belief or practice that differs in some important way from those authoritatively established, it is often defined as not mere deviance but **heresy**. This kind of dissent implicitly challenges the existing authority structure of the group. It suggests that the entire group should consider a different basis for its core beliefs and practices. The labeling and condemnation of heresy is the official leadership's assertion of its authority in the face of the challenge. The vigor with which labeled heresies have historically been prosecuted illustrates the seriousness of their challenge to established religious authority.

Conflict over authority is exemplified by the issue raised in the Roman Catholic church of "Americanism," which Pope Leo XIII condemned as heretical in 1899. Americanism generally referred to the idea that religion should be adapted to the individual cultures in which it is practiced; specifically it meant that the American culture required a different brand of Catholicism than European cultures. The origin of this perspective is attributed to missionaries to the Americans in the mid-1800s. They sought to adapt their missionary appeals to the peculiar characteristics of American culture, especially its democratic and pluralistic organization. Latent in the papal condemnation of Americanism, however, was the assertion of one particular system of church authority (Rome/pope/curia-centered) over another system (national/bishops/collegia-centered). The possibility that any national church authorities might develop separate strains of teaching authority was perceived as a direct challenge to the existing system of authority (Cross, 1958; McAvoy, 1957).

The label of "Americanism" continued to suppress much internal dissent in the American Roman Catholic church in the first half of the twentieth century. These same issues of authority were central in the deliberations of the First and Second Vatican Councils (1869 and 1962). The politics of the Americanist "crisis" have not been eliminated and continue to reoccur in clashes between Vatican authorities and some Catholic scholars and clergy (see Kurtz, 1986). The Roman Catholic church provides clear-cut examples of the definition of heresy because it has developed specific measures for defining and dealing with it. But Protestant history also involves numerous instances of heresy, schism, and clash over authority. While the particularist worldview of many Christian groups promotes this kind of internal conflict, non-Christian religions (e.g., Hinduism, Islam, and Shinto) have also known many schisms and internal divisions.

Internal conflict over heresy is a dynamic process. It can mobilize members of the group for action against the heresy and indirectly promote internal changes, revising its teachings, shifting organizational arrangements, or developing other innovations [e.g., see a case study of the 1953 Boston heresy case (Pepper, 1988)]. Or the group can absorb the alternative beliefs and practices, changing in the direction of alternative authority. Even in the suppression of heresy, the group impels adherents of the heresy to form an alternative social movement. The history of Christian sects illustrates this dynamism.

Religion is thus an important factor in social conflict, both within the group and with outsiders. This potential for conflict results both from qualities of religion and religious groups themselves and from the nature of the society as a whole. Nevertheless, the aspect of conflict is basically the obverse of social cohesion; a certain amount of conflict is part of the very structure that holds groups together. And because religion is one important way by which groups express their unity, it is also a significant factor in conflict. This dual relationship is illustrated in the Extended Application section of this chapter: an examination of recent conflict in Northern Ireland.

## Extended Application: The Conflict in Northern Ireland

Recent civil strife in Northern Ireland illustrates religion's capacity to promote both cohesion and conflict. The history of divisions in Northern Ireland is long, complex, and confusing to outsiders. This brief essay focuses on the religious dimension of this struggle, recognizing that other important historical, political, economic, and social factors are also involved.

### Religious Conflict?

One receives the impression from television and newspaper coverage that the strife in Northern Ireland is purely religious, with two antagonistic camps— Roman Catholic and Protestant—pitted against each other in senseless, deadly violence. Between 1970 and 1990, the sectarian strife claimed 2787 lives (Prokesch, 1990); tens of thousands more have been injured (Whitney, 1988). Neighborhoods and business districts have been bombed or burned, and a generation of children has grown to adulthood knowing no other way of life than intermittent warfare in their streets.

The image is puzzling to most Americans, who see the antagonists as so similar to each other. Unlike our national experience with conflict over segregation and discrimination, the opposing sides in Northern Ireland appear to share the same racial stock, language, and social class. These apparent similarities are, however, based on only a casual glance. Participants themselves apply finer distinctions; they are quick to identify someone as "one of us" or "one of them" on the basis of simple items of information—name, address, or school attended.

Nevertheless, the only differences apparent at a superficial level of analysis center around religion. Many Americans find it difficult to believe that religion is so important to anyone as to be worth fighting over. Others note the relative ease with which both Catholic and Protestant Irish immigrants adjusted to the heterogeneous and pluralistic religious scene in America (MacEoin, 1974b:1, 2). To participants in the Northern Irish conflict, however, the salience of religion as a genuine source of antagonism is unquestionable. In 1970 a British Broadcasting Corporation (BBC) interviewer asked a Belfast Protestant, "What do you have against the Roman Catholics?"—to which the response was, "Are you daft? Why, their religion of course" (cited in Rose, 1971:247). Americans and other outsiders ask: Is the conflict in Northern Ireland indeed a religious one?

This analysis will suggest that the answer is yes, but with qualifications. The conflict is not over theology or doctrine. It does not center on Protestantism or Catholicism, as most adherents of these religions in the United States, Canada, or elsewhere in Europe know them. The conflict in Northern Ireland could perhaps best be understood as the outcome of two mutually exclusive *civil religions:* one "Protestant" (or, more accurately, "Orangeism") and the other "Catholic" (strongly identified with "Republicanism"). Although the concept of civil religion is controversial, it appears useful in explaining the Northern Irish situation of religiously focused political strife.

### Conflicting Civil Religions

As suggested earlier in this chapter, American civil religion appears to be one of several forms of religious tie between citizens and the nation. In the United States, civil religion is separate from both state and church (i.e., particular religions). American separation of church and state makes possible (and, some sociologists would argue, necessary) a civil religion that *transcends* particular religions. Northern Ireland, by contrast, represents a situation in which religion (Protestant) is not separated from the state. Religion has been a significant consideration in political decision making, in applying social policy, in the actions of police and other officials, in the curriculum of national schools, and in the content of mass media. Relatively few persons in the country doubt that religion should be linked with the state; there is, however, profound dissension over *which* religion should shape the nation.

The nation-building potential of civil religion is evident in the intense conflict between these opposing views. Both Protestant and Catholic civil religions in Northern Ireland entail visions of nationhood. Whatever their political thrust, both national visions are also intensely religious in both content and style. Each group builds its own version of national identity imbued with religious significance, and each group engenders a strong sense of "us" against "them."

While particular religions (i.e., Roman Catholic and various Protestant groups) contributed to the creation of these opposing civil religions, they cannot control them. The civil religions are *separate religions,* with dynamics of their own.[2] Thus, religious leaders in both camps who attempt to quell the violence typically have little power over the strong sentiments set into motion in the name of religion. Similarly, the distinctiveness of Northern Irish civil religions from particular religions helps to explain why adherents are sometimes strongly critical of church leaders who are interested in ecumenism (i.e., interfaith tolerance and activities). Rather than follow their church leaders into interfaith activities, many believers use the civil religion as a basis for denouncing them.

The opposing civil religions of Northern Ireland illustrate that social cohesion is often the obverse of social conflict. Each group is held together largely by its sense of opposition to the other group. Both groups—especially the Protestants—are thrust into enclaves of tribal togetherness out of fear of the other. Reformation and Counter Reformation symbols and myths that are several hundred years old shape these fears and lock believers into nationalistic ideologies that cannot accommodate full religious pluralism.

### The Social Context

Northern Ireland was created in the 1920s after rebellion against Britain, when twenty-six counties of Ireland became a self-governing dominion (eventually recognized as an independent republic). The other six counties, an area of 5200 square miles in the northeast corner of the island, continued their union with Britain—that is, the United Kingdom. The part of the island retained by Britain included most of the industry and was economically and strategically important. Partition of the island in the 1920s encompassed as much area as possible in the British-held part without upsetting Protestant electoral dominance.

Roman Catholicism is the largest single denomination in Northern Ireland,

---

[2]Further corroboration of the distinction between particular and civil religions in Northern Ireland is found in survey data that show that commitment to particular Protestant religion is only weakly related to "Protestant" political attitudes about the conflict; commitment to Catholic religion is unrelated to "Catholic" political stances (McAllister, 1982).

accounting for roughly 35 to 45 percent of the 1.5 million population (the accuracy of the 1981 decennial census was severely disrupted, due to massive refusal to respond out of political fears; see Rowthorn and Wayne, 1988). The remainder of the populace is Protestant, particularly Presbyterian and Church of Ireland (i.e., Episcopalian). The Republic of Ireland, by contrast, is predominantly Catholic, so that on the island as a whole Catholics outnumber Protestants approximately three to one (MacEoin, 1987; Prokesch, 1990).

Irish Protestants and Catholics tend to be more fundamentalist and theologically conservative than their counterparts in the United States and England. Both groups exemplify religious particularism, the belief that one's group is the only legitimate religion. The self-righteousness and sense of opposition characteristic of particularistic worldviews make both groups strong sources of intolerance. Religious polarization in Northern Ireland has produced a social and psychological split in which there are no neutrals; even unbelievers are identified as Protestant or Catholic unbelievers. One of the most important products of this religious polarization is that the public does not believe (indeed, cannot conceive of) groups that claim to be nonsectarian (Beach, 1977).

All aspects of life in Northern Ireland are divided by these religious poles. There are separate Protestant and Catholic neighborhoods, playgrounds, schools, social clubs, charitable organizations, political parties, youth activities, sports, newspapers, and cultural events. As suggested earlier in this chapter, religion is especially likely to be a source of conflict between groups when it is coextensive with other important sources of identity. In Northern Ireland, religion is coextensive with almost all other significant social divisions.

Much of the social segregation results from systematic discrimination against Catholics, especially since the 1920s partition. To protect their power and economic advantages, Protestants severely discriminated against Catholics in employment, housing, civil service, public board appointments, electoral districting, and representation. By restricting housing for Catholics to certain neighborhoods and then gerrymandering the boundaries of voting districts, Protestants were able to maintain strong majorities in the councils of communities where they were numerically the minority. But not all segregation of Catholics was imposed by Protestants. Segregated Catholic schools are supported by the Catholic church hierarchy, partly in order to retain greater control over the socialization of Catholic youths and to protest the (Protestant) religious content of national school instruction (see Menendez, 1973).

The net effect of such total segregation is that few of the cross pressures that could reduce prejudice and conflict operate in Northern Irish social life. Polarization makes it difficult, if not impossible, to socialize with persons outside one's enclave. "They" are not real persons; "they" are described only by group myths, not by personal contact. Indeed, polarization has produced strong social controls. In-groups punish members who overstep boundaries and are too friendly to the out-group. Members of both enclaves have been beaten, tarred and feathered, and even killed for crossing the social boundaries. Sometimes religious sanctions are brought to bear on people who do not stay in their religious enclave. One Catholic bishop refused to confirm children who were not sent to Catholic schools. Some Protestant ministers use their pulpits to denounce fellow Protestants who accept ecumenism.

These current polarizations are the result of a long history, which figures significantly in the myths, symbols, and legends of the two civil religions. This history, however briefly presented here, is critical for understanding contemporary sentiments.

### Some Historical Background

Although the history of England's involvement in Ireland may be traced to the twelfth-century Norman conquest, events of the seventeenth century were especially critical in shaping contemporary problems. In the sixteenth and seventeenth centuries, the chief objective of England's Irish policy was to prevent Ireland from becoming a center for English rebels or a stepping stone for continental enemies. Thus, under the Tudor monarchs, the English began a system of plantations in Ireland, substituting loyal English settlers for potentially disloyal Irish or Old English landholders. The plantation system was particularly important in shaping the religious composition of Northern Ireland (Beckett, 1966:38–63).

**Plantation and Insurrection.** In the seventeenth century, the policy of plantation was extended with vigor and increasingly religious overtones. The English wished to disempower the native Irish and Old English landholders (predominantly Roman Catholic) and substitute new English and Scottish settlers (Protestant). The most extensive and successful plantation was in the province of Ulster (northern Ireland). Large areas of Ulster (i.e., counties Armagh, Cavan, Derry, Donegal, Fermanagh, and Tyrone) were confiscated and given to immigrant English and Scottish landholders. Since the seventeenth century, there has been a major social class distinction between the English (generally Church of England, called "Protestant") settlers and the Scottish (Presbyterians, Covenanters, etc., called "Dissenters"). Some Scottish-Irish later identified with the native Irish because of their own previous experience of discrimination from the English. In today's Northern Ireland, however, both Anglicans and Dissenters generally view their Protestantism as "common ground" against Catholicism. Some of these seventeenth-century settlers removed the Irish tenants from their land and replaced them with British tenants, while many kept their Irish tenants. Only about 2000 British families lived on these Ulster plantations by 1628. Thus, the large-scale removal of native Irish that the government intended did not occur then (Beckett, 1966:64–74; Clarke, 1967).

Discontent of the native Irish festered, and in 1641 insurrection broke out, especially throughout the north of Ireland. Ulster natives (mostly Catholics) attacked the Protestant colonists and seized many towns and fortifications. The Ulster insurrection had special significance to the English, who were then involved in a civil war between Royalists (largely Catholics and Anglicans) and Parliamentarians (mostly Puritans and Dissenters). Exaggerated tales of the 1641 massacres served as fuel for the English reconquest of Ireland. In 1649 Oliver Cromwell (a Puritan) landed in Ireland with a force of 12,000 soldiers and began a campaign so ruthless that 330 years later he is still identified as perhaps the most hated symbol of English oppression. Cromwell saw his mission as not only to quell a royalist uprising but also to bring divine revenge for the 1641 massacres. Reporting on his conquest of Drogheda, after which some 2000 townspeople were put to the sword, Cromwell wrote, "I am persuaded that this is a righteous judgment of God upon those barbarous wretches, who have imbrued their hands in so much innocent blood" (quoted in Beckett, 1966:79, 80).

The settlement following this reconquest solidified English dominance. The settlement forced all "disloyal" landlords to forfeit their lands, and new "loyal" settlers took their places. Catholics were particularly affected by this policy because the English parliament felt that they were, by definition, disloyal to English interests. In 1641 the majority of Irish landlords had been Catholic; after the Cromwellian settlement, the majority were Protestant (Beckett, 1966:74–81). Catholics lost virtually all political power because representation in parliament was based on landholding.

A subsequent revolution in England further polarized the Irish situation. The English monarch James II (a Catholic) was supplanted by a Dutch (Protestant) ruler, William of Orange (after whom the Orange Order of Northern Ireland is named). Irish Catholics sided with James II, and he brought his army to Ireland, hoping to reconquer England from that base. The Protestant colonists of Ireland had sided with William of Orange, who then brought an expeditionary force to Ireland. James took Dublin and laid seige to Londonderry (i.e., Derry), the main source of Orange resistance, but the city held out for fifteen weeks until reinforcements came. In 1690 the two armies met in a decisive battle at the River Boyne, and James was beaten on July 12. William's victory not only secured his position as king of England but also established Protestant supremacy in Ireland (Beckett, 1966:90–95; Simms, 1967).

These events, which occurred over 300 years ago, are still enshrined in the civil religions of Protestants and Catholics in Northern Ireland. The Catholic civil religion celebrates the heroes of the revolutionary uprising, and the Protestant civil religion commemorates the Orange defenders. In contemporary Derry, an annual (and sometimes violent) Protestant celebration parades the walls of the old city and proclaims to the Catholics living below the walls the seventeenth-century slogan "No Surrender" (see vivid descriptions in MacEoin, 1974b:87, 227, 232). In Belfast July 12 is a major Protestant holiday with bonfires, parades, speeches, and a mock battle to commemorate William's victory at the Boyne. The defeat is remembered in the Catholic civil religion as merely one of a long string of Protestant and English acts of oppression and injustice (the myths and images of this civil religion figure significantly in Irish literature, especially of the era immediately preceding the 1916 uprising; see Thompson, 1967). Thus, the history of "our nation" is remembered differently by the two groups. Events centuries past are the basis for two completely opposed sets of myths, legends, and heroes.

**The Deepening Split.** Events following 1690 further deepened the rift between Protestants and Catholics. Catholics were effectively excluded from

Parliament, and a number of "penal laws" excluded Catholics (the great population majority of the island) from Parliament, army and militia, positions in municipal corporations, all civil service, and the legal profession. Laws of inheritance and land tenure were changed so that it was virtually impossible for Catholic landholders to leave land to a Catholic heir. The laws forbade Catholics from sending their children abroad for education, and Catholics were forced to pay tithes to the (Protestant) Church of Ireland. These laws, in effect for over 100 years, had the desired result of suppressing Catholic political and economic power. The first widespread nationalist movement, arising in the first half of the nineteenth century, focused on the issue of "Catholic emancipation," especially from strictures on political representation and voting. This history laid the basis for the distinctively Roman Catholic character of modern Irish nationalism (Beckett, 1966:96–144).

Meanwhile, the economy and political situation in Ulster (the north) developed further apart from the rest of Ireland. Ulster included the only developed industrial sector in the country, and Ulster business owners and industrialists feared nationalist separation from Britain—their primary source of markets and supplies. Ulster was also the locus of the only large Protestant enclave, which felt threatened by the prospect of a Catholic nationalist movement in power. Some English politicians deliberately manipulated these fears in their fight against Irish home rule (i.e., allowing an independent Irish parliament). They "played the *Orange card*"—that is, encouraged religious fears and sectarian strife, advising violent resistance if the British Parliament should pass a home-rule bill. Lord Randolph Churchill visited Belfast in 1886 to encourage Protestant fears of home rule and left the slogans "Home Rule Is Rome Rule" and "Ulster Will Fight; Ulster Will Be Right"—mottos kept alive in contemporary Protestant civil religion in Northern Ireland (Beckett, 1966:146–157).

In the nineteenth century, both Protestant and Catholic national visions grew. Sometimes they were embodied in organizations and movements, but more generally they developed as an attitude or part of a worldview of the respective communities. The development of these nationalisms illustrates the distinction between *particular* religions (i.e., Roman Catholic, Presbyterian, and Church of Ireland) and Catholic and Protestant *civil* religions. The Roman Catholic hierarchy frequently opposed the nascent Irish nationalism, even though the movement was heavily Catholic. As the twentieth century approached, church leaders found their interests more closely allied to those of the establishment. They were particularly unhappy with the socialist strain of the pro-independence labor movement. Thus, while the Catholic hierarchy encouraged a general antagonism toward Protestants, it discouraged the growing Catholic nationalism. Similarly, especially in the North, the Orange Order and related Protestant groups developed their power and beliefs independent of particular Protestant churches, yet with their approval. Both Protestant and Catholic religious leaders generally encouraged distrust toward each other, but Protestant and Catholic civil religions developed independently of the churches.

The nature of the Irish Revolution (1916–1921) itself laid the grounds for the present conflict. The fighting was primarily between two relatively small, irresponsible armed forces, neither of which was controlled by its government. The British government was unable to exercise control over its special force, the "Black and Tans," and the Irish Republican Army (IRA) was not responsible to the Dail (the newly established Irish independent parliament). Never actually militarily "won," the revolutionary war did prepare both sides for a compromise: the division of the island into a politically independent part and a unionist part (see Beckett, 1966:157–166). The significance of irresponsible paramilitary forces in that conflict laid the groundwork for contemporary paramilitary violence in Northern Ireland; and the IRA (Republican nationalist) and Ulster Volunteer Force (Orangeist) are offspring of the armed forces of the Irish Revolution.

**Since Partition.** The 1921 treaty that followed the Irish Revolution established partition of the country. Northern Ireland consisted of six counties, which together had a Protestant majority of approximately 65 percent, but the Catholic minority was large and numerically dominant in several localities. These six counties had a parliament in Belfast and remained in "union" with England—thus the political tag "Unionist," retained even today. The other twenty-six counties (overwhelmingly Catholic) were given an independent dominion parliament in Dublin that eventually proclaimed the country a republic. The border was arranged to keep as much economically advantageous area for Northern Ireland as possible while retaining Protestant political dominance. Slogans of Unionists during border disputes (1922–1925) figure in their contemporary civil religion: "Not an inch" and "What we have we hold" (McCracken, 1967).

The fears of Protestant Unionists in Northern Ireland erupted in violence during the treaty years. Armed mobs in Belfast viciously attacked Catholic neighborhoods, forcing Catholics to depend on the IRA (i.e., the branch of the revolutionary army that is now an illegal paramilitary secret society) to protect them. Often, Protestant fears were deliberately exploited by (Protestant) employers to prevent Protestant and Catholic workers from uniting in labor disputes. With unemployment at 30 percent during the 1930s depression in Belfast, Catholic and Protestant workers organized a bipartisan demonstration. In response, Orange Order leaders stepped up their appeals to religious sectarianism, suggesting that the dearth of employment was all the more reason to discriminate against Catholics. The grand master of the Belfast Orange Lodge (and subsequent member of the Northern Ireland Senate) said in 1933:

> It is time Protestant employers of Northern Ireland realized that whenever a Roman Catholic is brought into their employment it means one Protestant vote less. It is our duty to pass the word along from this great demonstration, and I suggest the slogan should be: "Protestants, employ Protestants!" (quoted in MacEoin, 1974b:66).

The social and political pattern that emerged from increasing sectarianization severely discriminated against the Catholic minority in housing, jobs, law enforcement, and elections. Proportional representation in Northern Ireland's

Parliament was an early casualty of the Protestant struggle to retain power. Abolition of proportional representation guaranteed that minority parties could not coalesce to counterbalance the Unionist party. The Northern Ireland government also approved a systematic gerrymandering of electoral districts so that Catholic voters were isolated and underrepresented (see McCracken, 1967). Immediately prior to the present strife in Northern Ireland, gerrymandering resulted in Derry's 20,000 Catholic voters being able to elect only 8 councillors, compared to 12 councillors elected by only 10,000 Protestants. Similarly, in Omagh, Catholics comprised 61 percent of the population but elected only 9 of the town's 21 councillors; in Dungannon, Armagh, Enniskillen, and other large towns, Catholics were systematically deprived of a fair proportion of power in county and town government (London Sunday Times Insight Team, 1972:34, 35; MacEoin, 1974b:58–59).

Catholics were also discriminated against in housing. This feature was politically significant because until 1969—when the British government forced the Northern Irish government to change its electoral practices—only householders and their wives could vote. Thus, policies that prevented Catholics from becoming householders also kept down their voting strength. City and county councils have controlled the allocation of much housing, apartments and single-family dwellings alike. These councils gave strong preference to Protestants in public housing, and they further restricted Catholics by reducing the number of new houses built in their neighborhoods. In predominantly Catholic Derry, the Protestant-controlled city council built only 136 houses between 1958 and 1966 and none after 1966. Partly as a result, over 1000 single-family housing units in the city were occupied by more than one family and sometimes by seven or eight. Over 1500 families (almost all Catholic) were on the waiting list for housing, with an average of ten years waiting time (London Sunday Times Insight Team, 1972:35–37; MacEoin, 1974b:59).

Widespread job discrimination was the result of preferential treatment of Protestants in both private and public employment. Belfast's largest employer, the shipyards, had in 1971 only 400 Catholic employees out of 10,000 (London Sunday Times Insight Team, 1972:36). Northern Irish public officials systematically granted more and superior jobs to Protestants. In 1961 Catholics accounted for only 13 of 209 officers in professional and technical grades of Northern Irish civil service, and there was only 1 Catholic out of 53 people at the top administrative grade of civil service. Catholics were also greatly underrepresented on the twenty-two public boards in charge of such services as housing, tourism, hospitals, and electricity. Disproportions created since partition have not been substantially altered by reforms introduced after the British resumed Direct Rule in 1972. In 1973 Protestants retained 95 percent of the 477 top civil service positions (MacEoin, 1974b:67–69).

The religious segregation of entire occupations and industries also meant that Protestants had an effective stranglehold on large areas of vital services. When in 1974 the British got several government groups to agree to share some political power with Catholics, the plan was destroyed by a brief strike of Protestant workers. These workers controlled entire occupations so strategic that their strike paralyzed the country. Protestant extremists have held demonstrations and threatened similar work actions to destroy the 1985 English-Irish pact on Northern Ireland.

Despite the widespread discrimination, however, the Protestants are not overall much better off economically than Catholics. Because both groups are largely working class, the conflict is not class conflict; Protestants of all classes unite against Catholics of all classes (cf. O'Brien, 1974; Rose, 1971). The economic discrimination has direct political implications, however. There is higher unemployment among the Catholic working class than among the Protestant working class, thus encouraging workers to emigrate (usually to England) to find jobs. The higher emigration rate of Catholics had, for many years, counterbalanced their higher birth rate, preventing them from becoming the population majority (see London Sunday Times Insight Team, 1972:29–31); in the economic recession of the 1970s and 1980s, however, decreased employment opportunities in England reduced incentives to emigrate and exacerbated economic problems in the Catholic areas of Northern Ireland.

Protestants also controlled "law and order." Before the British intervention, the two main police forces in Northern Ireland were the RUC (Royal Ulster Constabulary) and the B-Specials (Ulster Special Constabulary). The RUC was established at partition and was supposed to be bipartisan. Catholics, however, were never proportionately represented on the force and, when recent strife broke out in the 1960s, were less than 10 percent. The B-Specials was a militia, almost exclusively Protestant. Indeed, the prime minister of Northern Ireland boasted in 1922 that "it is also from the ranks of the Loyal Orange Institution that our splendid B-Specials have come" (quoted in MacEoin, 1974b:62). Catholics had a reasonable basis for their suspicions that police forces used their power against the interests of the Catholic community. There is evidence that in some conflicts, the B-Specials openly joined Protestant mobs against Catholic civilians (see London Sunday Times Insight Team, 1972:132–142).

Besides possessing an arsenal (e.g., armored cars and automatic weapons), both forces were allowed discretionary powers by the Special Powers Act (introduced as a "temporary" measure in 1922 and eventually made permanent). This act, a source of considerable embarrassment to the British government, abridged civil freedoms by allowing any police officer to search, arrest, and imprison suspects without warrant, charge, or trial. A 1935 British inquiry concluded that "the Northern Ireland government has used Special Powers toward securing the domination of one particular political faction and at the same time toward curtailing the lawful activities of its opponents" (cited in MacEoin, 1974b:61–62).

Grave dissatisfaction with these features of Northern Irish life led to the 1964 formation of the Northern Ireland Civil Rights Association (NICRA), patterned after the American civil rights movement. Its demands were modest. They were (1) "one-man-one-vote" in local elections, (2) removal of gerrymandered boundaries, (3) laws against discrimination by local government and provision of machinery to deal with complaints, (4) allocation of public housing on a points system, (5) repeal of the Special Powers Act, and (6) disbanding of the B-Specials

(London Sunday Times Insight Team, 1972:49). The confrontations that resulted from the generally nonviolent demonstrations of NICRA occurred largely because the Northern Ireland government refused to recognize the legitimacy of the movement's grievances and equated it with Republicanism—the movement for national independence from Britain. Protestant enclaves similarly saw the civil rights demonstrations as "them" organized against "us." The stage was set for a long period of sectarian violence: Protestants versus Catholics and both Protestants and Catholics versus the British.

After more than two decades of conflict, several of the problems addressed by the Northern Irish civil rights movement continue to be major sources of tension and of serious social problems. One of the most pressing of these issues is inequality of socioeconomic opportunity. Comparing socioeconomic status of Catholics and Protestants in 1968 and 1978, one study found very little improvement in the relative status of Catholics (Moxon-Browne, 1983). Outright discrimination, together with workers' fears of hostility and unfair practices in the workplace, have perpetuated historical patterns of greatly disproportionate Catholic unemployment and underemployment. Despite nearly comparable attainment of educational qualifications, Catholics continue to be virtually absent from entire sectors of the economy, including better-paid manufacturing jobs, law enforcement, banking, insurance, and big business. In the expanding public service sectors, Catholic employees tend to be females, working in education, health services, and clerical jobs; except for police and prison work, public sector jobs for males decreased in the 1980s, providing little or no opportunity for Catholics (Eversley, 1989).

Male unemployment among Northern Ireland Catholics is particularly severe. A 1983–1984 survey found that 35 percent of economically active Catholic males, compared to only 15 percent of Protestant males, were unemployed (the rate for white males in Great Britain at that time was 11.6 percent). Furthermore, this study concluded that employed Catholics were greatly overrepresented in low-paid jobs with little job security. Interestingly, the study noted that the strife itself had artificially raised rates of employment, especially among Protestant males, through the provision of thousands of jobs in the various police forces and prisons (Rowthorn and Wayne, 1988). Not until 1990 was a tough equal opportunity act put into effect in Northern Ireland, and it remains to be seen whether this legislation will be effective, not only in opening jobs to Catholics but also in reducing discrimination in matters of promotions and job security (Prokesch, 1990).

In the area of housing, major changes occurred since the early civil rights protests. To eliminate the discrimination in public housing, the newly created Northern Ireland Housing Executive assumed responsibility for building, managing, and allocating public housing (about 40 percent of all new housing is in this public sector). By 1985 most of the worst housing had been replaced, and overcrowding was greatly reduced. At the same time, however, neighborhood segregation had increased. Many previously integrated neighborhoods became all-Protestant or all-Catholic, as families who were in the minority in each neighborhood left in fear or were actively driven out. Such segregation may have subsequently affected employment and occupational segregation, however, by narrowing the pool of workers who live near (and would not need to cross a presumably dangerous area to get to) a place of employment (Eversley, 1989).

Police actions in the same twenty years have, if anything, generally increased Catholics' sense of injustice. Throughout the 1970s, the various police forces in Northern Ireland were expanded, while the proportion of Catholics on the forces remained extremely small (usually less than 5 percent). Simultaneously, the scope of police powers was expanded, and civil liberties were curtailed. Police detained several thousand people without ever bringing charges, much less conducting a hearing or trial. Amnesty International investigated conditions in Northern Ireland in 1977 and found evidence of frequent maltreatment of suspects and political prisoners. These police actions were perceived as discriminatory against Catholics. A survey conducted shortly after the Amnesty findings were reported in the media found that some 63 percent of Protestants, but only 20 percent of Catholics, believed that reports of prisoner maltreatment were unsubstantiated (Moxon-Browne, 1983:147). Although in the 1980s levels of street violence were lower and police were better disciplined than in the 1960s and 1970s, the failure to integrate Catholics into their ranks proportionate to their numbers in the communities leaves police forces open to plausible suspicion of biases and outright partisanship in the administration of justice.

This capsule history of strife in Northern Ireland illustrates how two groups, defined by their religions, became so utterly segregated in their associations that cross pressures that could have prevented polarization never developed. The combination of this segregation with the religious particularism of each group makes the situation especially volatile. Indeed, if anything, the years of strife have further crystallized the separation of the two groups (Moxon-Browne, 1983:124–136). Systematic political and economic suppression of a large religious minority of natives by a colonial power made the resulting conflict reflect nationalistic as well as religious sentiments.

Now let us examine some of the chief characteristics of the opposing civil religions of Catholics and Protestants.

### Catholic Civil Religion
The national vision identified as Catholic in Northern Ireland is one of a united republic, encompassing all counties of the island. Thus, it is necessary to view the conflict in the context of the religious and political situation of the Republic of Ireland, as well. Although some Catholics in Northern Ireland do not favor union with the Republic of Ireland because of its lower standard of living (especially its social welfare benefits), they are nonetheless likely to identify themselves as Irish and feel a strong tie to the Irish cultural heritage (Beckett, 1966:175). A more militant version of this civil religion draws on the centuries-old Irish revolutionary myth; this version envisions the (Catholic) natives rising up to overthrow the oppressive (Protestant) colonial power. This militant nationalism does not accept the validity of the settlement that partitioned Ireland. A 1968 survey showed that 33 percent of the Catholics in Northern Ireland approved "on balance" the constitution of Northern Ireland, a

proportion probably reduced by subsequent polarization. The study concluded that Northern Ireland is a state that is governed without consent (Rose, 1971:189). A later survey (after various intervening constitutional changes had been tried unsuccessfully) found that about 25 percent of Catholics still favored a United Ireland as the "most workable and acceptable solution" (Moxon-Browne, 1983:103). Refusal to accept the Northern Ireland government's right to rule further feeds Protestant citizens' notions that Catholics are disloyal or traitorous.

An important factor in the Catholic civil religion of Northern Ireland is its readiness to accept the church–state relations of the Republic of Ireland as normative. Thus, Protestant fears are fed by what they see as the effects of the Catholic church on political and social policies in the neighboring Republic of Ireland. Although the influence of the Catholic church in the Republic is indirect, it is still potent (see Fulton, 1987a; MacEoin, 1987, 1974a, 1974b; O'Brien, 1974; Schmitt, 1973; Sheehy, 1969; Whyte, 1980). The Roman Catholic Church is the established church in the Republic of Ireland (Eire), including in its membership 95 percent of the Republic's population. The 1937 constitution of Ireland declared, until it was abridged in 1972, "a special position of the Holy Catholic Apostolic and Roman Church as the guardian of the Faith professed by the great majority of the Citizens" (Article 44 quoted in Whyte, 1980:24–61). Although the Protestant minority in the Republic is seldom discriminated against, the Catholic church has influenced legislation and social policy with which Protestants are unhappy. The Catholic church's considerable influence in national (i.e., public) schools affects Protestants, especially in areas where they are not numerous enough to support alternative schools. Church-influenced legislation in areas of "private" morality (e.g., divorce, contraception, and censorship) is similarly seen as unfairly restrictive by many Protestants. Thus even though Northern Ireland is politically distinct from the Republic of Ireland, both its Protestant and Catholic nationalisms refer directly to beliefs and practices of the Republic.

The salient imagery and ritual in the Northern Irish Catholic civil religion are symbols of independent Ireland: the flag ("tricolor"), the Irish language, national anthem and ballads of independence, celebration of heroes, and events of revolution. These symbols are always near the surface of interactions with the opposing group. Even nonsectarian events (e.g., civil rights demonstrations) are often the occasion for invoking the symbols of this civil religion; despite organizers' plans, demonstrators sometimes break into singing nationalistic ballads and hymns or unfurl a tricolor flag. These are natural expressions: Catholic militance on behalf of their rights has long been equated with nationalism. Catholic civil religion in Northern Ireland likewise often celebrates—with even more vigor than in the Republic—the heroes and events of Irish revolution: Emmet, Pearse, Connolly, Tone, Father Murphy, the Easter Insurrection, the Fenian Rebellion, Catholic Emancipation, Land League, Gaelic League, United Irishmen. All these elements mesh together in a national vision.

This nationalism is both anti-British and anticolonial (i.e., against Protestant "settlers"). The British "peace-keeping" forces in Northern Ireland have been embattled from both sides and are intensely disliked by both Protestants and Catholics; both groups believe the troops to be siding with their opposition.

Even moves by the British government to pressure the Protestant powers in Northern Ireland to recognize some Catholic rights are distrusted by the Catholic community. Britain has a long history of political maneuvers resulting in detriment to the Irish.

The Catholic civil religion in Northern Ireland is generally more tolerant of Protestants than the Protestant version is of Catholics. This acceptance is largely because Protestants, too, were prominent in the revolutionary heritage of Ireland. Protestants such as Wolfe Tone and Robert Emmet, inspired by the ideals of the French Revolution, led Irish revolts and are celebrated as martyrs in the civil religion of Catholic nationalists.

### Protestant Civil Religion

The Protestant civil religion of Northern Ireland is, by contrast, virulently anti-Catholic. Both a 1968 survey and a 1978 comparison found Northern Irish Protestant antipathy to Catholics and Catholicism to be considerably greater than Catholic antipathy toward Protestants and Protestantism (Rose, 1971:256; Moxon-Browne, 1983:94–95, 124–136). The religious form of Protestantism as a civil religion in Northern Ireland is particularly evident in the power and influence of the Orange Order. Orangeism is essentially the state religion of Northern Ireland. The Unionist party, which has dominated the government since its inception, has close, overt links with the Orange Order and is almost exclusively Protestant. Lord Craigavon, prime minister of Northern Ireland from 1921 to 1940, proclaimed, "I am an Orangeman first and a politician and a member of this parliament afterwards. . . . All I boast is that we are a Protestant parliament and a Protestant state" (cited in MacEoin, 1974b:53). More explicit about the link between the controlling party (i.e., Unionist) and Orangeism is the opinion of Brian Faulkner, prime minister of Northern Ireland before Britain stepped in to rule the province directly in 1972. Faulkner stated:

> There is no alternative to the invincible combination of the Orange Order and the Unionist Party. . . . The Unionist Party relies upon the Orange Order and likewise we in the Order trust the party. That vital faith must never be jeopardized by either partner (cited in MacEoin, 1974b:243).

The Orange Order, founded in 1795, has been especially strong in Northern Ireland in this century. The grand master of the Belfast Orange Lodge (the largest of the several lodges in the country) described the Order as "basically religious and only coincidentally political, a fellowship of all who embrace the Reformed faith, founded to safeguard the interests of the Protestant people against the aggressions of the church of Rome which historically claims to have power over princes" (cited in MacEoin, 1974b:31). The salvation theology of Orangeism requires a sense of opposition: It fights for salvation not only from sin but also from Catholicism. The grand master further stated, "We have to fight the pretensions of the church of Rome. . . . I am convinced that it has never stopped its efforts to obscure the gospel, and I believe we must contend for the way of salvation through Christ."

The Orange Order maintains nearly monolithic control over Protestant opinion. It includes in its membership one-third of Protestant men in Northern Ireland, and its strength is estimated at 100,000. The Order includes all social classes, Church of Ireland and Dissenters alike. Indeed, historically it was the main vehicle for establishing Ulster Protestant unity despite significant religious and social class differences, especially between members of the Church of Ireland and the more fundamentalist Free Presbyterians (Bruce, 1987). Together with active women's and youth auxiliaries, it involves more than 90 percent of the Protestant community in some places (MacEoin, 1974b:62). Not all Protestants believe in Orangeism, but Orange ties are essential to gaining many jobs, political offices, appointments to legal and medical positions, and housing.

The anti-Catholic stance of Orangeism is directly relevant to the Protestant national vision in Northern Ireland; a fundamental premise of that vision is that Catholics must not be allowed power. When some Catholics attempted to work within the Northern Irish political system in the 1960s, Unionist leaders tried to prevent their participation on grounds of their religion. One of the foremost Unionist leaders stated:

> I would draw your attention to the words "civil and religious liberty." The liberty we know is the liberty of the Protestant religion. . . . it is difficult to see how a Roman Catholic, with the vast difference in our religious outlook, could be either acceptable within the Unionist party as a member, or bring himself unconditionally to support its ideals. Furthermore to this, an Orangeman is pledged to resist by all lawful means the Ascendancy of the Church of Rome (cited by London Sunday Times Insight Team, 1972:37).

According to Northern Irish Protestant civil religion, Catholicism is incompatible with the Protestant conception of democracy, and the Protestant nationalist vision is, by definition, opposed to any Catholic power. One DUP (Democratic Unionist Party, the most extreme Unionist group) member of Assembly, interviewed in 1986, said:

> The Christian beliefs held by the vast majority of the DUP clearly influence their political decisions. We believe that Ulster's future is best secured in a society where the gospel message can be preached freely, and that is not under Rome. Where the gospel light in many little towns throughout the Irish Republic has died, it has gone out because of the control of Rome. . . . I'm in politics to preserve this province so that the gospel can be preached freely and the Christian message on any street corner and that cannot be done in a Catholic dominated state . . ." (quoted in MacIver, 1989:367–368).

This anti-Catholic stance is celebrated in the fiery speeches of Orange parades and commemoration of their history, Protestant heroes, and legends. One popular Protestant tune, called "Croppies (i.e., Catholics) Lie Down," goes:

> *Poor Croppies, ye know that your sentence was come,*
> *When you heard the dread sound of the Protestant drum.*
> *In memory of William we hoist his flag,*
> *And soon the bright Orange put down the Green rag.*

Although this national vision is inimical to the prospect of union with the Republic of Ireland, it also includes strong anti-British sentiment. Northern Irish Protestants are keenly aware of their inferior status in the British class structure and resent being treated as colonials. The militant version of the Protestant civil religion holds that loyalty to the British Crown is contingent upon the British enforcing the Protestant ascendancy of the seventeenth-century "Revolution settlement" (MacEoin, 1974b:40, 285). Some observers suggest that the British intervention may have unintentionally exacerbated the contemporary conflict, especially by creating conditions for an upsurge of Protestant populism (Bew and Patterson, 1983).

### Nationalism Without Nation Building

The Protestant and Catholic national visions of Northern Ireland appear to fit our model of civil religion, albeit in a different form from that of civil religion in the United States. Both Protestant and Catholic versions are mutually exclusive images of the nation. Both entail extensive myths, legends, rituals, and symbols of a long history of growing separation of peoples. Both civil religions entail imagery of the "chosen people." The Protestant version defines themselves as God's chosen people struggling to maintain the truth in a land of paganism and idolatry, whereas the Catholic version describes the Gaelic peoples as the chosen children of God, oppressed by the Protestants. Commenting on this Old Testament imagery, one astute observer remarked, "One could say that Ireland was inhabited, not really by Protestants and Catholics, but by two sets of imaginary Jews" (O'Brien, 1974:288).

The efforts of nonsectarian political groups, concerned clergy on either side, and various ecumenical peace groups to defuse the religious antagonisms in Northern Ireland are typically thwarted because the conflict is less between particular religions than between two civil religions. The differences in belief and practice between the particular religions—Protestant and Catholic—are important but are only a small part of this conflict. The division is essentially over national identity and vision, which are strongly shaped by a long history of experiences and cultural differences between the two groups.

As suggested earlier in this chapter, civil religion in many countries has considerable potential for modern nation building, forging unity from diverse peoples of a state and bridging barriers of language, particular religion, tribalism, and regionalism. In the case of Northern Ireland, however, neither civil religion has this potential. The state governs without consensus, and the people operate from opposing senses of national identity. Another model of civil religion or of church–state relations would serve the interests of nation building better, but the issue of what nation to build is still a political issue.

## Summary

Religion is a source of both social cohesion and social conflict. It contributes to the cohesion of a group and, at the same time, expresses its unity. Durkheim suggests that the moral unity of the group itself is the source of the sense of religious force or power experienced by participants. Thus, religious rituals incorporate members into the group by reminding them of the meanings and obligations of being "one of us." The civil religion thesis is an attempt to understand the expressions of moral unity of modern societies. It suggests that the nation may have a religious expression distinct from the particular religions of the people.

Religion contributes to social conflict because it is one basis of social cleavage. In societies where other loyalties cut across religious ties, religion is less likely to be a focus of serious conflict. Nevertheless, religion's capability for defining the boundaries of "us" against "them" makes it an inherently potential source of conflict. Particularist worldviews especially tend to intolerance and conflict. Religious groups are also subject to internal conflict over issues of deviance and control, authority and heresy.

The aspects of religion that promote both conflict and cohesion are illustrated by the strife in Northern Ireland. While particular Protestant and Roman Catholic religions have historically contributed to the active bigotry and intolerance there, a broader religious division—that between opposing civil religions with national visions, one Catholic and one Protestant—may also exist.

## Recommended Readings

### Articles

Robert N. Bellah. "Religion and Legitimation in the American Republic." *Society* 15 (4), 1978:16–23.

Robert N. Bellah. "Civil Religion in America." *Daedalus* 96, 1967:1–21; reprinted in McNamara (1984) reader and in Richey and Jones ed., 1974.

James S. Coleman. "Social Cleavage and Religious Conflict." *Journal of Social Issues* 12, 1956:44–56; also Bobbs-Merrill reprints S-47.

Phillip Hammond. "Religious Pluralism and Durkheim's Integration Thesis." In *Changing Perspectives in the Scientific Study of Religion*, Allen Eister, ed. New York: Wiley, 1974, pp. 115–142.

### Books

Nancy Tatom Ammerman. *Baptist Battles: Social Change and Religious Conflict in the Southern Baptist Convention.* New Brunswick, N.J.: Rutgers University Press, 1990. Ammerman analyzes the deep divisions—doctrinal, organizational, moral, and social cleavages—in one denomination, illustrating the conflict posed by the coexistence of factions of opposing orientations: one sectarian, the other denominational.

Robert Bellah and Phillip E. Hammond. *Varieties of Civil Religion.* San Francisco: Harper & Row, 1980. These essays focus on the usefulness of the concept of a civil religion in understanding the American situation.

Emile Durkheim. *Elementary Forms of the Religious Life.* New York: Free Press, 1965. Especially the introduction; Book 1, Chapter 1; Book 2, Chapters 6, 7; Book 3; and conclusion. Durkheim's classical study uses illustrations from the religion of the Arunta of Australia to explicate his theory of the social foundations of religious beliefs and practices.

Will Herberg. *Protestant-Catholic-Jew*. Garden City, N.Y.: Doubleday, 1960. Herberg's description of the "American way of life" as a quasi-religious belief system and his descriptions of Protestant, Catholic, and Jewish history in America are especially useful. While somewhat dated, this essay on religion in American life during the 1950s is still provocative.

David Kertzer. *Ritual, Politics and Power*. New Haven, Conn.: Yale University Press, 1988. A lively analysis of how ritual promotes cohesion and conflict, liberally illustrated with examples from many different cultures and historical periods.

Roger N. Lancaster. *Thanks to God and the Revolution: Popular Religion and Class Consciousness in the New Nicaragua*. New York: Columbia University Press, 1988. With rich ethnographic description, the author compares and contrasts the change-promoting and change-inhibiting religious expressions of Nicaraguan peasants and recent rural immigrants to cities: official Catholicism, folk religion, the Popular church, and evangelical Protestantism.

Charles S. Liebman and Eliezar Don-Yehiya. *Civil Religion in Israel: Traditional Judaism and Political Culture in the Jewish State*. Berkeley: University of California Press, 1983. This volume gives both an excellent contrast of civil religion in another sociopolitical setting and a helpful basis for understanding contemporary Israeli political action.

John Obert Voll. *Islam: Continuity and Change in the Modern World*. Boulder, Colo.: Westview, 1982. This highly readable account shows the variety of Islamic responses to modernization and the political situation of both Arab and non-Arab Muslim societies.

# CHAPTER SEVEN

# The Impact of Religion on Social Change

The process of social change highlights the relationship between religion and other aspects of the social system, especially the economic and political spheres. This chapter examines some of the key ideas about the interrelationship between religion and social change. Because the issue of religion and social change is so complex, its division in this chapter is largely for analytical purposes. We will first examine the ways in which religion supports the status quo and inhibits change, then follow by examining how religion promotes social change. The concluding analysis poses the questions: Under what conditions does religion have the greatest impact on society, and under what conditions is religion's influence likely to inhibit or promote change? Discussion of these questions is followed by an Extended Application to the case of Brazil, where religion has been an important force for social change.

## Factors in Social Change

The processes involved in social change are complex. It is difficult to isolate a single aspect such as religion in a chain of events that result in social change. For analytical purposes, we can speak of aspects of religion that promote or inhibit social change, but in most cases both are occurring simultaneously. As the example of millenarianism (discussed in Chapter 2) shows, a religious movement that arises in response to social change may itself help bring about social change.[1] A further complexity is the meaning of religion. Do we mean religious ideas, religious personages, religious movements, or religious organizations? All these aspects of religion need to be considered; yet, in any given situation, the relationships are often very complex.

[1]In framing the topic as an interrelationship, we are rejecting theories that treat religion as *nothing but* an epiphenomenon to other social processes, though we will discuss some of these theories later in this chapter. Any effort to present this complex relationship in the brief space of a chapter necessarily entails great simplification. Suggested sources at the end of the chapter and in footnotes can direct interested readers to further examination of this important theme.

We need to remember that change itself is neither necessarily good nor bad. **Social change** refers to any alteration in the social arrangements of a group or society. Of particular sociological interest is change that results in basic structural rearrangements (e.g., a new basis for social stratification or a change in a group's fundamental mode of decision making). Although the process of change itself is neutral, we often evaluate specific changes and potential changes by culturally established criteria. For example, we might consider the development of modern technology to be good or bad, but the criteria by which we make this judgment are not inherent in the process of change itself.

The effect of religion on a social development is not necessarily intentional. A religious idea or movement may become transformed into something very different from what its originators intended, and the influence of religion is often indirect. The Society of Friends (i.e., Quakers) became one of the foremost reformist sects in Christianity, even though their initial thrust was millenarian and somewhat mystical. As the movement developed, the Quakers' values of hard work and rejection of "worldly" amusements contributed to the wealth of numerous families in the sect. Their other values (e.g., a strong personal conscience on social and political issues) stimulated their active participation in political and social reforms. They were prominent in abolitionism and the underground railroad, established self-help projects during and after the famine in Ireland, and were active in campaigns for prison and legal reform. Thus, the ascetic, inward ideals of the original movement were transformed—first into monetary philanthropy and later into social activism (Isichei, 1967; Wilson, 1970:178–181). The impetus for its monetary philanthropy was the *unintentional* economic change produced by religious values, and the resulting social activism exemplified an *intentional* change orientation, or deliberate stance of opposition to the dominant society (Westhues, 1976).

## Religion Supports the Status Quo

There is an inherently conservative aspect to religion. Religion can evoke a sense of the sacred precisely because of believers' respect for tradition and continuity. Religious symbols link the believers' present experience with meanings derived from the group's tradition, and religious beliefs that are taken-for-granted truths build a strong force against new ways of thinking. Practices handed down through tradition as the god-approved ways are highly resistant to change. Although other aspects of religion promote social change, important elements in religion maintain the status quo.

A central theme in the sociology of religion is the relationship between religious ideas and the nature of the social groups that hold them. Social stratification, in particular, appears closely correlated with religious belief. **Stratification** refers to the differential distribution of prestige and privilege in a society according to criteria such as social class, age, political power, gender, or race. Religion has a different significance to various strata in a society, and different religious ideas will probably appeal to different groups in the society's stratification system. In the United States, Episcopalians and Congregationalists (i.e., United Church) draw members disproportionately from the upper classes, while Holiness and Pentecostal groups draw relatively large proportions of members from the lower classes (Demerath, 1965). Similarly, religious ideas of a social group often reflect social caste, as illustrated by black religion later in this chapter.

### Interest Theories: Marx and Engels

The connection between stratification and religious ideas has often been explained by interest theories such as Marx and Engels's explanation of religion as ideology. **Ideology** is a system of ideas that explains and legitimates the actions and interests of a specific sector (i.e., class) of society. The classical Marxian approach applies the concept of ideology only to ideas that embody the vested interests of the dominant classes; thus, the term has come to have a negative evaluative connotation. The concept of ideology applies more broadly to religion, however, if it is used neutrally. In this usage, ideology also refers to belief systems defending the interests of a socially subordinate group, thus justifying reform or revolution (Geertz, 1964; Lewy, 1974). From this perspective, recent sociology of religion has begun to explore some forms of popular religion as class-based production of religious meanings and practices (Aman, 1988, 1991; Gismondi, 1988; Lancaster, 1988; Maduro, 1982).

According to the dominant strain of Marxian analysis, the fundamental basis of social action is **material interests**, referring to considerations that give economic benefits and power to a person or group. Religious and philosophical ideas are seen as mere epiphenomena, after-the-fact explanations that justify or mask real motivations for behavior.[2] Another school of Marxian thought has, especially in recent years, emphasized the relative autonomy of religion. This school strives to study religious belief systems in themselves, locating them in a larger social-historical context. These theorists retain the insights of Marx and especially Engels but focus on the complex *reciprocal* influences between religion and social structure. By treating the relationship as complex, this latter Marxian approach identifies both the passive and active, conservative and revolutionary elements in religion (this strain of Marxian thought developed from the interpretations of Gramsci, Lukacs, and the Frankfurt school, among others; see Fulton, 1987b; Maduro, 1977; Mansueto, 1988; Mayrl, 1976).

Another concept explaining the change-inhibiting aspects of religion is the idea of **alienation**, which is central to the Marxian definition of religion. Marx used this concept to analyze the false consciousness that he believed religion engendered. Marx (1963:122) asserted:

All these consequences follow from the fact that the worker is related to the *product of his labor* as to an *alien* object. For it is clear on this presupposition that

---

[2]This dominant strain—"dialectical materialism"—is represented by the interpretations of Plekhanov, Bernstein, Kautsky, and later Stalin. Marx himself did not develop a careful, unified theory of religion; the writings of Engels and Kautsky were more fruitful for further development of Marxian theory of religion. For a lucid and critical examination of these early theorists, see McKown, 1975.

the more the worker expends himself in work the more powerful becomes the world of objects which he creates in face of himself, the poorer he becomes in his inner life, and the less he belongs to himself. It is just the same as in religion. The more of himself man attributes to God the less he has left in himself.

Religion, according to Marx, is the projection of human needs and desires into the realm of the fantastic. Religious alienation is thus the understandable reflection of the false consciousness inherent in this social system. Marx's insight into the change-inhibiting aspects of religious consciousness centers on the sense of domination by an alien force. Religion obscures real sources of conflict of class interests. Only a society that destroys such illusions can establish itself on the "truth of this world."

In this context, Marx stated that religion is the "opium of the people." He considered the distress that people expressed in their religion to be real but religion itself as an illusion preventing people from doing anything effective to remedy their condition. Religion soothed their distress, but any relief was illusory (Marx, 1963:43, 44). Often religion draws off dissent and zeal that might otherwise promote revolution. Thus, a Marxian reading of the history of the English working class suggests that the development of Methodism precluded revolution by harnessing nineteenth-century workers' dissent and fervor in a religious movement. (This thesis was first approvingly suggested in 1906 by British historian Elie Halévy; it was re-explored as a Marxian critique of English class relations by Hobsbawm, 1959, and Thompson, 1968.)

Although some evidence supports Marx's contention, studies of the correlation between religion and sociopolitical action generally suggest that the relationship is more complex, varied, and unpredictable (see Gerlach, 1974; L'Alive D'Epinay, 1969; Levine, 1990; Marx, 1967; Smith, 1982). Religion does relieve the tension of economic deprivation by substituting the value of religious achievement for economic achievement, and this substitution may indeed have an opiate effect because pressure for change is defused. At the same time, however, religion offers greater self-esteem by informing believers that they are superior, according to these alternative values. Such sense of superiority has transformative power, as exemplified by the zeal of the Puritans (Coleman, 1956:52).

### Interest Theories: Weber

Weber, too, noted the extent to which religious ideas served to legitimate existing social arrangements, especially the stratification system. Religion has historically explained and justified why the powerful and privileged should have their power and privilege. The wealthy might justify their privilege as a sign of God's approval of their hard work and moral uprightness. Weber observed that most religions provide **theodicies** both of privilege and of disprivilege.

Theodicies are religious explanations that provide meaning for problematic experiences—in this case, the discrepancies of stratification. For example, the Hindu doctrine of rebirth simultaneously justifies the privilege of the upper classes and gives meaning and some hope for the conditions of the lower classes. This belief explains that one's present condition is the result of behavior in one's former life. A favored situation (e.g., being born a man rather than a woman) is justified as the result of appropriate behavior in a former incarnation. Prescribed action for both the privileged and disprivileged is therefore to behave appropriately in one's present social situation in order to obtain a more favorable situation in the next life. This theodicy justifies and explains the social status of both privileged and disprivileged and deters the disprivileged from trying to change the existing arrangement—in this life, at least (Weber, 1946:253ff.; 1963).

Weber's interpretation of stratification was more complex than that of Marx. Weber distinguished between "class situation" (stratification based on economic factors) and "status situation" (stratification based on factors such as life-style, honor, and prestige). Thus, for Weber, the linkage between religion and social arrangements went beyond simple ideological justification for the material interests of a social group; it included ideas that served as a means of **status distinction**, as a focus for group cohesion, and as a basis for controlling economic opportunities (Bendix, 1960:86, 87).

Weber's conceptualization may be useful for understanding the change-promoting and change-inhibiting aspects of the "New Christian Right" in the United States. This movement (and similar movements in America and other societies) is pressing for change, albeit reactionary change. Reactionary movements typically fight the encroachments of "modernity" and urge the cultural return to values and norms of an earlier, more "pure" era. The "New Religious Right," for example, presses for traditional patterns of family life; this goal translates symbolically into crusades against the Equal Rights Amendment to the Constitution, abortion, sex education in the schools, gay rights, and other nontraditional family life-styles. A number of studies have shown a strong correlation between religious involvement (e.g., as measured by church attendance) and antifeminist sentiments and activism (see Himmelstein, 1986).

Some observers suggest that the appeal of such movements is a form of status politics, by which people whose status (i.e., relative prestige or honor) is threatened by changing cultural norms assert their values politically in order to re-establish the ideological basis of their status (Gannon, 1981; Lipset and Raab, 1981). The differences between supporters and nonsupporters of the "New Christian Right" are less matters of economic class than of cultural life-style (Johnson and Tamney, 1984; Shupe and Stacey, 1982; Tamney and Johnson, 1983). One study among U.S. Protestants found that people who were dissatisfied with the amount of social respect given to groups representing traditional values (e.g., people who were involved in churches, people who worked hard and obeyed the law, people like themselves) were especially likely to support the agenda and organization of the Christian right wing (Wald et al., 1989). Another analysis suggested that religion is a major factor in these political stands *both* as a source of traditional images of women and the family *and* as a network from which political traditionalist movements can be mobilized (Himmelstein, 1986). A study of Christian Patriotism, a sometimes-violent, radical right-wing Christian identity movement, found that fundamentalist beliefs powerfully shaped the extremist actions of members through their world-images: dualistic conspiratorialism, patriarchal misogyny, inner-worldly asceticism, and dispensationalist millenarianism (Aho, 1990:166–175).

Several researchers have hypothesized that the "New Religious Right" may be a political defense of a set of life-style criteria that form the bases of supporters' social status (Bruce, 1987; Harper and Leicht, 1984; Lorentzen, 1980; Page and Clelland, 1978; Wallis and Bruce, 1986). This political assertiveness may be all the more plausible because of the social and political turmoil of the 1960s and 1970s; many in the New Right interpreted such events as evidence of the failure of modernity and modern values, as well as the vulnerability of the United States as a nation (Simpson, 1983). As the previous description of the abortion debate shows (see Chapter 3), life-style aspects of status are important. Women whose identity and social status (i.e., relative honor) are bound up in traditional definitions of motherhood and women's proper roles have much to lose when the society shifts to a different set of cultural values (Luker, 1984). Thus, religion may serve to support both socioeconomic class interests and status-group interests. Whether its role was change-promoting or change-inhibiting depends on the social location of the interest group. Furthermore, beyond their purely ideological functions, religious ideas often shape or define the very criteria by which status in a given group is evaluated.

### Religious Legitimation

Religious legitimation of the status quo is sometimes the result of direct collusion between the dominant classes and the dominant religious organizations. Religious organizations and their personnel frequently have vested interests, which they protect by alliance with dominant groups in political and economic spheres. At the same time, dominant groups often try to manipulate religion to serve their purposes. Machiavelli recognized the legitimating power of religion and recommended that lawgivers resort to divine authority for their laws. He proposed:

> It is therefore the duty of princes and heads of republics to uphold the foundations of the religion of their countries, for then it is easy to keep their people religious, and consequently well conducted and united (*Discourses*, Book I, Chapter 12, as quoted in Lewy, 1974).

Through religious legitimation, wars have been justified as "holy wars," obligations have been justified as "sacred duty," and domination has been explained as "divine kingship." The theory of the "divine right" of kings (a Western political theory legitimating monarchies and reaching its apex in the seventeenth century) explained that the monarch's right to rule is God-given and that the monarch's ultimate responsibility for conduct of the state is to God. This theory interpreted the process of monarchical succession and the coronation/ anointment of kings as the working-out or expression of that God-given right. By implication then, a ruler who gained dominance by other means (e.g., by revolution, usurping the throne, or popular choice) was not legitimate in the eyes of God. The monarch had both a civil and sacred role. Divine right made the king priest-mediator of God's will to the state. Belief in the divine right of kings both legitimated the rule of monarchs and suppressed dissent.

In the United States, the constitutionally mandated separation of church and state reduced the direct religious legitimation of political power, but religion has been a significant source of persuasion in political and socioeconomic spheres. Religion has been used to legitimate slavery and racial segregation, industrialization and antiunionism, warfare and international policy—all of which generally served the interests of the dominant sociopolitical groups. The historical processes by which religion was drawn into legitimation of these interests is, however, complex (see sociological studies of specific historical situations such as Yinger's 1946 analysis of church responses to America's participation in World War II; Pope's 1942 classic study of religion's role in quelling a major mill strike; and Hadden's 1969 study of tensions over clergy activism, especially civil rights).

Similarly, religion has been used to legitimate changes favoring a dominant group (e.g., the wealthy or politically powerful). Imperialism and crusades have generally been supported by religious or quasi-religious belief systems. President McKinley explained the decision to wage the expansionist war against Spain by which the United States acquired Cuba and the Philippines as follows:

> I am not ashamed to tell you, gentlemen, that I went down on my knees and prayed Almighty God for light and guidance more than one night. And one night late it came to me this way. . . . There was nothing left for us to do but to take them all and to educate the Filipinos and uplift and civilize and Christianize them and by God's grace do the very best we could by them, as our fellow men for whom Christ also died (quoted in Ahlstrom, 1972:879).

Religious legitimations and institutional support have also been historically used to promote new economic arrangements such as industrialization. A South Carolina minister, supporting the industrialization of his region, said in 1927:

> To those who can read history it is unthinkable that any one fail to see in it all the hand of God bringing the many thousands from the bondage imposed upon us by social and economic forces which of ourselves we were powerless to control. . . . It is imperative that we think of Southern industry as a spiritual movement and of ourselves as instruments in a Divine plan. Southern industry is the largest single opportunity the world has ever had to build a democracy upon the ethics of Christianity. . . . Southern industry is to measure the power of Protestantism, unmolested. . . . Southern industry was pioneered by men possessing the statesmanship of the prophets of God. . . . I personally believe it was God's way for the development of a forsaken people (quoted in Pope, 1942:25).

### Religious Socialization

Religion legitimates not only the social system but also specific roles and personal qualities appropriate to existing structures. By promoting certain character types (e.g., the "hard-working individualist" or the "fatalistic happy-go-lucky" type) that are appropriate to a socioeconomic system, religion further legitimates that system. In socialization the individual internalizes these roles (as discussed in Chapter 3), which then come to exert an influence often greater than external social controls. Religious socialization often indirectly supports the socioeconomic status quo by teaching attitudes and values that adapt to that system.

The norms of existing social relations may be inculcated indirectly through religious rituals and socialization. In the nineteenth century, the social relationship of the Latin American noble-landowner to the peasants who tilled his land was that of the *patron*. The patron granted to the peasant farmers political and economic protection, with the use of a portion of land. In exchange, the peasants were honor-bound to serve the landlord and give him their loyalty. This relationship of personal submission was concretized and given moral strength by the religious practice of the "god-parent." The nobleman was linked to his peasant dependents as god-parent to their children at the time of their baptism. Even veneration of the saints in popular religiosity replicated this relationship of dependence relative to the patron. The attitudes and values transmitted through the institution of god-parent and the popular emphasis on patron saints indirectly supported the norms of the existing socioeconomic structure (Ribeiro de Oliveira, 1979:314, 315).

By observing the moral norms encouraged by a religious group, the individual will come to "fit" better into the existing social arrangement. This effect of religion for suppressing dissent was clearly recognized by a mill official who said: "Belonging to a church, and attending it, makes a man a better worker. It makes him more complacent—no, that's not the word. It makes him more resigned—that's not the word either, but you get the general idea" (quoted in Pope, 1942:30, 31). A restudy of this community found these attitudes still prevalent. Mill managers continue to think that the most valued service of local congregations is developing traits in workers of punctuality, hard work, sobriety, and obedience (Earle et al., 1976:26).

Many sectarian religious groups opposed to "the ways of the world" also teach their members values and behaviors that make them more successful in "the world." Holiness groups, for example, resocialize their members into the values of sobriety, hard work, and forgoing of present pleasures for future rewards (Johnson, 1961). Similarly, Black Muslims internalized a disciplined and ascetic religious ethic, indirectly producing socioeconomic benefit by enabling them to obtain jobs and run small businesses more readily than other blacks (Lincoln, 1989; Mamiya, 1982; Mamiya and Lincoln, 1988). Some 1960s religious movements helped their counterculture members "get straight," according to the values of the dominant society (Robbins and Anthony, 1972). By legitimating "appropriate" roles and personal norms, religion helps provide motivation for individuals to participate in the existing socioeconomic system (Robertson, 1977).

### Social Control

Religion is a potent force for social control. Although social control can be exercised for social change, it is typically change-inhibiting. External forms of religious social control are most evident. Within a religious group, sanctions for deviance from group norms are all the more potent because of their reference to the sacred. The idea of being judged not only by other humans but also by the gods is a powerful deterrent from deviance. Salvation religions such as Christianity exercise control in urging that believers conform to norms in order to assure their future salvation.

The forms of social control exercised by religious groups include informal sanctions such as ostracizing, shaming, or shunning the offender. Other measures include confession and exclusion or excommunication (Turner, 1977). Some religious groups have highly formal controls such as laws and religious courts. The social control exercised by religious organizations varies according to their power in the society. Societywide religious controls (e.g., the Inquisition and witch-hunts) are less likely in a pluralistic society. Nevertheless, America's modern equivalent of witch-hunts, McCarthyism and related right-wing extremism, was supported largely by members of religious groups that did not (then) favor pluralism—Roman Catholics and fundamentalist Protestants (Lipset, 1963). Even in relatively pluralistic societies such as the United States and Canada, religion forms a significant basis of the laws and formal order, as well as of informal social control.

Another relatively informal aspect of religious social control is the power of group loyalty to ensure conformity. The same loyalty can also be utilized for change-promoting action. When the group supports the status quo, however, religious loyalty ensures that the individual does nothing that would upset the group or work counter to its interests. The greater the accommodation of the religious group to the larger society, the more does members' loyalty to "our kind of people" inhibit change.

Internal religious controls are especially important. A person who is socialized into a religious perspective internalizes religious controls. Although socialization is never "perfect," most people internalize much of the normative content of their upbringing. Socialization has a built-in conservative thrust. When a person does something to have a "good conscience" or refrains from certain behavior to avoid a "guilty conscience," the belief system is exercising its control. Internal social controls are likely to impede change-oriented behavior because the individual feels guilty in breaking away from the learned norms.

## Religion Promotes Social Change

While certain aspects of religion inhibit social change, other aspects challenge the status quo and encourage change. In some circumstances, religion is a profoundly revolutionary force, holding out a vision of how things might or ought to be. Historically, religion has been one of the most important motivations for change because of its particular effectiveness in uniting people's beliefs with their actions, their ideas with their social lives. Religious movements such as the Great Awakening (i.e., several waves of revivals in the United States beginning in the latter part of the eighteenth century) have had tremendous impact on society, though the outcome may not have been part of the religious goals of the movement itself. These revivals were important sources of the abolitionist movement as well as of later temperance and prohibition movements. They also had an impact on the democratization of the American polity, making way for popular participation in what was largely an oligarchy of economically prosperous citizens (Ahlstrom, 1972:349, 350; Hammond, 1979; Jamison, 1961).

To understand recent developments in modern societies, sociologists have re-examined their classical foundations.[3] Social change issues were central in the works of Marx, Simmel, and especially Weber. These classical theorists have provided explanations for current developments and have inspired contemporary sociologists to new understandings. One product of this new thinking has been a more profound understanding of Weber's analysis of religion and modernization, going considerably beyond the relatively superficial "Protestant ethic" studies of the 1950s. We will discuss this new approach to Weber in more detail later in this chapter.

### Religion and Social Dynamics: Beyond Marx

Another product of new perspectives to the classics has been a rethinking of the Marxian approach to religion and social change. A number of neo-Marxian sociologists insist that an appropriately complex empirical approach to the scientific study of religion cannot be substituted for or anticipated by theoretical constructs. Marxian theory is thus used as a perspective from which study can be conducted rather than as a static assumption (Maduro, 1977:366).

This new Marxian approach to religion results in a more complex conclusion than classical Marxian interpretations about the relationship of religion to social change. Starting from some of Engels's later ideas, these sociologists view religion as being relatively autonomous from the economic substructure. These theorists come to a more rich, complex understanding of religion by focusing on its functions as a partially independent variable in social change. In contrast to classical Marxism, neo-Marxian approaches suggest that:

> (1) Religion is not a mere passive effect of the social relations of production; it is an active element of social dynamics, both conditioning and conditioned by social processes. (2) Religion is not always a subordinate element within social processes; it may often play an important part in the birth and consolidation of a particular social structure. (3) Religion is not necessarily a functional, reproductive or conservative factor in society; it often is one of the main (and sometimes the only) available channel to bring about a social revolution (Maduro, 1977:366).

These various new approaches to classical sociological interpretations of religion press toward a more complex understanding of religion and social

---

[3]Among sociologists of religion, this recent revitalization of classical issues is best represented by Berger, Luckmann, Fenn, Martin, and Robertson. Robertson's 1977 and 1979 analyses are particularly incisive statements of key themes (see also Beckford, 1989; O'Toole, 1984). Neo-Marxian thinking on religion has been especially lively in parts of Eastern Europe, Latin America, Germany, France, and Italy; see Birnbaum, 1973; Desroche, 1969; Houtart, 1980, 1988; Houtart and Lemercinier, 1984; Kersevan, 1975; Maduro, 1975, 1982; Mansueto, 1988; Thompson, 1986; Turner, 1983.

change. The question is no longer "Does religion promote social change?" but rather "In what ways and under what conditions does it promote rather than inhibit change?"

We will consider the change-promoting aspects of religion under three headings: religious ideas, religious leadership, and religious groups. In practice, of course, these frequently overlap (e.g., when a religious leader proclaims a new religious idea to a group of followers).

### Religious Ideas and Meanings

Ideas themselves do not directly effect change. Ideas indirectly influence society through people whose interests (i.e., all those things that could benefit them) lie in pursuing those ideas and applying them to social action. Religious ideas therefore effect social action in two ways: They may form the *content* of what a group of people try to do; and they may shape people's *perception* of what their interests are. The movement to abolish slavery in the United States, for example, had important religious impetus. Religious ideas explained the evils of slavery; religious movements created a pool of adherents receptive to abolitionist ideas; and religious interests (e.g., desire for salvation) motivated them to put the ideas into action. For such reasons, it is appropriate for sociologists to examine the emergence of new religious ideas in history.

### Religious Breakthrough: Weber.

By "ideas," we mean a broad concept of meanings, such as described in Chapter 2, rather than merely the formal ideas of doctrine or theology. Weber used the yet broader notion of **ethic**, referring to the total perspective and values of a religious way of thinking. Weber (1951, 1952, 1958b) examined several world religions for those aspects of their orientations to their god and their social worlds that inhibited or promoted certain socioeconomic changes, especially the process of "modernization." In these studies, Weber examined the social location of religious ideas and innovation in each society. He sought the sources of motivation for individual action, the relationship of the individual to the larger society, and the religious ideas that shaped how individual actors in that society perceived their social world. Weber was especially interested in locating historical points of **breakthrough**. These were periods in the development of a society when circumstances pushed the social group either toward a new way of action or toward reaffirmation of the old way. The movement toward innovation in the social system constituted a breakthrough. Weber noted that religion has been historically prominent in these breakthrough developments.

Like Marx, Weber held that people's vested interests, considered in the context of certain structural conditions, explained their actions. By contrast, he felt that these interests need not be purely economic ones; people could be motivated, for example, to protect their religious standing. Although religious ideas do not determine social action, according to Weber, they are significant in shaping actors' perceptions and interpretations of their material and ideal interests. Weber (1946:280) stated:

Not ideas, but material and ideal interests, directly govern men's conduct. Yet very frequently the "world images" that have been created by "ideas" have, like switchmen, determined the tracks along which action has been pushed by the dynamic of interest. "From what" and "for what" one wished to be redeemed and, let us not forget, "could be" redeemed, depended upon one's image of the world.

Weber considered religion one of the foremost sources of these ideas that shaped the direction of social action and were "switchmen" in the course of history.

In *The Protestant Ethic and the Spirit of Capitalism* (1958a), Weber analyzed one particularly important breakthrough—the development of the capitalistic mode of socioeconomic organization. He hypothesized a link between the rise of the Protestant worldview and the subsequent emergence of capitalism in Western society. To Weber, capitalism was a significant factor in modernization, a process he viewed with almost prophetic misgivings. The key characteristic of modernization is its *means–end* (i.e., functional) *rationality*. Weber emphasized that capitalism is not mere greed or acquisitiveness but the rational, systematic investment of time and resources toward the expectation of future chances of profit. Furthermore, the expansion of this perspective to a societywide economic system required socially available roles for performing these tasks: specifically, the role of the entrepreneur (Weber, 1958a:17–27; excellent brief treatments of Weber's thesis include Eisenstadt, 1969, and Robertson, 1977).

**Ideas and Individualism: Weber.** Weber argued that creation of a pool of individuals with the necessary values and characteristics to perform as entrepreneurs was essential to the emergence of capitalism. This development was made possible by a new form of individualism, an unintentional by-product of the main variant of Protestantism at that critical moment in history. Weber reasoned that the social role of the entrepreneur called for someone who valued hard work and considered deferred gratification as almost a virtue in itself. Early Protestantism, especially Lutheranism, contributed to this ethic by its interpretation of work as a vocation. The work of the layperson was thus viewed as a virtue, fulfilling a special call from God.

Capitalism further required persons willing to deny themselves rationally and systematically for the sake of achieving a future goal. The dominant strain of Protestantism (e.g., Calvinism, Puritanism, Pietism, Anabaptism) produced such qualities because of its **inner-worldly asceticism**, encouraging members to be active "in the world"—indeed, to prove their salvation by their socioeconomic actions. At the same time, these forms of Protestantism expected their members to forgo the pleasures "of the world"—not to spend their money on luxuries, drink, gambling, or entertainments. The "Protestant ethic" therefore became one of hard work, sobriety, financial care, and deferred gratification. Even though capitalistic gain was far from the goal of these Protestant values, according to Weber, the initial development of capitalism was made possible by the available pool of persons who shared these qualities.

Simultaneously, Protestantism resulted in a powerful new source of motivation for such economic action as capitalism. Protestant beliefs produced a new form of individualism—that is, a new mode of individual-to-society relationship. Under Roman Catholicism, the individual had experienced something of a "blanket" approval. The individual's standing before God and fellow believers was certified by belonging to the church and receiving its sacraments; thus, belonging to the *group* legitimated the individual member. Protestantism, especially its Calvinist varieties, offered no such security. Weber argued that the Protestant ethic made it necessary for individuals to legitimate themselves. This need for self-legitimation, combined with inner-worldly asceticism, produced a strong motivation for the kind of socioeconomic action that capitalism entailed.

Weber linked these values and motivations produced by the Protestant ethic with the emergence of capitalism, especially the critical role of the entrepreneur. Once established, however, capitalism no longer needed the Protestant ethic; mature capitalism, according to Weber, can be self-sustaining. Some sociologists, failing to recognize this point, have tried to apply Weber's terms to current attitudinal differences between Roman Catholics and Protestants in developed countries. These studies have produced conflicting results, but there is little consistent evidence of important differences between Protestants and Roman Catholics in attitudes toward work and consumption.[4] More relevant to Weber's thesis are studies of the impact of other-worldly or inner-worldly religious attitudes in developing nations (see Eisenstadt, 1968).

Numerous scholars since Weber have contested his Protestant ethic hypothesis. Some have argued that the constellation of attitudes and ideas that Weber called the Protestant ethic (if such a mind-set existed) coincided with more important socioeconomic changes such as technological developments, the availability of potential laborers, and the influx of new capital resources (e.g., from colonial holdings). They suggest that these latter factors, among others, were more significant than the Protestant ethic in the development of capitalism. Probably Weber himself would have modified his thesis had he finished his projected sociology of Protestantism before his death. *The Protestant Ethic* was among his earliest writings in the sociology of religion, and his later essays emphasized the change-promoting potential of religious sects as organizations more than the Protestant ethic per se (Berger, 1971; Weber, 1946:302–322).

**Religious Imagery.** Another aspect that contributes to social change is the capacity of religious meanings to serve as symbols for change. Religious symbols frequently present an *image* of future change. They create a vision of what could be and suggest to believers their role in bringing about that change. Change-oriented symbolism is often directed toward the social realm, as exemplified by the ideas of the "heavenly city," the "new Jerusalem," and the "chosen people." Many new religious movements, for example, have articulated visions of the

---

[4]See Greeley, 1964; see Bouma, 1973, for a detailed critique of specific empirical investigations purporting to test Weber's Protestant ethic hypothesis; the best-known study exemplifying these misunderstandings is Lenski, 1963; see counterarguments in Greeley, 1963, and Schumann, 1971; a somewhat different application is made by McClelland, 1961, with counterarguments in Robertson, 1978. Systematic critiques of the Weber thesis include Green, 1959; Samuelsson, 1964; and Tawney, 1926.

"kingdom of heaven on earth" and of how they ought to proceed to realize that
new order (Barker, 1988b).

Religion provides symbols of tradition and continuity, but religious idealization also gives symbols to a group's desire for change. An example of such
change-promoting religious imagery is liberation theology's sacralization of the
community of the poor and identification of this-worldly activism with Christ's
struggle to liberate humanity (Lancaster, 1988). For example, in the Nicaraguan
*Misa Campesina* (peasants' Mass), the "Kyrie" is sung:

> *Christ, Christ Jesus*
> *identify yourself with us*
> *Lord, Lord God*
> *identify yourself with us*
> *Christ, Christ Jesus*
> *solidarize yourself*
> *not with the oppressor class*
> *that wrings out and devours*
> *the community*
> *but rather with the oppressed*
> *with my people*
> *thirsting for peace.*

(lyrics by Carlos Mejia Godoy, 1981; quoted and translated by Roger N. Lancaster in *Thanks
to God and the Revolution* © 1988; reprinted by permission of Columbia University Press)

These symbols enable people to conceptualize their situation and to manipulate change-promoting elements of their social world. Sometimes the specific
symbols evoked are not very effective. The future sought by a group may be
impossible to realize; but genuine social structural changes are unlikely to occur
unless people have new ways of thinking about their social world—a set of
symbols that evoke images of change (see Maduro, 1977:366).[5]

### Religious Leadership

Social change often requires an effective leader who can express desired
change, motivate followers to action, and direct their actions into some larger
movement for change. Religion has historically been a major source of such
leaders largely because religious claims form a potent basis of authority. The
prototype of the change-oriented religious leader is the **prophet**, whose social
role is especially significant.[6]

A prophet is someone who confronts the powers that be and the established
ways of doing things, claiming to be taken seriously on religious authority.
There are two types of prophetic roles. One is the **exemplary prophet**, whose
challenge to the status quo consists of living a kind of life that exemplifies a
dramatically different set of meanings and values. The Buddha is a good example of this kind of prophet, whose very way of life is his message. The other

---

[5]Maduro's theories utilize the seminal work of Pierre Bourdieu, whose theories of symbolic
production are useful to a sociology of religion; some of Bourdieu's works are beginning to be
available in English (1977).

[6]This discussion of prophets and charismatic authority follows Weber's theories (1946;
1947:458ff.; 1963).

type is the **emissary prophet**, who confronts the established powers as one who is sent by God to proclaim a message. Most of the Old Testament prophets were of this type. The emissary prophet has historically been an important source of change because the message proclaimed offered a new religious idea (i.e., ethic) and a different basis of authority. The prophetic message was often one of judgment and criticism. Whether the message called the people back to a previous way of life or directed them to some new way, it nevertheless called for *change.*

The role of the prophet is the opposite of the priestly role. A **priest** is any religious functionary whose role is to administer the established religion—to celebrate the traditional rituals, practices, and beliefs. Most clergy and church officials perform priestly roles. The basis of priestly authority is priests' location in the religious organization as representatives of that establishment, and their actions mediate between its traditions and the people. The prophet, by contrast, challenges the established way of doing things, not only by messages of criticism but also by claiming an authority outside the established authority. Thus, the role of the prophet is essentially a force for change in society.

Weber proposed that **charisma** was the authority basis of leaders such as prophets. Charisma refers to "a certain quality of an individual personality by virtue of which he is set apart from ordinary men and treated as endowed with . . . specifically exceptional powers or qualities . . . [which] are not accessible to the ordinary person" (Weber, 1947:358, 359). The authority of the charismatic leader rests on the acceptance of his or her claim by a group of followers and on the followers' sense of duty to carry out the normative pattern or order proclaimed by the leader. Thus, the charismatic leader is a source of both new ideas and new obligations.

The charismatic leader says, in effect, "It is written . . ., but I say unto you . . . ." By word or deed, the charismatic leader challenges the existing normative pattern, conveying to the followers a sense of crisis, then offering a solution to the crisis—a new normative order. Charismatic leaders may arise either outside or within the institutional framework. Charismatic authority is *extra-ordinary.* It breaks away from everyday bases of authority such as that of officials or hereditary rulers. Thus, in certain periods and social settings, charismatic authority is highly revolutionary.

The authority of the Islamic leader, Khomeini, in the 1979 revolution against the Shah of Iran illustrates this quality of charisma. Khomeini stood in opposition to the rationalization of Iranian society, the modernization of which had upset large segments of the populace. Counterposing a religious charismatic figure against the political leaders "fit" a traditional Shiite model of religious action (Arjomand, 1984, 1988). The combination of Khomeini's personal characteristics and his appeal to Shiite Islamic dogma was powerful. Although Khomeini disclaimed his revolutionary function, he was viewed as the leader who, according to prophecy, would deliver the people; the prophecy itself was, in turn, reinterpreted as support for his charismatic authority (Kimmel, 1989).

Pure charismatic authority is, however, unstable. It exists only in *originating* the normative pattern and the group that follows it. By the process of **routiniza-**

**tion**, charismatic authority is transformed into a routine or everyday form of authority based on tradition or official capacity, and the new religious group comes to serve the "interests" of its members. Although this routinization usually compromises the ideals of the original message, it is a necessary process for the translation of the ideal into practice. Routinization frequently dilutes the "pure" ideals and sometimes results in a comparatively static organizational form; however, it is also the process by which charisma comes to have its actual impact on history (see M. Hill, 1973b; Weber, 1963:60, 61).

The history of the Christian churches is a series of charismatic innovations, routinization, and stable organization, followed by new appearances of charisma. Saint Francis of Assisi was a charismatic leader, an exemplary prophet whose message was also one of judgment—that Christian life must return to its pristine ideals of poverty, community, and simplicity. The early stages of the Franciscan movement were a serious threat to the established church and the sociopolitical order with which the church was intertwined. There was little to distinguish Franciscanism from other threats of the day—Albigenses, Waldenses, and various other pietistic movements, each with its own charismatic leaders. The other movements were eradicated, often militarily by the sheer might of the established powers. The Franciscan movement, by contrast, was encapsulated as a religious order, and its subsequent routinization made it no longer a serious threat to the church. Nevertheless, the Franciscan ideals had great impact and were periodically raised as a critical standard against the complacency of the church and the order itself.

Similar cycles of charismatic innovation and routinization into organizational stability have occurred in other religious groups (and, indeed, in nonreligious institutions, such as political groups). The Mormons moved from a highly charismatic form of authority under Joseph Smith and Brigham Young to a stable, routinized form under church officials called "bishops" (many of whom are laymen). Similarly, Black Muslims have undergone several phases of charisma and routinization, each time altering the group's ideology, tactics, and form of organization (Lincoln, 1989; Mamiya, 1982).

The development of mass communications, especially those using image technologies, such as television and mass advertising, has added a new and complex dimension to the generation of charisma. Mass media possess much more flexible and powerful means to shape the presentation of a leader (or group) to generate the image of exceptional powers or qualities for which the audience might impute charismatic authority. Modern movements may use these mass media to create and sustain a sense of urgency, crisis, or imminent disaster to appeal to a mass constituency (Barkun, 1974).

Not only political figures but also religious leaders can frame their images and their messages for maximum media impact. The television coverage of the journeys of Pope John Paul II resembled a serialized narrative, evoking recurring themes. In repetitive short news clips, the television viewer "saw" the pope set in images of conquest (e.g., with the keys of the city, honor guards, red carpets, in palatial settings). Other media images included symbolic confrontations with death (e.g., located by cemeteries, concentration camps, soldiers with machine guns) and triumphal images (e.g., cathartic, expressive images of attentive, cheering people, flagwaving, applauding, attempting to touch the pope) (Guizzardi, 1989).

Charismatic figures portrayed by the media are thus transformed by the selection process inherent in those media: Images mediate the reality to the audience. Some would-be charismatic figures also deliberately manipulate mass media to enhance their position by creating a sense of awe and power unlikely to be attributed in face-to-face interaction with the leader.

Because it is the followers who impute charismatic authority to their leader and who put the charismatic ideal into practice, any analysis of the impact of religion on social change must examine the role of the religious group or community in that change.

### The Religious Group
Whether the religious group is a small band following a charismatic leader, a growing religious movement, or a staid and established religious organization, it is a potential force for social change. This potential exists because religion—especially the religious community—is a source of power, which is a fundamental category in the sociology of religion. The same power can be and usually is, however, used to support the status quo. Religion is not only an experience of power but often also results in the sense of being empowered. Thus, the followers of the charismatic leader may experience a sense of power in their relationship with the leader and with fellow believers that enables them to apply the new order to their social world.

This sense of power, especially when identified with the leadership of a charismatic figure, can give a developing religious movement great dynamism. In 1525, for example, Thomas Muntzer issued this call to battle in the German Peasants' War:

> Do not despair, be not hesitant, and stop truckling to the godless villains. Begin to fight the battle for the Lord! Now is the time! Encourage your brethren not to offend God's testimony because, if they do, they will ruin everything. Everywhere in Germany, France, and neighboring foreign lands there is a great awakening; the Lord wants to start the game, and for the godless the time is up. . . . Even if there were only three of you, you would be able to fight one hundred thousand if you seek honor in God and his name. Therefore, strike, strike, strike! This is the moment. These villains cower like dogs. . . . Have no concern for their misery; they will beg you, they will whine and cry like children. Do not show them any mercy. . . . Strike, strike, while the fire is hot. . . . Have no fear, God is on your side! (quoted in Lewy, 1974:115).

Another potential is the religious group's capacity to unite previously disparate segments of a society. Religious sentiment can bridge barriers of tribe, family, nationality, and race. Enthusiastic religious movements (e.g., the Jesus People or Pentecostals) often experience intense egalitarianism in their early stages. Similarly, in some developing nations, religious groups are sometimes the only forces capable of uniting people steeped in a tradition of tribalism. The followers of Simon Kimbangu in Zaire and Jehovah's Witnesses in Kenya and

Zambia exemplify this unifying dynamism. Religion often provides a bridge enabling people of different social class interests to work together, as illustrated by the early stages of the Solidarity movements in Poland and Catalonia (Johnston and Figa, 1988). The religious community can produce a sense of group consciousness and solidarity that focuses a group's awareness of its needs and interests and strengthens its efforts to achieve its goals (Levine, 1990; Lewy, 1974:217–220).

Another change-related feature of religious groups is that they are microcosms of political participation. Throughout history, members of religious groups often got a "first taste" of political agency through the polity of their religious group, rather than their state. Long before they had the opportunity for democratic participation in their governments, many Presbyterians, Baptists, and Congregationalists, among others, experienced a measure of political agency in their church organization. These experiences may have shaped their eventual image of a good government and prepared them to be more active and effective than people who had no such experience. Similarly, Latin American Basic Christian Communities (CEBs, described further in this chapter) have changed members' expectations for political participation and action (Levine and Mainwaring, 1989; see also Bruneau, 1988; Hewitt, 1990; Ireland, 1989; Mainwaring, 1986b).

Certain forms of religious collectivity typically have greater change-oriented potential. Sects and cults are, by definition, in greater negative tension with society than denominations or churches (see Chapter 5). Sects are particularly likely to mobilize efforts for change because of their characteristic collective orientation and insistence that religion be applied to all spheres of members' lives (see Yinger, 1963). Churches and denominations, by contrast, generally have more substantial social bases for action but are less likely to challenge the status quo. As Chapter 5 demonstrates, different kinds of religious collectivities have different potential for change-oriented action. Furthermore, these collectivities experience organizational transformations that reduce or promote their dissent.

## Factors Shaping the Interrelationship of Religion and Social Change

The earlier parts of this chapter explored those aspects of religion that are change-inhibiting, change-promoting, or both. A more complex question now arises: Under what conditions is religion likely to be change-inhibiting or change-promoting? The circumstances surrounding a change-oriented effort often determine the degree to which the effort is effective. Thus, we also need to ask: Under what conditions is religion most likely to be an effective source of social change? And what are the social factors that maximize or minimize religion's influence in a society?

### Qualities of Beliefs and Practices

Certain qualities of some religions' beliefs and practices make them more likely to effect change than other religions. A sociologist would seek to learn some of these qualities by asking the following questions.

*Does the belief system contain a critical standard against which the established social system and existing patterns of interaction can be measured?* Those religions that emphasize a critical standard (e.g., a prophetic tradition or a revolutionary myth) pose the potential of internal challenge to the existing social arrangements. The prophetic tradition of the Israelites was a basis for subsequent religious challenges to the established way of doing things (Weber, 1952, 1963). Similarly nations such as the United States, France, or the Republic of Ireland, where the civil religion embodies a revolutionary myth, have a built-in point of reference for internal criticism (Stauffer, 1974). This prophetic aspect of American civil religion was central to the appeal of the civil rights movement. Kennedy's inaugural address was similarly built on prophetic aspects of American civil symbols.

Ethical standards also provide a basis for internal challenge to existing social arrangements. The social critic can say, in effect, "This is what we say is the right way to act, but look how far our group's actions are from these standards!" Many Americans judged their nation's conduct of the war in Vietnam as immoral. The emphasis of ethical standards within certain religions provides a regular ground for social action and change (Nelson, 1949; Weber, 1963). Furthermore, the content of norms and ethical standards influences the kinds of resulting social action. Some religious groups emphasize personal or private moral norms such as strictures against smoking, drinking, gambling, premarital and extramarital sexual activities, divorce, homosexuality, abortion, and contraception. Other religious groups put greater emphasis on public morality, focusing on issues such as social justice, poverty, corporate responsibility, ethics of public policy, and war. Both approaches may result in efforts for social change. The temperance movement of the 1920s and the antiabortion movement of recent years both exemplify movements based on personal moral norms; the civil rights movement of the 1960s and the sanctuary movement of the 1980s were efforts to apply moral judgments to the public sphere (Crittendon, 1988).

The existence of such value standards within a group's belief system sometimes provides a basis of legitimacy for leaders who try to move the collectivity to social action. For example, in the 1950s and 1960s, priests ministering to Hispanic farmworkers in California could refer to the Catholic church's social teachings (e.g., papal encyclicals) regarding social justice and workers' right to organize (Mosqueda, 1986). Despite objections from local clergy and bishops whose congregations (and financial resources) included the grower/employers, these leaders gained legitimacy from the existence of strong standards and ideals taught by their religion at an international level.

*How does the belief system define the social situation?* Individuals' perceptions of the social situation are shaped largely by how their belief system defines that reality. If a religion informs believers that their misfortune is part of God's plan to test their faith, they are not likely to challenge that misfortune. Believers are unlikely to try changing a situation that the belief system has defined as one that humans are powerless to change. Belief systems that embody this kind of fatalism are not conducive to social activism. Some belief systems promote fatalism as a response to the expected imminent end of the world. One Seventh Day Adventist explained why she did not vote or get involved in Hispanic political efforts in her community: ". . . this world is going to be destroyed anyway. We should be focusing our interest on heavenly things. I fear that if we become too involved in these things, we may overlook important things in life" (quoted in Hernández, 1988).

Similarly, belief systems that are voluntaristic are unlikely to result in concerted efforts for change because they define problems in the social situation as the sum of all the individual shortcomings. Accordingly, social ills can be overcome only by converting all individuals to the right way of life. Certain religious definitions of the situation (e.g., fatalism and voluntarism) are less likely than other religious definitions to result in concerted efforts for structural social change.

*How does the belief system define the relationship of the individual to the social world?* This question is a generalization of the Weberian theme described earlier in this chapter. Weber pointed out that certain belief systems encouraged different kinds of individualism and that this individual-to-society relationship is critical to social action. Weber also distinguished between religions that promote a "this-worldly" as compared with an "other-worldly" outlook. Buddhism's interpretation of the material world and aspirations as illusion discourages this-worldly action. By contrast, many strains of Protestantism emphasize one's "working out" of salvation in this world and one's "stewardship" (i.e., responsible use) of God-given worldly resources. While "this-worldly" religious perspectives are generally more oriented toward social action, an "other-worldly" focus can serve to withhold legitimacy from the dominant society (Fields, 1985; Goldsmith, 1989). These aspects of the belief system determine the likelihood and direction of action for social change.

### The Cognitive Framework of the Culture

Relevant to any discussion of religion's role in social change is the cognitive mode of the culture. What are the main ways that people in a culture think about themselves and their actions? While these qualities of the culture may be related to specific items of religious belief, they are more general. Discovering the cultural aspects that shape the influence of religion on social change would include the following factors.

*Is the religious mode of action central to the cognitive framework of the culture?* And are other modes of action (e.g., political) foreign to the way people think? Religious modes of action may be the *only* channel people have for affecting their world; other modes of action may be not only foreign but literally inconceivable. In much of Latin America, most people's worldview is saturated by religion, and religion has the potential to be a viable and vital mode of social action.

Economic dissatisfaction and political dissent may be expressed in religious terms and resolved in religious modes of action (Maduro, 1977; Worsley, 1968).

In cultures where the religious mode of action is pervasive, people are unlikely to conceive of other ways of organizing for change. If another mode of action is introduced, people are likely to be highly suspicious because the idea of a group with no religious identity is impossible to them; they cannot conceive of a group for whom religion is irrelevant to mutual activities. The People's Democracy, a radical independent political movement arising in Northern Ireland during the late 1960s strife, was utterly anomalous because, in that society, a nonsectarian social group was inconceivable to most persons (Beach, 1977). In the United States, by contrast, nonreligious modes of action are both common and conceivable partly because of the ideological separation of church and state but also because of the high degree of institutional differentiation in modern society. Religious institutions are separated from other institutional spheres such as economic and political ones. The difference between these two cultural settings is not only in the social structure but also in the characteristic ways the people think about themselves and their actions.

*Are religious roles and identification significant modes of individual action in the culture?* Does it make sense in the culture for an individual to claim a religious identity or a special religious role? And are religious roles understandable forms of leadership? For Joan of Arc to make religious claims for her leadership made sense to the people of her culture, whereas similar claims in a modern society would probably result in derision or labeling as symptomatic of mental illness. In cultures where religion is important in determining how one thinks of oneself or how others identify one, religion is probably a more significant vehicle for change than in cultures where religious roles are less acceptable.

These aspects of religion suggest that not merely the belief content of religion determines its influence on social change but also the underlying cognitive framework of people in that culture. Sociological analysis of the change-promoting and change-inhibiting characteristics of religion must take into account these cognitive modes, which shape people's ways of thinking about themselves and their world.

### The Social Location of Religion

Most sociological analyses of religion's impact on social change have focused on the social location of religion in various societies. They refer to the structural relationship between religion and other parts of society (i.e., where does religion "fit" into the total pattern of how things are done?). Also, the internal structure of religious groups is related to their larger social location. The following criteria suggest some of the structural aspects of religion in relation to society that are relevant for understanding the extent of its impact on social change.

*Is religion relatively undifferentiated from other important elements of the society?* In most simple societies, religion is diffuse. It permeates all activities, social settings, group norms, and events. Religion, not a separate institution, is undifferentiated even in people's ways of thinking. The person who plants seed with a ritual blessing does not think of the blessing as a separate religious action;

it is simply part of the right way to plant seed. One of the definitive characteristics of modernization is *differentiation* of various spheres of social life. Relatively complex modern societies are characterized by structural (and often spatial) segregation of different parts of social life into identifiable separate institutions.

We can envision societies as falling somewhere on a continuum between these two polar types:

← ——————————————————————————————————————————————— →

Highly undifferentiated                                    Highly differentiated

(religion diffuse)                                         (religion segregated)

In societies where religion is relatively undifferentiated, any change-oriented action is also likely to be religious action. Some of the millenarian movements described in Chapter 2 exemplify religious actions that are also sociopolitical. In relatively undifferentiated societies, religious dissent is often a way of expressing political and economic dissatisfactions. By contrast, although religious action *can* express other dissatisfactions in relatively differentiated societies such as the United States, members of such societies often think of each institutional sphere as separate and of religious action as irrelevant to the political or economic sphere.

*If religion is relatively differentiated from other institutional spheres, do strong ties exist between religious institutions and other structures—especially political and economic ones?* The more linkages that exist between religion and other institutional spheres, the more likely that religious movements for change are expressions of dissatisfaction with other spheres as well. Historically, dissatisfaction with political and economic spheres has been prominent, but dissatisfaction with other differentiated institutions is also possible. For example, a number of contemporary religious movements show strong currents of dissatisfaction with the dominant medical system in American society.

Even in highly differentiated social structures, linkages often exist between institutional spheres. Such linkages range from formal ties (e.g., the state church in England) to highly informal connections (e.g., the "coincidence" of closely overlapping membership between a congregation and other social and political associations in a community). Earlier in this chapter, we saw how religion is often used to legitimate the status quo and serve the interests of the dominant group. Religious organizations themselves often have vested interests such as lands, wealth, and power to protect. These kinds of linkages with political structures and the stratification system are precisely why religion is often used as a vehicle of dissent *against* the political or stratification system.

In societies where religion is closely linked by formal or informal ties with the state, religious movements are especially likely to have political (though not *only* political) significance. The Protestant Reformation was a highly political movement, but religious ideas were also very important. In the United States, where the official linkages between religion and the state are minimized, there is still a considerable informal religiosity (e.g., Congressional prayer breakfasts) and an entire set of beliefs and practices related to civil religion (see Demerath and Williams, 1990). The strong connection between religion and America-love-it-or-leave-it in the 1960s made plausible a specifically religious mode of political reaction. Precisely because the dominant political system was supported by mainstream religiosity, religion was a significant force for political dissent.

Similarly, in situations where religion is closely linked with the stratification system, religious movements are likely to be expressions of socioeconomic dissatisfaction. Development of the array of black denominations and sects in the United States was partly in protest against the parent groups' complicity in a white-dominated caste system of economic rewards, status, and power. The close ties between dominant religious groups and the existing stratification system—in which the black caste was treated as categorically inferior—meant that religious protest could be a simultaneous expression of dissatisfaction with the stratification system (see Frazier, 1974; Lincoln, 1974; Lincoln and Mamiya, 1990; Washington, 1972; Wilmore, 1983).

Many theories about the development of religious movements focus on economic or status deprivation (as discussed in Chapter 5). A relationship clearly exists between religious discontent and disprivilege, but it is not so simple or direct as some of these theories imply. The utterly poor and powerless rarely form change-oriented movements because of their fatalism, lack of resources, and the sheer struggle for basic subsistence. At the other end of the socioeconomic ladder, the powerful and privileged rarely form such movements because the established ways of doing things in the society serve their interests well. The broad spectrum of persons between these two poles, however, are difficult to characterize in their likelihood of forming movements for change. Nevertheless, it is possible to generalize that where religion is structurally linked with socioeconomic privilege, religious dissent and movement for change are likely to express dissatisfaction with the existing distribution of privilege and power. Where such dissatisfactions are not adequately handled by the existing religious groups, fertile ground exists for new change-oriented religious movements.

*Do other modes effectively compete with religion for the expression of human needs, the development of leadership, and organization of effort toward social changes?* In some societies, religious groups may be the most effective vehicle for social change because they are better situated to mobilize change-oriented action. In others, other modes of action such as political or legal action may be better developed and more effective. Thus, in the early years of the U.S. civil rights movement, the black churches were well situated to be vehicles for change. In 1958, 400 black ministers in Philadelphia launched a "selective patronage" campaign (i.e., boycott) against industries that depended heavily on black patronage but discriminated against blacks in hiring. They used the pulpit for announcing target employers and succeeded in stopping some of the blatant job discrimination. The segregation in churches may have aided this effort because none of the "opposition" was in the audience (Lincoln, 1974:9). Similar organizational efforts by the churches were important in the Montgomery bus boycott, sit-ins, and freedom rides (see McCoy, 1964).

In some situations, religious organizations and leaders become virtually the only available voices for change, due to the cooptation or repression of other avenues, such as political parties, legal avenues, the press, universities, and labor organizations (see Levine, 1990). In South Africa, church leaders (except those of the Dutch Reformed churches) protested apartheid from its inception in 1948. By the 1980s, the government had imprisoned, banned, or exiled most secular antiapartheid leaders, so religious leaders (taking considerable risk) became the foremost spokespersons and activist leaders. In 1988 Anglican bishop Desmond Tutu (who was awarded the Nobel Peace Prize in 1984), together with twenty-five other church leaders from sixteen denominations, openly defied government bans and called on all Christians to boycott the election for segregated municipal councils. They declared, "By involving themselves in the elections, Christians would be participating in their oppression and the oppression of others" (Thompson, 1990:239).

*What is the social location of religious leaders?* The potential of charismatic or traditional religious leaders to effect change largely depends on their social location: ties with a pool of potential followers, links with resources that can be mobilized in a movement, and connections with networks of related movements or leaders. Leaders of any movement require these kinds of ties. Recognizing the importance of these linkages, social organizations threatened by dissent often isolate potential leaders in a social location where they are cut off from the people and resources necessary to establish a change-oriented movement. Church officials, for example, might assign a potential "troublemaker" to a remote and tiny parish. Although charismatic leaders are less restricted by such official pressures than other leaders, their social location also influences the likelihood of their gaining a sufficient base of support. Whether or not religious leaders or movements succeed in their efforts for social change depends on their location relative to social resources, organization resources, and a pool of potential followers.

Religious leaders have greater influence in the black community than is characteristic in most white communities. Just as the black church was often the primary institution in the community, so too the preacher was the leader (Hamilton, 1972). Although this role is declining in recent years, many blacks still look to their preachers for leadership on social and political issues. Studies have shown that blacks are more likely than whites to approve of political activism from their preachers and of using the pulpit to discuss political issues (Lincoln and Mamiya, 1990; Nelsen and Nelsen, 1975). One study of urban black laypersons found that 68.5 percent rated qualities of aggressive political leadership as important in their clergy; 78 percent rated expressions of active concern for the oppressed as important (Harris, 1987).

The black clergy activists of the 1960s civil rights movement used their influence as religious leaders, and their style of leadership and exhortation was characteristic of the preacher. Clergy such as Martin Luther King, Jr., Leon H. Sullivan, Fred Shuttlesworth, Wyatt Walker, and Ralph Abernathy did not distinguish between their roles as religious leaders and civil activists. In the civil rights movement, as in some other 1960s protest movements, dissent itself had a religious quality. The importance of the civil rights movement was not only its

specific achievements in gaining voting rights, desegregating educational opportunity, and eliminating discriminatory legislation but also in making blacks
conscious of their power to effect change.

Part of the effectiveness of leaders like Martin Luther King, Jr., and Jesse
Jackson derives from their ability to combine a mode of sociopolitical action with
a religious style to which people could respond enthusiastically. They skillfully
wove together political and religious images, symbols, and modes of expression
(Wilmore, 1983; see also Marable, 1984; Spillers, 1971; Washington, 1985). The
leadership style of the preacher "made sense" in the black community, enabling
clergy activists to move the people to action.

One important factor in the relative influence of religious leaders is the
impact of pluralism and institutional differentiation on the salience of religious
authority. In a pluralistic religious situation, competing worldviews each have
their own spokespersons who can exercise influence that is not necessarily
recognized, however, outside their immediate group. A Mormon bishop's pronouncement on a social issue may be printed in national newspapers and read
with interest by non-Mormons, but it is not likely to be considered authoritative
by persons outside the Mormon fold.

Similarly, institutional differentiation has often led to the idea that religious
leaders should speak out only on "spiritual" matters. One survey found that 82
percent of the laity agreed that "clergymen have a responsibility to speak out as
the moral conscience of this nation." Nevertheless, 49 percent felt that "clergy
should stick to religion and not concern themselves with social, economic, and
political questions" (reported in Hadden, 1969:148, 149). This opinion reflects
the sense that religious leadership is not relevant to other institutional spheres
and that "moral conscience" is generally a matter of private-sphere activities
(e.g., marriage and family). Pluralism and institutional differentiation, important features of some modern societies like the United States, considerably
diminish the impact of religious leadership and religious movements.

### The Internal Structure of Religious Organizations and Movements

In societies where religion is relatively differentiated with distinctive
religious organizations, members' access to religious power is an important
variable in whether that religious group can be change-promoting. A centralized
priestly hierarchy that controls religious "goods" (e.g., salvation) is in a strong
position of power to shape the direction and force of that group's efforts. The
American Roman Catholic church was able to racially integrate its congregations
and schools long before most Protestant denominations or public schools
because it used an authority that superseded local authorities and opinion
(see Campbell and Pettigrew, 1959, for a study of the tension between local
opinion and organizational security in ministers' involvement in civil rights
activism).

Centralized religious organizations typically support the societal status quo
because they also have vested interests in the economic and political arrangements of that society. In some cases, however, a religious organization identifies

with those out of power and privilege (e.g., when it is the religion of a colonial people). In either situation, the access to religious power can be the source of social power, whether for change or stability. Similarly, religions that incorporate a more democratized sense of religious power (e.g., pentecostalism and other forms of experiential religiosity) have the potential to convert that sense of power to social action. Typically, these groups do not organize themselves for action, however, because other elements of their belief systems do not encourage social action. While the potential for change-oriented action is present in such groups, the loose internal structure of their organizations would lead to a very different form of action than that of a centralized organization.

Another feature of a religious organization that sometimes strengthens its effectiveness in social change is the degree to which a group has larger organizational support outside its immediate situation. Roman Catholic clergy in Poland were more able to challenge those in power in the country because theirs is an organization with a supranational base (Mojzes, 1987; see also Chodak, 1987; Chrypinski, 1989). If their religious organization had only a local base of power and authority, the clergy would be more vulnerable to suppression or cooptation by the established national political-economic powers (Westhues, 1973).

In some situations, the larger organization tries to foster change in the face of local resistance. In South Africa, the National Council of Churches has been at the forefront of confrontation with the state over the issues of human rights and apartheid (i.e., the government policy since 1948 of rigid segregation and stratification of racial groups). The Dutch Reformed Church of South Africa (the denomination of the majority of the dominant white group, the Afrikaners) has generally supported apartheid and provides scriptural justification for the policy (Cowell, 1985; Jubber, 1985). When they refused to alter their proapartheid stand, two white South African Reformed churches were suspended in 1982 by the international denomination, the World Alliance of Reformed Churches (Barkat, 1983). Finally, in 1986, the Dutch Reformed Church of South Africa rescinded its support of apartheid. While not actively condemning it, as did the international denomination, the church statement did affirm that the "use of apartheid as a socio-political system . . . cannot be accepted on Christian ethical grounds . . ." (quoted in the *New York Times*, October 23, 1986).

The national centralized denomination served a similar role in supporting ministers speaking out against child labor and other working conditions in mills (Pope, 1942:195–198). A study of twenty-eight American religious groups found that the ability of church leaders to carry through controversial policies (such as racial integration) is dependent on the degree of authority given them by a central formal authority. Ministers were in especially weak positions in groups with congregational polity, where their congregations had power to dismiss them without reference to a higher authority (Wood, 1970, 1981). At the same time, however, the larger organization can stifle or suppress local initiatives for change, as illustrated by the tension between the (local) Latin American bishops and the pope at the 1978 Puebla Conference in Mexico, as well as Rome's later pressures to curb Liberation Theology in Latin America.

## Extended Application: Religion and Social Change in Brazil

As the preceding section suggests, many factors contribute to the likelihood that religion will be an effective factor in promoting social change in a given historical situation. In recent years, some Christian churches—primarily the Catholic church—have been vigorous agents for social change in several Latin American countries. Brazil is one of the most interesting and well documented of these cases.

When we say that religion has been a potent force for social change in a particular situation, that does not mean that all aspects of religion have been involved, or equally involved. Nor does it mean that the activist religious groups share the same goals. In Brazil, for example, the Catholic church has generally pressed for greater civil freedoms and democratization of political and religious structures. At the same time, there are some groups within the church that are urging greater authoritarianism and return to "traditional" forms in family, economic, and political spheres.

Overall, however, the Catholic church in Brazil is promoting general social structural changes that will probably have significant long-term impact. Indicative of its stance are the hierarchy's strong statements on human rights and socioeconomic conditions, the development of Liberation Theology, and the development of grass-roots organizations, Basic Ecclesial Communities (CEBs). The term used within the church for this change-oriented direction is "the preferential option for the poor," meaning that, by design, the church is identifying with the masses of poor, trying to represent their needs and interests, and structurally opening itself to their fullest participation in church and society (Adriance, 1985a).

### Historical Background

The contemporary approach to religious influence taken by the Catholic church in Brazil cannot be appreciated without reference to the history of the church, the Brazilian state, and the changing relationships between them. Brazil was colonized by Portugal in the sixteenth century, and from the outset the Catholic church was tied to the colonial government in a unique arrangement because the Pope gave the Portuguese crown control over the colonial ecclesiastic structure. Accordingly, the church had no separate colonial organization and was nearly totally dependent on the state for its organization and influence, but both were engaged together in the goal of domination of the entire colony: to bring in resources for the colonial power and souls for the church. The ideal was the establishment of a Christendom—the coextension of religion and rule (Bruneau, 1982:11, 12).

Although the church was successful in gaining members (more than 90 percent of the entire populace became Catholic), its actual influence was weak and disconnected. There was a perennial shortage of clergy, and many of those who served were poorly trained and supervised. Churches in rural areas (i.e., most of the country) were under the indirect control of land-owning families.

Popular religiosity only vaguely reflected formal church teachings or Catholic theology (see Bruneau, 1982; Ribeiro de Oliveira, 1979).

A new period in the life of the Brazilian Catholic church began in 1889 when the state became a republic and modeled parts of its constitution after those of France and the United States. The church was abruptly separated from the state, and many of its prerogatives were removed: control over marriages and education, government subsidies for religious education and clergy salaries, and so on. The church responded with political efforts to regain formal power or influence. It also adopted a European model of internal organization; this major change has been called the "Romanization" of the Brazilian church. The effects of this reorganization were strong linkages to the central church in Rome (providing both resources and ideas) and almost exclusive attention to the development of urban bourgeois parishes. The vast majority of the population, nominally Catholic but poor and often remote, was ignored (Bruneau, 1982:17–19).

By the 1930s, the Brazilian church had become internally well organized and held an important political base among the urban upper and middle classes. Simultaneously, the basis of socioeconomic power in the entire society was gradually shifting. The economy had been dominated, well into the nineteenth century, by a landholder/peasant farmer structure. In the twentieth century, urban industrialization and rural plantations (with wage laborers) became predominant (see Ribeiro de Oliveira, 1979). Under the pressure of a powerful bishop, the revolutionary government of dictator Getúlio Vargas granted the Catholic church substantial privileges. His mandated new constitution effectively guaranteed the Catholic church considerable political power. Between 1934 and 1964, the church cooperated intimately with the Brazilian state (Adriance, 1985a, 1986; Bruneau, 1982).

Within the Catholic church there were some seeds of change. Pope Pius XI had begun a movement called Catholic Action to try to mobilize laypeople to restore the influence of the church in society. In the 1950s, the French model of Catholic Action (less other-worldly and more social reform-oriented than its Italian counterpart) became influential in Brazilian Catholicism. The generally conservative hierarchy was initially receptive to Catholic Action because it feared three threats to its own influence: Communist movements in Brazil, Protestant missionary successes, and the spread of synchretic religions such as Umbanda. Although Catholic Action and its related submovements for students, factory workers, and farmers ceased to exist by the mid- to late 1960s, Catholic Action influenced the "preferential option for the poor" in the 1970s through its development of a critical social consciousness, emphasis on the role of the laity, and structural development of local communities of believers (Adriance, 1985a, 1986).

Simultaneously, there were in several Latin American countries the beginnings of the articulation of a "Liberation Theology." Brazilian bishop Dom Helder Camara was identified with the early developments of such ideas, and Brazilian theologian Leonardo Boff has been a prominent exponent of Liberation Theology. In general, these theologies of liberation share the broad principle that the goal of spiritual salvation is not incompatible with—indeed, often necessitates—this-worldly action for the liberation of the oppressed. Liberation Theology is not just a new set of theological interpretations. According to one of its formulators, it "does not stop with reflecting on the world, but rather tries to be part of the process by which the world is transformed" (Gutiérrez, 1973). Thus, within the church there was the development of a set of *ideas* that made possible a whole new way of relating to the social world; the ideals and ideas of Liberation Theology specifically pointed to a transformative approach to religious action.

### The Political and Socioeconomic Situation: The Bishops' Response

The "preferential option for the poor" grew out of a repressive political situation. The military coup of 1964 resulted in imprisonment, torture, and in some cases murder of persons opposing the military regime. As the bishops became aware of the repression, they began to take a stand for human rights. The bishops, too, were subject to repressive measures, but—unlike many other spokespersons (such as newspaper editors or labor leaders)—they had support and some protection from the outside: the Conference of Latin American Bishops and the Vatican itself.

In 1968 the military regime began to impose even more repressive measures to suppress the growing political opposition. The congress, courts, and press were severely curtailed and controlled; legal protections for civilians were curbed; and the use of torture became widespread. In May 1970, the conference of bishops published its first major public criticism of the government, denouncing the use of torture and the absence of fair trials for political prisoners. In 1972 the Episcopal Conference authorized the establishment of Justice and Peace Commissions at the national and diocesan levels. One of the most active diocesan commissions published two books in 1975, demonstrating the link between private death squads and security forces and the inequitable distribution of income and economic resources in that area. The commission also obtained the support of church groups from other nations for its programs for political prisoners and their families. Under the auspices of the national Justice and Peace Commission, efforts began in 1975 to help the Indians and small landholders protect their land claims against the encroachment of government "development" projects and large-scale agribusiness developments. Large numbers of participants in these legal-aid and land-claim programs were harassed, arrested, or murdered (Smith, 1979).

Two related aspects made the Brazilian political-economic situation particularly problematic: the ideology of "national security" and the proclamation of the "Brazilian miracle." The Doctrine of National Security was invoked to legitimate the total control by the military. Brazil has fought no wars with its neighbors in this century; its military buildup was almost exclusively for the maintenance of internal control. The ideology of national security was based on the premise that there were internal enemies working together with international communism to try to subvert the state. It further held that Brazil would become a world power through full economic development. Such economic development, according to the Doctrine of National Security, requires that the nation be secure from "unreliable" elements within it. Thus, the military considered itself to be justified in using repressive measures in all situations of potential unrest, such as dissatisfaction among industrial workers and evicted peasants, newspaper editorial criticism, student activism, and so on. Even political opposition to the military's rule was deemed suspect, and opposition parties were suppressed (Bruneau, 1982; Calvo, 1979).

The Catholic bishops have spoken out against this policy, protesting that:

> inspired [by the national security doctrine], the military regimes, in the name of their battle against communism and for economic development, declare an "antisubversion war" against all those who do not agree with the authoritarian vision of the organization of society (Episcopal Conference of Brazil, 1977, as quoted in Calvo, 1979:150).

In the same statement, the bishops argued that one of the chief problems with this ideology is that it attempts to equate the needs of the government with the needs of the nation—the people. The bishops stated that this ideology leads to the overvaluing of the security of the state and the debasement of individual security: "This is to reduce the people to silence and a climate of fear" (quoted in Calvo, 1979:147).

Concomitant with the Doctrine of National Security, the military government proposed far-reaching economic changes, resulting in what was called the "Brazilian miracle." Two of the main features of this economic change were the wooing of investment capital, especially from multinational corporations, and restructuring of much agricultural production toward greater profitability and export income. "National security" was invoked to make these changes possible, for example, by suppressing labor organizing and enforcing wage freezes (Fernandes, 1984).

The "miracle" was Brazil's accelerated economic growth between 1968 and 1975; Brazil moved from twenty-first to fourteenth in rank among developing countries, based on per capita GNP. The "miracle" did not help most Brazilians, however. The imbalances in the distribution of wealth were made yet worse. The lower half of the entire population saw its share in the total national income drop from 14.5 percent in 1960 to 13.9 percent in 1970 (and further to 13.5 percent in 1976). Meanwhile, the top 5 percent of the population increased its income dramatically, from 27.4 percent of the national total in 1960 to 36.3 percent in 1970 (cited in Bruneau, 1979). Other shifts, such as the expansion of agribusiness production of export crops, made subsistence (such as the cost of the staple beans) disproportionately expensive for the poor. More recently, Brazil has suffered from the worldwide economic crisis, affecting its export markets, inflation rate, and production growth rate. The country has considerable foreign indebtedness, and the economic slowdown has produced further hardships, especially for the unemployed or underemployed working class (see Bruneau, 1979; see also Riding, 1988).

The Brazilian bishops openly denounced the "Brazilian miracle" for the poverty it engendered. They attacked the economic policies that pushed

thousands of peasant farmers off the lands their families had farmed for generations, and they questioned development projects (such as the exploitation of the
Amazon) that displaced the native Indians and poor farmers but brought them
no benefit. Indeed, one observer concluded that "The church has become the
primary institutional focus of dissidence in the country" (Bruneau, 1982:151).
Through its influence and ongoing strength, the church effectively de-
legitimated the military regime (Bruneau, 1985). Beyond the leadership and
formal statements of the bishops, however, a more far-reaching religious influence has been the development of grass-roots organizations, Basic Ecclesial
Communities.

### Basic Ecclesial Communities
The social form of the *Comunidade Eclesial de Base* (or CEB) has spread rapidly
in at least one-half of the countries of Latin America. In Brazil there are some
80,000 such communities (Bruneau, 1988). CEBs represent a fundamental
reorientation of the church in both its internal organization and its ability to
influence popular initiatives for social change (Adriance, 1984, 1986; Bruneau,
1982, 1985, 1986, 1988; Hewitt, 1987; Mainwaring, 1986a, 1986b; Smith, 1979).

Basic Ecclesial Communities are small (about thirty people) local groups of
Christians organized within a neighborhood, workplace, or profession. In principle, they are "basic" in the sense they are organizations of and for ordinary
people, the masses, the "base" of the society and the basis of the church. They
are "ecclesial" in that they are to be in communion with the church, worshipping
through sacraments, prayers, and liturgy, and founded in the continual reflection on the Scriptures and their application to the concrete lives of members. By
emphasizing "community," the CEBs involve members in interaction, equality,
and collective consciousness (Bruneau, 1986). They are the primary structural
way by which the Brazilian church is trying to exercise its "preferential option
for the poor."

CEBs have had significant influence in consciousness raising and in giving
their members experience and self-confidence in democratic modes of participation (Bruneau, 1988; Levine and Mainwaring, 1989). Even after the political and
economic conditions that spawned them have passed, the CEBs may continue to
influence Latin American society through these grass-roots learning experiences. CEBs borrowed some of Catholic Action's methods, including going
beyond simple literacy education to a process the Catholic Action movement
called "see-judge-act." CEBs called the learning process "conscientization," by
which ordinary participants learned to evaluate the realities of their everyday
lives and problems, search for the social structural roots of those problems, and
consider solutions in the light of their readings of Scripture.

These communities were organized primarily as groups of laypersons,
maintaining ties with the church through "pastoral agents." This principle of lay
organization was partly necessitated by the church's massive shortage of clergy;
in 1975 the priest-to-population ratio was 1 to 8,528. Although nuns have
assumed many pastoral duties in Brazil, their numbers are also decreasing by
about 1 percent per year. The church needed to involve laypeople to achieve the

desired contact with the masses (Bruneau, 1986). At the same time, the leadership role of the pastoral agents should not be underestimated.

Pastoral agents (usually nuns, deacons, priests, or trained lay organizers) are typically the driving force in initially organizing the CEB and giving it direction and focus. Pastoral agents teach local leaders necessary leadership skills, suggest materials for discussion and learning, and encourage local members to move into deeper levels of reflection, understanding, and often, social activism (Adriance, 1984, 1985b; see also Doimo, 1989). One study found that the degree of social activism of CEBs was not affected by the social class of the membership (middle-class CEBs were activist, too) but was related to the leadership of the pastoral agents affiliated with the group (Hewitt, 1987).

The spiritual and social activism aspects of the CEBs are closely intertwined. In contrast to the other-worldly and fatalistic approach of most popular religiosity in Latin America, the religiosity promoted by the CEBs emphasizes the individual believer's application of his or her faith to the concrete realities of everyday life. They teach that poverty and suffering are not God's will or the individual's fate but the result of specific social and economic forces. A study of residents of an Amazonian area, for example, found that CEB members were far less likely than non-CEB members to believe that various problems had no solution or could be resolved only by God. Furthermore, CEB members were likely to look to the community (rather than the individual) as the proper agent for response to problems (Bruneau, 1986). CEBs urge their members to apply their faith to action in this world. Thus, CEBs are deliberately promoting a very different definition of the situation and a different individual-to-social-world relationship.

The situation in Brazil is continually changing, and it is difficult to predict the future role of the hierarchy and Basic Ecclesial Communities. After 1985 the government came under the control of a civilian government, and many repressive measures were lifted. The church has continued to speak out against the elitist politics and gross inequalities of wealth and productive resources; it has pressed for agrarian reform, income redistribution, civil rights for indigenous peoples, and broad democratic reforms (Levine and Mainwaring, 1989). At the same time, however, the Vatican has appointed very conservative new bishops and has ordered numerous other changes in the Brazilian hierarchy, resulting in reversal of many social policies and programs for reform (Brooke, 1989; MacEoin, 1991).

While direct political repression became less of a threat, by 1985 the country was deep in an economic crisis, mostly due to the impossibility of repaying Brazil's massive foreign debt, incurred during the years of the economic "miracle" of industrial development. The country instituted numerous austerity measures, partly to satisfy foreign lenders in hopes of restructuring the debt. These austerity measures were particularly damaging to the urban poor, but the fledgling middle classes also suffered and dwindled (Walton, 1989). These economic crises, with the concomitant uncertainty of triple-digit inflation and inability to plan for the future, have both created the popular demand for nontraditional responses from their religious and political leaders and disrupted the small sense of security on which nascent popular democratic participation was based (Maduro, 1989).

As other forms of expressing dissent become legal and viable, the church's role as the voice of dissent may disappear or weaken. Relations with the supranational church organizations (notably the Conference of Latin American Bishops and the Vatican itself) are tense and continually shifting, precisely because of disagreements about the proper role of the church in political and economic matters (see Mainwaring, 1986b:237–253). The CEBs are hardly a unified force within the church, and it is difficult to predict their future direction. Nevertheless, any future development within the Brazilian church or government will have to take into account the values raised by the "preferential option for the poor" and the democratic modes of participation learned by a large proportion of the populace in their experience in CEBs (cf. Dodson, 1990).

## Summary

Certain qualities of religion tend to support the existing socioeconomic arrangements; at the same time, however, religion also promotes social change and has considerable potential for change-promoting action in certain social situations. Interest theories explain that religion frequently supports the vested interests of the dominant social classes, legitimating their dominance, socializing believers to comply with it, and utilizing religious social controls for deviance from it. Religious elements can be potent forces for social change; religious groups, leadership, ideas, and images can promote change-oriented action.

Because the relationship between religion and social change is complex, we have focused on the conditions under which religion is likely to be change-inhibiting or change-promoting. Social-structural aspects of religion are important variables that influence the impact of religion on social change in any given situation. The social location of religion, the relationship of religious institutions to other institutions in society, and the internal structure of religious organizations are all criteria in determining whether religion is change promoting or change inhibiting and its degree of effectiveness in social action.

These factors that shape the interrelationship of religion and social change are illustrated by the religious situation in Brazil. When the church decided to take a "preferential option for the poor," it was uniquely situated to influence certain changes. The cognitive framework of the culture and the social location of religion made religious leadership and organization plausible forces for human rights and socioeconomic change efforts.

## Recommended Readings

### Edited Collections

Jeffrey K. Hadden and Anson Shupe, eds. *Prophetic Religions and Politics: Religion and the Political Order*. New York: Paragon, 1986.

Daniel H. Levine, ed. *Religion and Political Conflict in Latin America*. Chapel Hill: University of North Carolina Press, 1986.

Thomas Robbins and Roland Robertson, eds. *Church–State Relations: Tensions and Transitions*. New Brunswick, N.J.: Transaction, 1987.

### Books

Liston Pope. *Millhands and Preachers*. New Haven, Conn.: Yale University Press, 1942. A classic sociohistorical analysis of the role of religion in a famous mill strike.

Max Weber. *The Protestant Ethic and the Spirit of Capitalism*. New York: Scribner, 1958. The most readable of Weber's classical studies of religion.

Peter Worsley. *The Trumpet Shall Sound*. New York: Schocken, 1968. A Marxian analysis of Melanesian cargo cults. Worsley's critique of Weber's social change theories is the weakest part of this highly readable, well-documented study of religion's change-promoting impact in a concrete historical setting.

# CHAPTER EIGHT

# Religion in the Modern World

Major changes have occurred in the social location and significance of religion in Western societies over the last several centuries. Indeed, it appears that no developed or developing country (including non-Western ones) has remained untouched by these processes. The situation of religion in the modern world is different; the linkages of religion with societal and individual life have changed. Any examination of the nature of religious change must be grounded in an understanding of other major changes in the structure of society—or, indeed, the world.

The secularization thesis is an interpretive paradigm used by many sociologists to try to "make sense" of these historical processes. There is profound disagreement over the meaning of this concept, but most interpretations of *secularization* refer to a historical development by which religion has lost (or is losing) a presumed central place in society. The secularization thesis poses the issue: Does modernization inevitably reduce or eliminate the influence and importance of religion? For classical sociologists and their heirs, the secularization thesis is essentially an attempt to *explain the emergence of the modern world* because many thinkers feel that modern society differs absolutely from what came before it (Luckmann, 1977:16, 17). In their examination of the social location of religion, these theorists have highlighted several processes of social change, which are useful to our understanding of the structure of modern societies: institutional differentiation, changing patterns of legitimation and authority, rationalization, privatization, and individuation.

The general concept of "secularization" is, however, imprecise and broad, lending itself to nonobjective polemics (such as arguing that the end point of the process is either desirable or undesirable). The concept of secularization itself is not very useful because it typically implies a unilinear historical development—inevitable decline of religion and/or religiosity. The nature of social change is, however, far more complex; for example, throughout this book, there are many examples of historical moments when one aspect of religion is ascendant precisely as another aspect is becoming less salient. Religion is thoroughly embedded in so many facets of society that unilinear interpretations cannot possibly portray the complex ways religion reciprocally influences society. Often the empirical data used to buttress the secularization interpretation are based on narrow, substantive definitions of religion that cannot encompass new forms that religion might take. In itself, it is not a particularly important *sociological* question whether religion, thus narrowly defined, does or does not decline. Far more important are questions about the social-structural changes that have produced the modern religious situation and what those structural changes imply for the individual, the society, and humankind.

## The Religious Situation in Modern Societies

Religion in the modern world *does* occupy a very different location from its place in various premodern types of societies. We should remember, too, that there have been profound changes in the structure of society even since the time of the classical sociologists, who formulated their theories during the early stages of industrialization. The pace and quality of social and technological changes in the last 100 years have resulted in new social structures and new modes of religiosity, especially in advanced industrial societies (Beckford, 1989). By trying to understand the changed situation of religion, sociologists also gain understandings of the nature of contemporary societies. For example, how has the nature of national or societal cohesion changed? How has the individual-to-society nexus changed in the face of such significant social-structural changes? How does religion reflect the globalization of economic, political, and legal/moral issues?

When we think about rapid social change, especially with regard to religion, we must avoid romantic notions about the "good old days." Our image of traditional "old-time religion" is often historically very limited (i.e., referring to a period of the heyday of church religiosity that, in the United States, encompassed several decades ending in the midtwentieth century). We conveniently forget that throughout much of America's relatively short history the majority of Americans were apparently unchurched, highly susceptible to "weird" assortments of nonofficial religious teachings, superstitious, or ignorant of church doctrines and practices. Much of what we identify as "traditional" religious morality, such as norms regarding gender roles, abortion, incest, and abstinence from undesirable substances, dates primarily from Victorian times, with the domestication of religion in the face of early industrialization.

Nor were the "good old days" necessarily all that "good." Yes, the quaint little church on the village green back then had the potential to be a place of neighborly warmth and cohesion, where everyone knew everyone else, and where religious belief, practices, and shared values were more fully supported in everyday life by all in the community. That same firm sense of tradition and community, however, also gravely restricted individual freedom: Choices of marriage partners, occupations, leisure-time activities, and political options were all controlled, sometimes subtly and often overtly. The traditional societies that so firmly supported traditional religion were generally authoritarian, patriarchal, highly stratified, and nondemocratic. Indeed, the very discovery of the individual, with emotional needs and human rights and prerogatives of choice, is a peculiarly *modern* feature of a society, as discussed further below. So, in trying to understand the location of religion in modern society, let us strive for a more neutral appreciation of social changes: Some qualities we may value have been lost or transformed; other valuable features have been gained.

This chapter first examines the various processes of societal change toward modern forms of (industrial) society, and for each process we explore the implications for the individual and for the society. Subsequently, we look briefly at the effects of additional changes wrought by globalization and by certain characteristics of advanced industrial societies. These interpretations of societal changes suggest major changes in the nature and location of religion but do not necessarily imply religious decline. They do, however, emphasize significant aspects of *change* not only of the institution of religion but also of politics, work, family, and other spheres of social life.

### Institutional Differentiation
Institutional differentiation refers to the process by which the various institutional spheres in society become separated from each other, with each institution performing specialized functions. For example, religious functions are focused in special "religious" institutions, separate from other institutions such as educational, political, and economic. The contrasting image behind the concept of differentiation is that in simpler societies the beliefs, values, and practices of religion directly influence behavior in all spheres of existence and religion is diffused throughout every aspect of the society. Workers might pray over their tools at the beginning of each workday, and an intragroup conflict might be expressed in a religious ritual. In complex societies, by contrast, each institutional sphere has gradually become differentiated from others. The division of labor in complex societies is similarly differentiated, with specialized roles for each different function. In a highly differentiated social system, the norms, values, and practices of the religious sphere have only *indirect* influence on other spheres such as business, politics, leisure-time activities, education, and so on (Parsons, 1971:101). This means that religion influences these other areas through the personally held and applied values and attitudes of people who are active in each sphere, rather than directly through specifically religious institutions such as the church.

Some theorists point to differentiation as evidence of the declining influence of religion. They interpret the facts that religion is not diffused throughout the society and that specifically religious institutions have limited control over other institutional spheres as evidence of religion's diminished strength and viability. Particularly important is the loss of control over the definition of deviance and the exercise of social control (Wilson, 1976:42). During the Middle Ages, the churches defined and prosecuted deviant behavior, exercising social control both through informal measures (e.g., confession or community ostracism) and formal measures (e.g., church courts). In most modern societies, however, the churches have no such direct control. Courts are more independent, and laws are the province of specialists. Medical institutions have similarly acquired much power over social control, defining deviance as "sick" and calling medical social control measures "therapy." The net effect of this differentiation is that separate institutional specialists *compete* for areas of control that previously were mainly church prerogatives. Although religious institutions still proclaim their definitions of deviance and use some measures of social control, their influence is limited in most modern societies.

Most of the evidence supporting the hypothesis of increasing institutional differentiation is from historical studies. Other evidence can be drawn from conflict between institutional spheres over control of a given function. For example, recent court conflicts over civil disobedience, euthanasia, and forced cesarean deliveries involved competing jurisdictional claims by representatives of several institutional spheres (Fenn, 1978; Irwin and Jordan, 1987; Willen, 1983). Different institutions also compete for control over the definitions of health, illness, and healing (Conrad and Schneider, 1980). These contemporary issues exemplify the process of delineating the boundaries of institutional differentiation.

**Implications for the Individual.** There are two important implications of extensive institutional differentiation: one for the situation of the individual and one for the larger society. For the individual, the process of differentiation involves a conflicting development. On the one hand, differentiation appears to go hand-in-hand with the discovery of self—the unique individual within society. Religion has been an important factor in the increased awareness of and modern emphasis on the individual (Bellah, 1964). On the other hand, differentiation results in segregation of the individual's various roles. A woman's role as mother is not considered relevant to her role as mayor; a man's role as religious believer is not considered relevant to his role as corporate manager.

Specialization in the division of labor extends not only to the separation of institutional spheres but also to specialization of roles within each sphere. One is not merely a "factory worker" but, more specifically, the one who turns the third bolt on the left side of the object coming down the assembly line. Each institutional sphere dictates its own expectations of the individual so that he or she should become an effective performer of that role. Requirements of institutional functioning, however, often conflict with the individual's personal goals, preferences, or needs. The role of advertising account executive may require the individual to wear certain kinds of clothes, to eat and drink in certain kinds of places, to have tact, to lie, to flatter, to be self-effacing, polite, or even obsequious in order to please clients and keep accounts (good examples of such role requirements are found in Berger, 1964; Hochschild, 1983; Terkel, 1975).

Role requirements of occupations vary, but all tend to exclude qualities that do not contribute to achieving the goals of the organization. Values such as moral qualms or self-realization are not necessarily negated; they are simply relegated to another institutional sphere and considered irrelevant to the job. We will discuss these developments as we find them implied in other processes. The key point is that the individual may experience a conflict between the needs and goals of the self and the demands of these social roles (see Simmel, 1959, 1971).

**Implications for Society.** Similarly, the processes of differentiation contribute to society's difficulty in mobilizing the commitment and efforts of its members. Values from one separate sphere do not readily motivate behavior in another. Why should one vote in an election, serve in the army, or work hard on the job? In societies with relatively little institutional differentiation, behavior in work, politics, and social or military service is often motivated by values from other spheres—family, community, religion, and tradition. In contemporary society, the main motivating force in public spheres appears to be the promise of certain levels of consumption—that is, a material standard of living (Fenn, 1974:148).

The process of differentiation is not inexorable. There appear to be limits to the effectiveness of specialization. Some workplaces are experimentally reducing job specialization, with hopes of increasing worker motivation. Even such experiments, however, are dictated by the criteria of the economic sphere; their goal is not to make work meaningful but to increase productivity. Nevertheless, the process of differentiation has important implications for the location of religion in contemporary society. The effective criteria of public institutional spheres—notably the economic—are separate from the values of the private sphere.

Religion is relegated to the private sphere, a development detailed further in this chapter. The individual's desire for meaning and belonging must be pursued in the private sphere. Apparently, the same differentiation that makes possible the "discovery of the self" also frees the institutions of the public sphere to ignore or counteract the autonomy of individuals under their control. The tension between the demands and needs of institutions at the societal level and the demands and needs of individuals makes this an important issue in both the sociology of religion and sociology in general.

### Authority and Legitimacy

**Legitimacy**, as defined in Chapter 6, refers to the basis of authority of individuals, groups, or institutions, by which they can expect their pronouncements to be taken seriously (Fenn, 1978:xiii). Legitimacy is not an inherent quality of individuals, groups, or institutions but is based on the *acceptance* of their claims by others. If an individual proclaims "it is absolutely imperative for Americans to reduce their consumption of gas and oil by 15 percent," on what basis does this person claim to be taken seriously? Such a statement by the president of the United States would be based on a different source of legitimacy than the same statement by a spokesperson for the Union of Concerned Scientists, the editor of *Newsweek*, the National Council of Churches, or, for that matter, a Chicago elevator operator.

The location of religion in contemporary society reflects societal changes in the bases of legitimacy. Relatively stable societies typically have stable sources of legitimacy. The key criterion in such societies is usually traditional authority such as the inherited authority of the patriarch or king. Institutional differentiation often produces a different kind of authority: the authority of the holder of a specialized role or "office." Claims to be taken seriously are based not on who one is but on what position one holds. The authority of a judge, for example, is based on the role for that office, rather than the person.

Religion legitimates authority indirectly in traditional societies by its pervasive interrelationship with all aspects of society. Myth and ritual support the seriousness of all spheres of life. The chief, priest, or matriarch can speak with authority because their roles correspond to or reflect the authority of divine beings. Historic religions legitimate authority more directly, as shown in Chapter 7. Such historic religions as Christianity, Islam, and Judaism have similarly given authority to pronouncements on education, science, economic policy, law, family life, sport, art, and music. Whether directly or indirectly invoked, the images and symbols of the sacred are a source of legitimacy (Fenn, 1978:xiii). The taken-for-granted quality of religion as a source of legitimacy characterizes all of these earlier societal situations.

**Conflicting Sources of Authority.** The main feature of legitimacy in contemporary society is that the differentiation process has resulted in competition and conflict among the various sources of legitimacy of authority. As Bellah (1964) suggests, historic and early modern religions experienced such competition to a limited extent. The beginnings of differentiation of religious from political institutions meant that religious institutions could come in conflict with political institutions. The potential to legitimate implies the corollary ability to delegitimate power, whether in the context of government, courts, workplace, or family (see Luckmann, 1987). Historically, the churches could authoritatively evaluate whether the state was engaged in a "just" war and could authoritatively criticize business practices they judged as "usurious" (i.e., charging unfair interest rates on loans). In contemporary society, by contrast, religious institutions must actively compete with other sources of legitimacy. Personal, social, and political authority are more uncertain.

This competition for legitimacy can be seen in a number of issues in recent years. Civil disobedience over civil rights, abortion, sanctuary for political refugees, and the wars in Vietnam and Central America was often based on a different source of authority than that of the prevailing legal or political ones. Civil disobedience often claimed religion or higher human values as the basis for conflict with civil authorities. Sometimes court events involve a clash of several different sources of legitimacy. In one court case, claims were made from legal, medical, parental, and religious bases of authority. A young person had been comatose for months and "as good as dead" in the commonsense view. Her body was kept alive by technological intervention, and eventually her parents sought legal permission to terminate the "extraordinary" measures of keeping her body alive. No single authority held uncontested legitimacy. The court case was complicated. Medical experts gave their opinions on the medical definitions of death. Legal experts raised issues of the legal rights and guardianship of comatose patients. Theological experts offered briefs on the borderlines of life and death, and the girl's father made a thoughtful personal statement about his request. The relevant issue is not merely the uncertainty of the outcome but that the court is the arena in which medical, legal, religious, and parental figures vied to have their statements taken seriously (Fenn, 1982; Willen, 1983).

**Pluralism.** One particular source of this uncertainty of legitimacy is **pluralism**, referring to a societal situation in which no single worldview holds a monopoly. This feature is an especially important characteristic of the American and Canadian religious scenes (see D. Martin, 1978:1–99, for a sophisticated description of factors related to secularization in nonpluralistic as well as pluralistic societies). Historic religions were characteristically monolithic: They established *the* worldview of their society and had a monopoly over the ultimate legitimation of individual and collective life. Where alternative views coexisted with the dominant one, they were absorbed and coopted or effectively suppressed and segregated. Examples of absorption include the incorporation of early monasticism into church-controlled monastic orders or the development of sects within Hinduism. The alternative worldviews of medieval Jews among Christians and Zoroastrians (i.e., Parsees) among Hindus were no threat to the monopolies of these dominant worldviews because the minority religions were effectively suppressed and segregated (Berger, 1967:134, 135).

Pluralism is sometimes used in a narrower sense to describe the political or societal tolerance of competing versions of truth. Highly pluralistic societies such as the United States, Canada, and Australia typically have several diverse groups with competing worldviews. By contrast, Spain and Russia exemplify a generally monopolistic situation. Between these two polar types, however, there is much variation. In England and Scandinavian countries, there is an official state church but relatively high tolerance of minority positions. A more rigid situation exists in Northern Ireland; although there is "official" tolerance at the societal level, there is little intergroup tolerance. In the Netherlands and Belgium, less volatile but religiously divided societies, most aspects of society have been split into mutually exclusive "pillars" or "columns," according to people's religion. Political parties, hospitals, mass media, youth clubs, social welfare organizations, and schools were all divided according to these religious lines, largely as a protective device for the separate religious enclaves. The divided columns were given political and societal recognition, although leaders discouraged intergroup mingling (Dobbelaere, 1989; Laeyendecker, 1972).

Berger emphasizes that pluralism, in both this limited sense and in the broader sense previously described, greatly affects the situation of religion. Where worldviews coexist and compete as plausible alternatives to each other, the credibility of all is undermined. The pluralistic situation relativizes the competing worldviews and deprives them of their taken-for-granted status (Berger, 1967:151). A farm family in Spain may take the Roman Catholic worldview for granted. It is the worldview of their friends, relatives, and neighbors, and it is a part of everyday life. Although members of the community may vary considerably in piety or religious activity, non-Catholic worldviews are probably literally inconceivable.

In a pluralistic situation, by contrast, no single worldview is inevitable. In the United States, a committed Roman Catholic's neighbors are Baptist, Jewish, Unitarian, Lutheran, atheist, and Zen Buddhist. The government and the society do not (formally, at least) give favorite legitimacy to any of these worldviews, nor does anyone's god accommodate anybody's quandary by sending down a lightning bolt to get rid of all the "wrong" believers. If people wish to protect the belief that their own worldview is uniquely true, they must isolate themselves from alternative worldviews. In U.S. society, especially in relatively urban settings, that is not easy to do. On the job, in school, through the media, working for a political party or a social "cause," or even playing softball—and increasingly in neighborhoods, social clubs, and parties—Americans are exposed to others who hold different worldviews from their own. The impact of the pluralistic situation is thus that the various worldviews in society also compete for legitimacy. No single view has such uncontested legitimacy that a person expressing it authoritatively could be certain of being taken seriously (Berger, 1967:151).

Religious pluralism, in the broadest sense of the term, is not like having a choice between Rotary and Kiwanis. The pluralism of worldviews is qualitatively different from a pluralism of affiliations or associations because the taken-for-granted quality of any one worldview is undermined by pluralism. Differentiation made it possible to conceive of religion as an entity separate from other institutional spheres. Pluralism, furthermore, made it possible to conceive of religions; the very concept implies a stance of some distance, a meaning system that one does not personally believe (Hammond, 1974:119, 120).

**Pluralism and Legitimacy in the United States.** The evidence supporting this interpretation of social change is fairly strong, at least in the United States; but we need to remember that pluralization is a process that is neither inexorable nor complete. In colonial America, the principles of pluralism and religious tolerance were gradually established, but colonial society was far from the pluralistic mode of contemporary society. It was narrowly Protestant and adamantly Christian, even though less than 10 percent of the populace actually belonged to churches (see Wood, 1988). Colonial Massachusetts, for example, was virulently anti-Catholic, with stringent laws against "popish" ideas and practices. Quakers and Baptists were also suppressed, sometimes violently. Only Pennsylvania, Rhode Island, and—for a period—Maryland and New York had some legal support for religious tolerance, but even there religious minorities were not treated equally before the law. Many of the colonies denied Catholics, Jews, and minority Protestant groups such rights as voting, holding public office, and conducting religious services in public (Ribuffo, 1988). Later, when the Constitutional Convention of 1787 prohibited any religious test for public office, it dramatically reshaped church–state relations in the new republic, eliminating preferential treatment not only for any one religion but also of religion over nonreligion (Wood, 1988). The principle of religious liberty, together with the growing multiplicity of religious views, pressed the government and especially the courts to resolve conflicts. The courts needed moral answers to resolve legal questions, yet the increasing ethnoreligious diversity of the country made it impossible for the courts to resort to the language and legitimacy of orthodox Protestantism. In 1879 the Supreme Court had to decide whether polygamy (i.e., plural marriage) would be legally permitted if it were part of a religious group's (i.e., Mormons') practices (*Reynolds* v. *United States*, 98 U.S. 145, 1879). The court decided against making "religious" exceptions to "the law of the land" (Hammond, 1974:129–130).

Similar conflicts arose over whether to allow exemptions on religious grounds from national ceremonies (e.g., saluting the flag) and military service, release time from public schools for religious instruction, prayer in public schools, taking prohibited drugs for religious purposes, and so on (see Manwaring, 1962; see also Demerath and Williams, 1987; Hammond, 1987). It would be a mistake to see these conflicts as merely church–state issues; the very *process* of their resolution reflects the pressure of an increasingly pluralistic situation, in which no single religious worldview is granted legitimacy. Note the change between the language of cases in 1892 and 1965:

> *Church of the Holy Trinity* v. *United States*, 143 U.S. 226, 1892:
> Mr. Justice Brewer wrote that events in our national life "affirm and reaffirm that this is a religious nation." Deciding that a statute prohibiting importing aliens for labor was not intended to prevent a church from hiring a foreign Christian minister, the Court quoted with approval two earlier judicial opinions stating that "we are a Christian people, and the morality of the country is deeply ingrafted upon Christianity" and "the Christian religion is a part of the common law of Pennsylvania."

> *United States* v. *Seeger*, 380 U.S. 163, 1965:
> In a conscientious objection case, the Court determined that "belief in relation to a Supreme Being" (thus exemption from military service) shall be determined by "whether a given belief that is sincere and meaningful occupies a place in the life of its possessor parallel to that filled by the orthodox belief in God of one who clearly qualifies for the exemption." Exemption is not limited to monotheistic beliefs. Mr. Justice Clark noted the prevalence of a "vast panoply of beliefs," thus Seeger's beliefs qualified as "religious" and he was exempted (cited in Hammond, 1974:130, 131).

One general impact of pluralism and differentiation is to create a problem of legitimacy for both the individual and the society. The problem of legitimacy at the societal level involves the society's very basis for authoritative decision making and its grounds of moral unity or integration. Such problems highlight the issues of church–state tensions and the nature of American civil religion, discussed further below. At the individual level, the problem of legitimacy makes the individual's meaning system more precarious, voluntary, and private.

**Problems at the Societal Level.** In a pluralistic situation, worldviews and authoritative claims compete; this results in the diffusion of sources of legitimacy among many agents in society. These competing claims may appeal to sacred or quasi-sacred sources of authority, even if not using explicitly religious symbols (Fenn, 1978:45). A good example is, again, the courts. Hammond suggests that the pluralistic situation creates problems of defining order (because no monolithic worldview exists to authoritatively define it), but the social situation still needs order. Legal institutions are then called on to establish and interpret a uniform order, and the result is a constellation of legal institutions with a "decided religio-moral character" (Hammond, 1974:129). Thus, the hypothesis of secularization is connected with the issue of civil religion (i.e., the national expression of unity, as discussed in Chapter 6). According to this interpretation, the courts are trying to articulate a basis for moral unity where no such foundation is given by specifically religious institutions. Some sociologists view the civil religion as attempting to overcome problems of legitimate authority and societal integration in U.S. society.

Similarly, the Indonesian national guiding principles, the "Panch Shila," represent a government-promulgated civil religion. These principles include belief in one god, national unity, guided democracy, social justice, and humanitarianism. They are derived from Indonesia's principal religious and cultural traditions but attempt to transcend the particular religions to unify conflicting ethno-religious communities and to link people divided by tribal, linguistic, and geographical barriers (see Geertz, 1973:193–233; Demerath, 1989). Fenn (1978) argues that development of generalized religious symbols or ideology is linked with larger societal problems of unstable sources of legitimacy of social and political authority.

The problem of legitimacy results from the collapse of a societal shared conception of order (Wilson, 1976:100; Wilson specified "transcendent" order, but the collapse of any shared worldview has equivalent results). What some people decry as "secularity" is not the result of lost belief followed by immoral behavior; first lost is the overall sense of moral community and consensus (literally "thinking and feeling together"). Without agreement on the way to live together, claims of moral authority make no sense (MacIntyre, 1967:54).

This problem affects both individual and societal decision making. How is it possible for human values to determine public policy in a pluralistic society? Is the role of religion in political decision making reduced to that of one more interest group vying with opposing interest groups? Or is it even possible for a pluralistic society to agree on human values at a societal level? The obvious examples of religious groups asserting themselves on a national level have typically been issues of legislating what might be called "private morality"—homosexuality, abortion, pornography, divorce, gambling, or drinking. Less publicized but still important have been the efforts to apply religious values to decision making on "public morality"—civil rights, the Vietnam War, arms proliferation, and the human environment.

The difficulties of reaching a consensus on values at the societal level in a pluralistic society are illustrated by the issues involved in developing a national energy policy for the United States. The energy situation could be viewed as merely the conflict of competing interests: producers versus consumers, producers versus environmental protection groups, voters versus major campaign contributors, the United States versus other nations, and so on. Policy decisions could reflect a juggling of these interests according to relative power, influence, technological expertise, financial muscle, or political pressure. Nevertheless, human values have also been raised: fair distribution of scarce resources, concern for the "have-nots," human health and safety, responsibility to future generations, and responsibility in international relations. None of these interests or values specifies the exact outcome of any policy decision, but the specific factors accepted as relevant in the decision-making process do make a difference. Is consensus on values necessary or even possible? And if so, does the society consider human values relevant or important to decisions in the public sphere?

Fenn (1972:27) argues that the instability of sources of legitimacy produces two seemingly disparate social responses: the *development of minority and idiosyncratic definitions of the situation*, together with increasingly *secularized political authority*. On the one hand is pressure to desacralize the political authority—for example, removing any ideological notions of what is "good" from decision making and replacing them with criteria such as due process and technical procedures (see also Hammond, 1974). On the other hand, challenging the civil religious synthesis results in spreading access to the sacred. Thus, individuals and groups develop their own particular (i.e., "idiosyncratic") views and symbols for which they claim the same seriousness as recognized religions. Individual and group claims to social authority multiply as the uncertainty of boundaries becomes evident. Fenn (1978:55) states, "Secularization increases the likelihood that various institutions or groups will base their claims to social authority on various religious grounds, while it undermines the possibility for consensus on the meaning and location of the sacred."

Court cases present some interesting examples of the ambiguities at this stage. The use of peyote (a hallucinogen) by some Native Americans has had a long tradition. The Native American church, instituted in 1918, ties together themes from Native American religions and incorporates peyotism in its ritual. In 1962 a group of Navajos using peyote in their religious ceremony was arrested and convicted under a California law prohibiting possession of peyote. The California Supreme Court overturned the conviction (*People* v. *Woody*, 394 P 2d813, 1964), however, granting Native Americans immunity under the First Amendment of the Constitution, which guarantees free exercise of religion. Despite the promulgation of federal policies to protect Native American religious freedom (see Michaelson, 1987), few court battles have upheld those freedoms. In 1990 the U.S. Supreme Court overturned a lower court's decision and refused to grant religious exemption from Oregon's laws against possession or use of peyote (*Oregon Employment Division* v. *Smith*, as reported in Greenhouse, 1990).

The courts were even more pressed on the religious freedom issue when individuals and groups in the larger culture began to use hallucinogens for religious purposes. A number of "psychedelic churches" (e.g., the Church of the Awakening and the League of Spiritual Discovery, both founded in the mid-1960s) have tried to establish their constitutional immunity for the religious use of drugs (mainly marijuana, peyote, and mescaline). The best-known case on this issue was in the mid-1960s, when Dr. Timothy Leary was convicted and sentenced to thirty years in jail and $40,000 in fines for illegal possession of marijuana. Dr. Leary, a researcher of psychedelic drugs, claimed that marijuana use was an integral part of his religious practice. On appeal, the court upheld the conviction, commenting that Leary's religious use of marijuana was occasional, private, and personal. The court declared a "compelling state interest" in drug prohibition:

> It would be difficult to imagine the harm which would result if the criminal statutes against marijuana were nullified as to those who claim the right to possess and traffic in this drug for religious purposes. For all practical purposes the anti-marijuana laws would be meaningless, and enforcement impossible. . . . We will not, therefore, subscribe to the dangerous doctrine that the free exercise of religion accords an unlimited freedom to violate the laws of the land relative to marijuana (*Leary* v. *United States* 383f.2d841, 1967).

This kind of conflict illustrates the lack of moral unity and problems of legitimacy in modern societies such as the United States. Similar cases involving idiosyncratic perspectives include conflict over whether Transcendental Meditation (TM) should be allowed in the schools, whether faith healers or Scientologists could be banned as fraudulent, or whether to allow movements with strong political overtones (e.g., Black Muslims) the prerogatives of religious organizations (see Alley, 1988; Burkholder, 1974; Pfeffer, 1974, 1987; Robbins, 1981, 1985; see also *The Annals of the American Academy*, Nov., 1979).

Legitimation problems are disturbing because they undermine the ability of society to maintain belief in a symbolic whole that transcends the separate identities and conflicting interests of society's component parts (Fenn, 1978:8). Pluralism and institutional differentiation are generally important factors in this process because they break down the overarching worldview—the symbolic whole. These processes make it impossible to achieve a new firm source of societal integration and legitimacy. At the same time, however, they increase the likelihood that people will need and seek this symbolic whole.

**Problems at the Individual Level.** Pluralism, as suggested, undermines the taken-for-granted quality of the worldview. The individual's own meaning system receives less social support and becomes precarious, voluntary, and private. This too can produce conflict for the individual. Pluralism increases personal ambiguity: What am I to believe? How am I to act? On what basis can I decide?

A good example of the widespread conflict and doubt created by religio-cultural pluralism is the issue of abortion. At the societal level, the main issue is whether to legally permit abortion and, if so, under which circumstances. Thus, societal resolution requires response to conflicting legal, biological, religious, and social claims. At the individual level, however, the issue includes personal decisions: Is it right for me to have an abortion? What shall I teach my children about abortion? And how shall I respond if my friend or daughter wants an abortion? Religious pluralism contributes directly to the quandary. Some religions teach that life begins at the moment of conception; others allow that intrauterine life (i.e., the developing fetus) is qualitatively different from life after birth. Accordingly, the former view would equate abortion with killing, while the latter might allow abortion but counsel that it is still not a desirable action. Many other religious groups hold still different interpretations in good conscience. Little consensus exists even within religious groups. The problem for many individuals, then, is that while the decision is often defined as a moral issue, there is profound disagreement over the "good" path. Whatever the individual chooses, there is no massive social support for that decision.

Personal value decisions are important, but a more critical issue at the individual level is the impact of the problem of legitimacy for personal *identity*, which (as developed in Chapter 3) is influenced and supported by religion. The individual's worldview is an important element of personal identity (Luckmann, 1967:70). But one need not go so far as Luckmann, who links personal identity with religiosity, to appreciate the connection between religion and identity. The individual's subjective meaning system legitimates that person's hierarchy of goals, values, and norms. What happens, then, if this key part of the individual's identity is undermined? This problem of legitimacy of the self is related to the process of "privatization" discussed later in this chapter.

### Rationalization

**Rationalization** is the process by which certain areas of social life are organized according to the criteria of means–ends (or functional) rationality. This linkage was a central thread in the works of Weber (1947; 1958a), who viewed a special form of rationality as the outstanding characteristic of modern society.

The key characteristic of modernization, according to Weber, is increasing emphasis upon functional (i.e., means–ends) rationality.[1] In traditional societies, for example, a farmer prepares the soil for planting in a certain way because "that's the way it's always been done" (i.e., traditional behavior); such orientations are extremely resistant to change. By contrast, the modern farmer is more likely to consider alternative and new methods, judging each according to rational criteria such as relative productivity, costs versus benefits, and appreciation of land values. Although Weber's concept of rationality may be useful in understanding the place of religion in contemporary society, we must avoid an overrationalized conception of human action.

**Rationality and Modernization.** According to Weber, modern Western society has a "rationalized" economy and a concomitant special "mentality." A **rational economy** is functionally organized, with decisions based on the reasoned weighing of utilities and costs. The **rational mentality** involves openness toward new ways of doing things (in contrast with traditionalism) and readiness to adapt to functionally specialized roles and universalistic criteria of performance. Although these forms of rationality originated in the economic

---

[1]Note, however, that Weber used the concept of rationality in several ways, giving rise to some ambiguities in his theories; see Luckmann, 1977:16, 17.

order, they have extended into political organization and legal order—the modern state. Weber argued that religious motives and legitimations played a central role in bringing about this form of organization and mentality—for example, by the development of universalistic ethics (i.e., the norm of treating all people according to the same generalized standards) and by the development of religious drive for rational mastery over the world. Nevertheless, this rationality, once a part of societal structure, became divorced from its historical origins and acquired an impetus of its own (Weber, 1947, 1958a).

Weber (1958a) closed his early essay on the Protestant ethic with an almost prophetic evaluation of possible outcomes from the process of rationalization. He observed that the rationally organized order of modern society presents itself to the individual as an overwhelming force. All people engaged in the extensive system of market relationships are bound by the norms of functional rationality. This is exemplified by the great difficulty of establishing alternative economic arrangements, as countercultural communes have discovered. The conditions of the modern economic order have become an "iron cage" of instrumentality, which in Weber's opinion was far from desirable:

> No one knows who will live in this cage in the future, or whether at the end of this tremendous development entirely new prophets will arise, or there will be a great rebirth of old ideas and ideals, or, if neither, mechanized petrification, embellished with a sort of convulsive self-importance. For the last stage of this cultural development, it might well be truly said: "Specialists without spirit, sensualists without heart"; this nullity imagines that it has attained a level of civilization never before achieved (Weber 1958a:182).

Weber's analysis suggests that the differentiation process alone does not account for important changes in contemporary society. Rationalization specifies the direction of differentiation: Differentiation proceeds according to the criteria of means–ends rationality. Businesses can (and many do) specialize tasks according to nonrational criteria. Nevertheless, the main thrust of organization in business (and increasingly in other spheres such as schools, government, churches, and other voluntary organizations) is to extend the criteria of functional rationality: effectiveness, efficiency, cost–benefit analysis, and specialization of tasks. While differentiation produced separate norms appropriate to each institutional domain, rationalization of public sphere institutions means that nonfunctional values (e.g., kindness, honesty, beauty, or meaningfulness) are generally irrelevant to action within these institutions. The ethical regulation of an impersonal rational organization is thus impossible (Weber, 1958a:331).

If modern society is indeed moving in the direction of increasing functional rationality, this process implies problems at two levels: the location of individual meaning and belonging and a conflict between corporate control and values versus personal autonomy and values. Personal meaning is not only relegated to the private sphere but is also undermined by the dominant rationality of other spheres. Rationalized medical practice, for example, does not deal with the "meaning" of childbirth. It is more functional to treat the woman's and child's bodies as objects to be manipulated. Personal meaning, satisfaction, and emotions in giving birth are not overtly denied by the medical process, but they are subordinated to "rational" criteria of efficiency and medical management. The individual seeking to apply meaning to personal experiences is in a weak situation relative to the powerful institutions for which individual meaning is irrelevant.

**Disenchantment of the World.** Another feature of rationalization undermines the individual's personal sense of meaning and belonging: Weber called it **"disenchantment"** of the world, referring to the process by which things held in awe or reverence are stripped of their special qualities and become "ordinary." Protestantism thus brought about much disenchantment of what Roman Catholicism had held in awe (Berger, 1967:111), emptying the believer's world of angels, saints, shrines, holy objects (e.g., candles, blessed water, and oil), holy days, and elaborate sacraments. Rational science also promotes disenchantment, explaining natural phenomena without reference to nonnatural categories of thought. Phenomena previously attributed to miracles are reinterpreted by rational science as natural (e.g., "spontaneous remission of disease"). The key feature of the rationalization process is not so much the particular explanations of phenomena but the *belief* that all phenomena *can* be rationally explained (Weber, 1958a:139).

Thus, rational science undermines other ways of knowing. The process of rationalization results in a dichotomy between "serious" and "nonserious" interpretations (rational explanations being serious and other forms of explanation nonserious). The cognitive style appropriate to modern bureaucratic structures in the public sphere and utilized in science, medicine, law, and so on, is not amenable to the cognitive style of religion—which, by contrast, allows reference to a transcendent, empirically nonverifiable realm and allows undifferentiated "experiencing" as a valid way of knowing. The rational cognitive style of modern public institutions has filtered down to everyday life, though it is not consistently or pervasively applied there. Consequently, not only is one's everyday knowledge more restricted to rationally knowable items but the world itself is indeed transformed. The way in which people think of the world becomes distinct from the way in which they think of themselves and each other (Gellner, 1974:196). The process of rationalization means that the rational mode of cognition applies to those institutional spheres that "really matter" (i.e., are serious); other modes of cognition (e.g., fantasy, play, religion) are treated as frills of private life (i.e., as nonserious).

Rationalization does not necessarily mean that people are generally more rational than in the past. Some evidence exists that superstition and folk religion are widespread elements in highly industrialized societies. A 1968 study in the area of London found a core of approximately 8 percent of respondents who strongly believed in superstitious rituals (e.g., avoiding ladders, knocking on wood, and throwing salt); approximately 23 percent strongly believed in astrology; some 60 percent believed in premonitions; and 30 percent believed in "ghosts or spirits that can sometimes be seen or sensed by people." The proportionately greater amount of superstitious interpretation of events among women and working-class persons apparently reflects these believers' relative lack of control over the events of their lives. Whereas rational modes of cognition are increasingly dominant among the strata of modern society that feel in control of their lives and their world, nonrational modes may increasingly reflect other strata's sense of loss of control (Abercrombie et al., 1970).

This situation on the individual level may therefore reflect societal conflict between the extension of corporate control and the degree of individual autonomy. The modern economic situation is characterized by the increasing power and scope of **corporate actors** in conflict with the power and rights of individuals. The corporate actor is one who fills a role in an institution; the role, rather than the individual, is part of the corporate world (Fenn, 1978:66), and the corporate role is functionally rational in the institutional area. There is little or no integration of the meaning of institutional roles into a subjective meaning system for the individual actor. This lack of integration does not disturb the business or political worlds; the individual is still controlled by the corporate norms. It does not matter in a teddy bear assembly line who (or why someone) attaches the eyes; all that matters is that the role is satisfactorily performed.

Increasing anonymity of functionally specialized roles makes the individual replaceable. The corporate actor is also less responsible as a person, able to claim, "I didn't do anything wrong; I was only doing my job." Individuals may consequently segregate "rational" institutional norms in their consciousness, compensating for their lack of autonomy in the public sphere by a somewhat illusory sense of autonomy in the private sphere (Luckmann, 1967:96, 97; also see Fenn, 1978:81, 82, on the legitimation of the corporate actor's nonresponsibility).

**Privatization and Individuation**

**Privatization** is the process by which certain differentiated institutional spheres (e.g., religion, family, leisure, the arts) are segregated from the dominant institutions of the public sphere (e.g., economic, political, legal) and relegated to the private sphere. This segregation means that the norms and values of the private sphere are irrelevant to the operations of public sphere institutions (Berger, 1967:133). It also implies that the functions of providing meaning and belonging are relegated to institutions of the private sphere. The hypothesis of privatization is directly related to the previously discussed processes of differentiation, pluralization, and rationalization. Differentiation creates separate institutional spheres, in which formerly pervasive institutions such as religion and family are compartmentalized. Pluralization contributes to the complexity of society and makes it difficult, if not impossible, to speak of cultural consensus. And rationalization of the economic (and by extension, the politico-legal) institutional spheres renders the functions of meaning and belonging irrelevant in public institutions.

Whereas differentiation, pluralization, and rationalization all refer to conditions in the larger society, privatization primarily describes the residual category of self. Privatization implies that the individual finds sources of identity increasingly only in the private sphere. Concomitants of personal identity are a sense of order, an interpretation of reality, a system of meaning, and the integration of oneself into a larger community (see Mol, 1976:9–15). The individual finds fewer sources of identity in public sphere institutions largely because the order obtaining in this sphere is functional rationality. All that contributes to making, maintaining, or changing personal identity is located in the private sphere. The individual's very self is privatized.

On the one hand, such privatization promotes some of the personal freedoms enjoyed by many in modern society. One freedom is the extent to which one's reference groups (i.e., those whose opinion of one "counts" in shaping one's behavior) are not imposed by kin or neighborhood but are more freely chosen. The very idea that one could freely choose one's religion is a peculiarly modern concept. Privatization produces greater individual autonomy because it carves out pockets of life in which individuals' motives and range of options are not fully controllable by institutions representative of the larger society. Ironically, American fundamentalism is as much a product of this feature of modernity as it is an antimodern campaign against religious pluralism and differentiation. American Christian fundamentalism's emphasis on voluntary association (e.g., as a freely chosen adult faith commitment) is a clearly modern feature, along with its emphasis on the value of the individual and on religious liberty (Lechner, 1990).

On the other hand, privatization apparently results in problems of meaning and integration for many individuals. These problems may be reflected in the relatively widespread quest for holistic worldviews, as expressed by many alternative health movements, agrarian communes, and contemporary religious movements. Holistic perspectives such as these express a desire for integrating all aspects of life, gaining a sense of wholeness in social, physical, psychic, and spiritual life.

Privatization implies problems in legitimating oneself. Identity becomes problematic. Sources of order, meaning, and community have been undermined; all have become increasingly voluntary and uncertain. Luckmann (1967:99) suggests that this voluntary quality contributes to a sense of autonomy in the private sphere, perhaps making up for the individual's lack of autonomy in institutions of the public sphere:

> Once religion is defined as a "private affair" the individual may choose from the assortment of "ultimate" meanings as he sees fit—guided only by the preferences that are determined by his social biography.
> An important consequence of this situation is that the individual constructs not only his personal identity but also his individual system of "ultimate" significance (used by permission).

This self-selected construction is, according to Luckmann, the contemporary social form of religion. While church-oriented religion continues to be one of the elements that some people choose for their constructions, other themes from the private sphere (e.g., autonomy, self-expression, self-realization, familism, sexuality, adjustment, and fulfillment) are also available in a supermarket of "ultimate" meanings (Luckmann, 1990, 1967:100–114; see also Fenn, 1978:xii, on identity and privatization). Luckmann identifies autonomy as one of the central themes of modern religiosity; however, Luckmann is less optimistic about its significance. He suggests that individual autonomy has been redefined to mean the absence of external restraints and traditional limitations in the private search for identity. Nevertheless, the institutions of the public sphere have real power over the individual; performance of one's roles in these spheres must conform to institutional requirements. Thus, autonomy may be limited to the private sphere (Luckmann, 1967:109–115).

Luckmann (1967:115–117) raises a critical note in a "postscript" when he observes the discrepancy between the subjective autonomy of the individual in modern society and the objective autonomy of the major institutions of the public sphere. He wonders if the removal of nonfunctionally rational considerations from these institutions has not contributed to the dehumanization of the social order. The irony, then, would be that privatized religiosity, by sacralizing the increasing subjectivity of the individual, supports this dehumanization; by withdrawing into privatized religious expressions, this new mode of religiosity fails to confront the depersonalized roles of the public sphere. If this is true, modern forms of religion do not have to legitimate society directly—they support it indirectly by motivating retreat into the private sphere.

Bellah's (1987) interpretation of privatization is similarly pessimistic. Drawing comparisons between modern Japan and the United States, he notes that both religion *and* politics appear to be increasingly relegated to the private sphere. Accordingly, the modern state could be legitimated neither by religious nor political discourse, but rather "managed" by bureaucratic and technological expertise, sometimes decorated by the (real or media-created) attractive personalities of leaders (cf. Williams, 1988).

While privatization refers to the location of religion in modern society, the process of individuation refers more to the development of the *form* of modern religion. **Individuation** is the process by which the individual and his or her concerns come to be seen as distinct from the social group and its concerns. Bellah (1964) views modern forms of religion as a new stage of religious evolution, in which religion as a *symbol system* evolves and develops increasing ability to adapt to its environment. Bellah describes several historical patterns of religion: primitive, archaic, historic, early modern, and modern—each characterized by a common pattern of symbol system, style of religious action, religious organization, and social implications.

Bellah's evolutionary scheme links the structural differentiation of religion with the process of individuation. Historic religions "discovered" the self; early modern religions enabled a greater acceptance of the self; and modern religion represents an even greater emphasis on the autonomy and responsibility of the self. Modern religion is characterized by an image of the dynamic multidimensional self, able (within limits) to continually change both self and the world. The mode of action implied by this image is one of continual choice, with no firm, predetermined answers, and the social implications of modern religion include the image of culture and personality as perpetually revisable. Bellah notes, however, that whether the freedom allowed by this perspective will be realized at the social rather than purely individual level remains to be seen (Bellah, 1964:374).

Robertson's synthesis of themes from Weber, Durkheim, and Simmel similarly emphasizes the twin processes of individuation (especially the quest for individual autonomy) and societalism (i.e., the growth in societal power over individual members). He suggests that the individuation process has proceeded so far that large-scale institutions of the public sphere cannot easily resist it. Some emerging religious movements of recent years may prompt a new mode of relationship between the individual and society (Robertson, 1977:305; 1978:180; 1979; cf. Dreitzel, 1981; McGuire, 1988). Indeed, the very privatized form of modern religiosity may make it more effective than traditional religiosity as a bridge between autonomous individual actors, on the one hand, and societal institutions and humankind, on the other (Riis, 1988).

Reviewing developments of modern forms of religiosity in France, Hervieu-Léger (1990:S22) comments that these phenomena are "vehicles of an alternative rationality which is as much in harmony as in contrast with modernity." Rather than weaknesses, she observes that features of privatized religion are particularly effective strengths in modern society. For example, the purely voluntary associative basis of religious networks makes them more flexible supports for the mobile modern individual. Similarly, the contemporary religious emphasis on personal experience and the individual's "right to subjectivity" are well suited to personal motivation and decision making in a situation where values and ends are not institutionally "given." Hervieu-Léger suggests that modernity produces not the "decline" of religion but a new place—a new work for religion to accomplish.

## Extended Application: Religion and Healing

Many of the processes of change described above have affected the relationship between religion and healing. In many respects, the belonging- and meaning-providing elements (not only of religion but also of family and other institutional providers) have been removed from the entire institutional area of healing. Medical treatment in the Western medical system is not related to any larger meaning system. The Western medical system, while allowing other sources to project meaning for health and illness, generally denies that meaning is related to medical problems or their resolution. The process by which the institution of healing has changed so thoroughly illustrates the broader concepts we have discussed above.[2]

### Differentiation
Western medicine has gradually differentiated itself from other institutions. Traditionally, healing was interwoven with numerous other institutional

[2]Portions of this discussion are adapted with permission from *Health, Illness and the Social Body* by Peter E. S. Freund and Meredith B. McGuire, 1991, Prentice-Hall, Inc., and from my essay, "Religion and Healing," pp. 268–284 in *The Sacred in a Secular Age*, P. Hammond, ed. (Berkeley: University of California Press, 1985).

domains, notably the family and religion. Long before there was any distinctive occupation of medicine, healing was the function of mothers and other nurses, herbalists, folk healers (e.g., persons with a special gift for setting bones), religious persons, midwives, diviners, and so on. Although the institutional differentiation of medicine proceeded especially rapidly in the last century, its roots can be traced at least as far back as early Greek and Persian medicine. Characteristics of this differentiation include the development of a distinctive body of knowledge, a corps of specialists with control over this body of knowledge and its application, and public acknowledgement (or legitimacy) of the specialized authority of medical specialists.

The specialized occupation of "doctor" did not develop until the Middle Ages. Healing in this period, however, was highly supernaturalized. The church attempted to control healing because the power to heal was believed to come from spiritual sources, either evil or good. Nonclerical healing was severely limited. The Lateran Council of 1215 (which represented the apex of church power in the Middle Ages) forbade physicians from undertaking medical treatment without calling in ecclesiastical advice. Healing that took place outside clerical jurisdiction was suspect of having been aided by the devil. The two main groups of healers thus suspect were Jewish doctors and "white" witches (typically women). Jewish (and, in small numbers, Moorish) physicians kept their ancient medical lore alive throughout the Middle Ages, while the dominant European medical knowledge was a stagnant form of late Greek medicine (i.e., Galenism). Therefore, Jewish physicians were sought by the wealthy, to the disgust of Christian clergy who proclaimed (for example) that "it were better to die with Christ than to be cured by a Jew doctor aided by the devil" (cited in Szasz, 1970:88).

The masses, however, were served by an assortment of folk healers such as white witches—members of the community who used herbs, potions, magic, charms, and elements of pre-Christian religions to cure disease and ward off evil influences (see Ehrenreich and English, 1973). They served as midwives and were often consulted for advice about personal problems as well as magic. These healers were a special target of the Inquisition in several countries. A papal bull (i.e., official writ), the *Malleus Mallificarum* (1486), became a manual for witch-hunts. The *Malleus Mallificarum* (Sprenger and Kramer, 1948) singled out women healers as especially dangerous:

> All witchcraft comes from carnal lust, which in women is insatiable . . . Wherefore for the sake of fulfilling their lusts they consort with devils . . . it is sufficiently clear that there are more women than men found infected with the heresy of witchcraft . . . And blessed be the Highest Who has so far preserved the male sex from so great a crime.

Similarly, witch-hunts in Protestant countries (e.g., England and New England) considered the "good" witch to be especially dangerous (Szasz, 1970:89).

The specialized occupation of doctor developed largely through the establishment of university medical schools and the later state regulation of credentials for physicians and creation of medical guilds. This specialization also had the effect of legitimating the increasing monopoly of physicians against the claims of other healers. Nevertheless, the formally recognized physicians were unable to command a monopoly over healing services, largely because the state of their craft did not inspire public confidence.

### Rationalization

The pre-eminence of the modern medical profession has its roots in the rationalization and further differentiation begun in the Renaissance and proceeding rapidly since about 1850. This rationalization occurred mainly in two spheres: the application of rational science to medical discoveries and the rational organization of the profession. The rationalization of medical knowledge meant increasing reliance on scientific methods of discovery and utilization of technology and technique. Careful observation and systematic recording, dissection and autopsies, and pharmacological experiments contributed to the formal knowledge of the field, whereas some of these methods had been previously forbidden by church authorities. Technological developments such as the invention of the microscope and stethoscope also were a part of this rationalization process. Especially important was the discovery, in the latter part of the nineteenth century, that specific agents (e.g., bacteria) caused specific diseases (Freidson, 1970:16). Access to specialized knowledge and specialized technologies uniquely "qualified" the physician.

The increasing rationalization of medical knowledge led to a focus on *disease,* a biophysical entity, rather than on *illness,* the complex social, psychological, and spiritual situation of the sick person (Kleinman, 1978). Rational medicine has a built-in tendency toward treating the human body as an object rather than a person. This narrow biological determinism promotes the image of the physician as active and powerful, the client as passive and objectlike. The sick person must give up control of his or her own body and depend on the benevolence and knowledgeability of the professional (Young, 1976:18, 19).

Rational organization within the medical profession went hand-in-hand with specialization of knowledge and technical skill. Simultaneously, physicians consolidated their control and prestige by annexing and subordinating a number of potentially competing occupations such as nursing, pharmacy, and medical technology. Other competitors such as midwives and bone-setters were effectively driven out of practice and their functions taken over by the medical profession (Freidson, 1970:52). In the United States, one approach to medicine (called "allopathy") gained a monopoly over medical practice, education, and licensing. Important competing medical approaches such as homeopathy, osteopathy, chiropractic, and naturopathy—each positing different interpretations of causes and treatments of illness—have been effectively subordinated or suppressed by the monopoly of allopathic medicine (Wardwell, 1972).

The dominant medical system has also effectively reduced the legitimacy of actions of other "encroaching" institutional areas such as religion and the family. In 1967 the pastor of First Church of Religious Science was found guilty of illegal practice of medicine for treating members' "emotional and weight problems" with hypnosis—judged by the court to be an appropriate procedure only for medical therapy (cited in Szasz, 1970:91). Other religious groups have experienced similar problems (see Burkholder, 1974). In 1990 a Massachusetts court convicted a couple who postponed medical treatment for their sick child while using Christian Science healing approaches. Although testimony suggested that the child's problem might not have been surgically reparable even if a medical doctor had been consulted, the parents were tried for manslaughter (see Beck and Hendon, 1990, for descriptions of this and related cases).

Various courts have similarly overruled religious or moral objections to medically "required" procedures. Jehovah's Witnesses believe that blood transfusions are forbidden by Scripture, but the courts have generally upheld the medical authorization of transfusions, even for unwilling recipients (*U.S.* v. *George*, 239 F. Supp. 752, 1965). Medical authority supersedes parental authority in decisions presumed to determine life or death (precedent cases in 1952, 1962, and 1964 are cited in Burkholder, 1974:41). Even in instances where medical ability to prevent death is doubtful, greater legitimacy is given to medical rather than parental authority. In 1977, for example, a Massachusetts court ordered that parents of a child dying of leukemia submit the child to medically prescribed chemotherapy, rather than to a less unpleasant (but also less respectable) naturopathic therapy.

### Legitimation and Social Control

The result of these processes of differentiation and rationalization (together with the efforts of the interested parties—the physicians themselves) is that the medical profession today has an officially approved monopoly over the definition of health and illness and the treatment of persons defined as "sick" (Freidson, 1970:5). A good case could, in fact, be made that the church authority of the Middle Ages was undermined first by legal institutions, which, in turn, were undermined by the medical institution. The medical profession has become a key legitimating agency in Western societies and thus has a major role in social control.

Society must deal with individuals who differ significantly from socially established norms. The concept of **social control** refers to all mechanisms a society uses to try to contain its members' behavior within those norms: deterrents, incentives, rewards, and punishments. Whether the society punishes the individual depends largely on its determination of the individual's responsibility for the deviant behavior. A person who fails to go to work for two weeks is likely to be fired, but if it is determined that the absence resulted from sickness, the person will probably be exempted from normal obligations of work. At the same time, however, a person who is permitted to take the "sick role" is obliged to try to get well, cooperate with medical help, and act appropriately "sick" (Parsons, 1972).

The key issues in understanding social control are definition of deviance, determination of responsibility, and administration of punishment. Religious, legal, and medical institutions have served as significant agents of social control in all three aspects. The definition of deviance is a particularly important aspect because the norms by which deviance is defined are themselves socially constructed. The definition essentially depends on the social group's idea of what is "normal" (Becker, 1963). If a society defines noncompetitive or unobtrusive behavior as the norm, any individual who is competitive or stands out is likely to be considered deviant. Deviance is a product of the group that labels it.

Religious, legal, and medical institutions have all contributed to the definition of deviance in society. For example, "Though shalt not steal" defines a religious norm; stealing is a sin. Legal systems define similar norms of behavior, and violation of the norms is a crime. Medical systems also define what is normal or desirable behavior; deviance is sickness. But the relative weight of religious, legal, and medical institutions in defining deviance has shifted. As the Middle Ages waned and these three institutions became increasingly differentiated, the religious organizations still held greatest influence in defining societal deviance. This pre-eminence continued into the eighteenth century, but in the United States and France (and later other European countries) the legal mode of defining deviance gained ascendancy. In the United States, the increasing pre-eminence of the legal definitions was promoted by pluralism (as previously noted) and by increasing rationalization of the nation-state.

The significance of legal definitions of deviance has waned somewhat in the twentieth century, and medical definitions of deviance have gained pre-eminence. This shifting balance is clearly reflected in the 1954 precedent-setting court case *Durham* v. *United States*, 214 F 2d, 863, which decided that "an accused is not criminally responsible if his unlawful act was the product of a mental disease or mental defect." Abe Fortas, the court-appointed defense in this case and later a Supreme Court justice, stated, "Psychiatry is given a card of admission [into the courtroom] on its own merits, and because of its own competence to aid in classifying those who should be held criminally responsible and those who should be treated as psychologically or emotionally disordered" (cited in Szasz, 1970:317). The shift in balance favoring medical definitions of deviance corresponds chronologically with the period of rapid professionalization of medicine, when medical discoveries and technology proceeded rapidly and public faith in science and medicine was increasing.

The appeal of medical definitions of deviance over legal or religious definitions is understandable. Religious definitions appear too nonrational and, in a religiously pluralistic country, they lack societywide acceptance. Legal definitions, while more rational, appear to hinge too greatly on human decisions—the judgment of twelve ordinary citizens on a jury, for example. Medical definitions of deviance, by contrast, appear more rational and scientific.

The concept of sickness, however, far from being a neutral scientific concept, is ultimately a *moral* one, establishing an evaluation of normality or desirability (Freidson, 1970:208). A wide range of disapproved behavior has been defined by the medical profession (especially its psychiatric branch) as sick: alcoholism, homosexuality, promiscuity, drug addiction, arson, suicide, child abuse, and civil disobedience (see Conrad and Schneider, 1980). The seeming rationality of medical diagnosis thus masks the evaluative process. Jurisdiction of medicine over these behaviors does not depend on medical *knowledge* of their causes or cures, for medicine is no better able to cure alcoholism or homosexuality than can religion or law. The pre-eminence of medical definitions is based on the popular and juridical *acceptance* of medical authority (Bittner, 1968; Freidson, 1970:251–253).

### Privatization

A foremost characteristic of the institution of medicine in modern Western societies is that it has divorced the function of curing disease from the function of providing meaning and belonging to the sick person. The medical institution has limited itself to the cure of disease (a biophysical entity) and the physical tending of the diseased person. The meaning- and belonging-providing functions of healing are treated as relatively unimportant and are relegated to the private sphere institutions of family and religion. Thus, if a person has cancer, the physician diagnoses a "malignancy" and uses surgery, radiation, or chemotherapy. The "disease" is an abnormal mass of cells in the body. The physician typically does not address, however, the person's *illness*: what cancer means to the person, how it is experienced, how it affects the person's life, feelings, and personal relationships (Kleinman, 1973, 1978). There is some recognition of problems of meaning for cases of "mental illness," but their treatment is also segregated in separate institutions with separate specialists. Most physicians, however, see the provision of meaning as totally unrelated to the cause or healing of medical problems.

Because a high percentage (roughly half) of patients' visits to physicians are for nonspecific complaints (i.e., for which no clear biophysical basis can be found), it appears that illness (i.e., the social and psychological problems of the person) needs as much attention as the disease or biophysical difficulties (Kleinman, 1978:63). For example, if a thirty-three-year-old mother of five children goes to a physician complaining of pains in her legs and increased difficulty with chronic asthma but no biophysical basis for the leg pain is found, what is the physician's response? The doctor may say there is nothing wrong with her legs and nothing to be done for her, or decide that her trouble is "all in her head" and prescribe placebos (i.e., pretend pills) or antidepressant drugs; or the physician may refer her to a specialist for illnesses that are "all in the head." The physician treats her physical problems as fundamentally unreal and inappropriate for medical attention.

Two recent social developments challenge the separation of healing from the provision of meaning and belonging. One is the increasing prominence of healing in new religious movements, both outside and within the churches. Many of the new religious movements emphasize physical, emotional, and spiritual healing. At the same time, quite a few established churches have renewed interest in their healing ministry. Faith healing previously carried the connotation of being invariably limited to uneducated, lower-class sectarians; now, however, it is prominent in numerous middle-class churches and movements, such as the Catholic Charismatic Renewal, Order of St. Luke (Episcopalian), and Women Aglow (an interdenominational Pentecostal association). Faith healing is typically used in addition to medical treatment. Few middle-class healing movements recommend forgoing secular medicine; rather, they generally emphasize the limitations of modern medicine—especially its inability to heal the whole person.

The second development is the parallel rise of a number of alternative therapies. Some of these alternative therapies draw heavily on the Christian tradition; others are adaptations of Eastern practices. Many more are eclectic combinations: Christian and Western occult traditions, Eastern thought and practice, Native American healing, elements of psychotherapeutic approaches, science and pseudoscience. Alternative therapeutic approaches are widespread among middle-class, well-educated persons—sometimes as the adherents' sole belief system and sometimes in combination with more traditional religious belief systems. A study in a relatively staid East Coast suburban community found alternative therapies widespread; hundreds of suburbanites in this catchment area were found to have employed nonmedical healing, and over eighty-five different healing methods were found in use (McGuire, 1988; similar indicators of the prevalence of these approaches are found in Gallup Poll, 1976; Wuthnow, 1976b, 1978a).

It appears that, while members of "modern" Western societies are utilizing secularized medicine, they are also seeking something more. A number of similarities between the new Christian healing movements and the other alternative healing systems emphasize the challenge of these movements to the secularization of health and healing. Both the Christian and alternative healing groups assert the need for *wholeness*. They emphasize that true healing depends on the recognition that social, emotional, and spiritual aspects are intimately intertwined with the physical.

Another parallel between Christian healing movements and other alternative therapies is that both proffer alternative *norms for healthiness*. Their definitions of health are far broader than those used by the medical profession. These broad definitions imply that healing is an appropriate response to a wide range of physical, emotional, spiritual, social, economic, and political "problems." The symbolic and metaphorical significance of health and healing are highlighted. Spiritual values, such as oneness with God, balance, and harmony with the universe, are critical. Movements such as these are contemporary dissent against the hegemony of the dominant medical system. They argue that the provision of meaning and belonging are central to the healing of illness.

The limitations of highly rationalized medicine are evident when we ask: What are the proper ends of modern medicine? The rationalized biomedical model operates on a highly interventionist stance against disease; accordingly, the conquest of death became a major goal of medicine (Comaroff, 1984). Although this model acknowledges that ultimately all persons will die, it also makes it difficult and sometimes impossible to *decide* that any given person will be allowed to die (see Kleiman, 1985; Muller and Koenig, 1988).

Due to technological developments, physicians can prolong lives, even though often they cannot actually cure the disease. Indeed, our cultural values turn technological intervention into an "imperative": If we have the technological capability to do something, then we *should* do it. In the context of medical institutions, once a technology becomes available its use becomes almost inexorably routinized and considered standard. A physician's failure to apply this standard care—no matter how inappropriate for the individual patient—would be considered reprehensible (Koenig, 1988).

The technological imperative is deeply embedded in many institutional responses to health crises, in large part because rationalized institutions give primacy to technological and expert knowledge. In recent years, serious ethical issues have developed regarding the ends of medicine (see Callahan, 1987, 1989). Should a terminally ill person be treated aggressively until the moment of death? What medical intervention should be used for an accident victim in a persistent vegetative state, for an irreparably damaged newborn, or for a medically unsalvageable patient (such as someone who has advanced untreatable cancer who subsequently has a heart attack)? Are there limits to the use of medical responses that themselves cause pain or other suffering? Under what conditions should a patient or patient's family be allowed to refuse a prescribed treatment, such as chemotherapy or surgery?

Increasingly, such issues are coming before the courts, but the courts have typically deferred to the medical profession itself as the proper authority on matters of defining life and death. On issues such as these, however, there is no technical (i.e., purely medical) answer. Callahan (1987:179) notes, "The 'sanctity of life' has to be the sanctity of personhood, not merely the possession of a body." Moral and legal rights adhere to persons, but there are no neat biomedical criteria to distinguish when or whether a body is a person. The choice of such criteria must come from outside the rationalized domain of scientific medicine. The effects of institutional differentiation, pluralism, and rationalization in modern societies, however, make it extremely difficult to engage in the necessary ethical discourse at a societal level.

Because Western medicine is focused on the curing of "disease" rather than healing of illness, the provision of meaning is privatized and undermined. Secularized medicine is generally unable to deal with illness, the complex of perceptions and experiences of the sick person. One critic, who speaks from both a medical and anthropological background, suggests that Western physicians should be trained to treat sickness in the context of the patient's culture and psychology (Kleinman, 1978:68). Although such attentiveness to the illness experience would probably produce a more humane treatment of patients, it is doubtful that it would satisfy the patient's desire for meaning. The Western medical paradigm is based on some presuppositions that cannot be readily reconciled with the belief systems of many patients, especially in a culturally and religiously pluralistic society. Although physicians could be taught to respect the beliefs and values of, for example, Jehovah's Witnesses (a worthwhile task), it is improbable that this would overcome physicians' evaluation that their own paradigm is superior.

The Western medical paradigm denies the validity of other paradigms, including the interpretive ones provided by religion, family, and the ethnic community. The Western medical system has come to conclude and defend that health, illness, or dying has no real meaning other than the biophysical realities determinable by empirical, rational means. Provision of meaning is allowed to institutions of the private sphere, as voluntary and purely subjective interpretations, only as long as that interpretive scheme does not interfere with the biophysical treatment of the disease.

## Religion, Power, and Order in the Modern World

World events in recent years have highlighted the extent to which religion is an appropriate focus for study of the continuing tensions between tradition and modernity. The resurgence of fundamentalist Islam in Mideastern politics, the application of Christian liberation theologies to political situations in Latin America, and the intensification of church–state issues in the United States exemplify the relationship of religion to political and economic life in modern and modernizing societies.

The first part of this chapter described some theories of social change as they apply to the situation of religion; that discussion focused on the implications of four processes for the society and for the individual member. Some social scientists suggest that it is necessary to look beyond the borders of any given society for the full impact of modernization and related social changes. They argue that the global interdependence of societies dramatically alters both their internal workings and the ways they relate to each other (Kokosalakis, 1985). Although these interpretations of global social change are relatively new and undeveloped, they appear to be a very useful point of departure for our understanding of the relationship between religion, power, and order in the modern world.

Global interdependence results from the process by which the entire world has gradually become part of a single economic system. Whereas in the not-too-distant past most economic activity was local or regional, increasingly the scope of economic interdependence has broadened—to whole societies, whole continents, and ultimately the entire world.

The scope of a global economic situation is difficult to imagine. It means that decisions made in distant places by unknown powerful people affect most aspects of one's own everyday economic situation. In traditional societies, most economic relationships are within the community, and transactions are handled face-to-face: producing, buying, and selling all of one's needs take place in the immediate community.

Modern Western societies have in this century been accustomed to a larger scope; societal economies are taken for granted. When a company opens a new factory in Texas and closes its older plants elsewhere, its employees in New Hampshire may be frustrated and angry about losing their jobs, but they can usually at least envision some of the societal factors influencing their plight. It takes a considerable leap of knowledge and sociological imagination for those workers to understand increasingly global operations, such as when the company closes all but the finishing operations of its manufacturing in the United States and has its basic production done in Brazil, the Philippines, and Mexico.

Other experiences of global economic interdependence facing Americans include the impact of the federal budget deficit, trade deficits, international commodity cartels (oil, coffee, cocoa, etc.), trade embargoes, massive foreign lending and borrowing, and vast multinational corporations with assets greater than the entire gross national product of many entire countries.

People of developing countries of the Third World have a different experience of this interdependence. Their economies were initially developed as colonies of First-World nations, and the colonial legacy often makes economic, social, and political independence extremely difficult. Furthermore, developing countries are often totally dependent on the international economic powers for markets, development loans, and imported technology. Often these countries have only a few export products (typically natural resources, such as minerals) and cannot even set their own prices for the sale of these in the international economy.

"Development" in these contexts frequently means only that the Third-World country allows multinational corporations to build factories, extract natural resources, affect the environment, and pay reduced taxes, in return for the employment of a number of that country's unskilled or semiskilled workers. The fragility and one-sidedness of this kind of "development" are linked to total dependence on a vast world economy in which the developing countries have little or no leverage.

Not only are economic considerations increasingly of a global scope, but also political aspects are necessarily intertwined with world politics. The politics (and consequent military decisions) of a country are closely connected to its position in the world economy and its economic relationships with other countries in that economy. This connection is not a simple one, however, because multinational corporations typically span several government units and have no necessary loyalty to any of them.

In general, recent history suggests that "national societies become related to religious cultures and traditions as the economic and power structure of these societies become interdependent at a global level" (Kokosalakis, 1985:368). Because the study of global interrelationships, especially as they are linked with religion, is only beginning to develop, this analysis can only suggest some relevant aspects. If one were to predict important themes for future sociological research and theory, they would include (1) the religion-politics interface in each nation; (2) various specific responses to human-existential issues, such as human rights, concerns for the environment, and the equitable distribution of resources for a "good life"; (3) the social significance of individuation (i.e., the social conception of individuals as persons, as relatively autonomous actors; and (4) issues of power and order at the global level, especially the problem of legitimation (Robertson, 1981, 1989).

The nature of the linkage between religion and politics depends partly on the nature of the religious belief system(s) and political system involved. Chapters 6 and 7 suggest some of the factors influencing the relative activism of a religion in sociopolitical spheres. The connection between religion and politics varies widely from one society to another. The United States is a particularly interesting arena because of its high degree of pluralization, its Constitutional specifications of a unique relationship between church and state, and its national identity and role in world politics.

## Church–State Tensions in the United States

As the degree of interdependence at the global level increases, church–state tensions at the societal level are also likely to increase. This is because church–state issues are the locus of societal struggles to articulate a national identity and moral character (Robertson, 1981:193; 1989:19). Two particular areas of tension in the United States include legal conflicts over what may be properly called "religion" and political conflicts over the religious stance of the American people and its proper role in public life.

**Religious Boundaries and the Law.** Two historical mandates shape the church–state relationships in the United States: (1) a firm Constitutional prohibition against the state establishing any religion as *the* religion of the land (later this was further articulated as a "wall of separation" between church and state) and (2) Constitutional protection for free practice of religion (i.e., "religious liberty"). These two central norms have been the focus for various church–state tensions throughout the history of the United States (for a compendium of Supreme Court decisions regarding the Establishment Clause and the Free Exercise Clause, see Alley, 1988).

Two societal factors made church–state relations more problematic, especially after World War II. One major change was the increased number, socioeconomic status, and influence of American Catholics; this was the first important challenge to taken-for-granted Protestant assumptions in the nation's concept of "religion." Subsequent proliferation of diverse new religious movements, both inside and outside of Christian traditions, has confronted the state with an even more varied array of beliefs and practices claimed as "religious." The second major change is the nature of the Supreme Court itself, which has been more willing to hear church–state cases in recent decades (Demerath and Williams, 1987). Furthermore, in attempts to reassert influence in the public sphere, numerous religious groups have turned to litigation and legislation, for example, over issues of public education, abortion rights, homosexual rights, and religious communication. The depth of U.S. pluralism makes it unlikely that either new laws or court decisions will satisfy these groups' wish for a decisive break with state neutrality on religion (Hammond, 1987; see Hunter, 1987a). Such politicization of religion reflects changes in the nature of the modern state (discussed below) and has parallels in other countries (such as Egypt and Israel) where orthodox religions, confronting the pluralism of the modern world, are attempting to reassert control through the law.

As shown in the first part of this chapter, numerous precedent-setting cases have arisen over whether a given practice (e.g., polygamy, taking of hallucinogenic drugs, exemption from military service, or refusal of blood transfusion) should be allowed under the protection of the principle of religious freedom. A key criterion in such decisions was whether the practice was part of what could be considered a valid religion.

To enforce the principle of religious liberty, therefore, the courts must define religion. In this century, in several cases regarding conscientious objection (to war), the Supreme Court adopted very broad definitions. Subsequently, however, various courts have been faced with cases requiring decisions as to whether deviant religious movements (such as the Unification Church or Children of God) are indeed "religion" and thus due protection of religious liberty (see Kelley, 1987; Pfeffer, 1987; Robbins, 1981, 1985, 1987; Robbins et al., 1986). Further complications arise as to which aspects of a religious group's actions deserve protection. Most modern religious organizations are extensively involved in activities that are not purely spiritual—running hospitals and schools; investing in real estate and stocks; operating TV, radio, and film studios; offering life insurance; and so on. Many court cases have had to delineate which aspects of religious organization deserve protection under the "free practice" clause.

Church–state tensions have been exacerbated by the steady increase of government regulation of a wide range of activities affecting religious groups. In recent years, religious groups have been subject to many new tax laws, labor regulations, immigration investigations, and licensing and accreditation regulations for institutions and personnel. The increasing internal power of the centralized state has made the boundary between church and state yet more problematic.

In 1985 the U.S. government charged eleven clergy and laypersons with violation of federal immigration statutes; the defendants made no secret of their intent to assist Central American refugees to enter the country illegally and to provide them "sanctuary." The Sanctuary Movement held strong religious motivations for their aid to refugees, who they believed were fleeing political oppression in countries like El Salvador and Guatemala. The movement argued that the U.S. government was breaking its own immigration law, which requires asylum to victims of political oppression in other countries. Whether the activities of the Sanctuary Movement could be defined as "religious practices" became a key issue in the court case (*United States* v. *Aguilar*) because the government's primary evidence was based on testimony from paid infiltrators. Is the free practice of religion compromised by government infiltration? Spokespersons for the government argued that the "bugged" gatherings were not really church services and prayer meetings because they included discussion of nonspiritual issues; the sanctuary workers argued that these were very much religious gatherings and that their religion was not limited to a purely spiritual realm (Wogaman, 1988).[3] In the context of increased power of the centralized state,

[3]The court decided to allow use of the tapes obtained by government infiltrators, but ironically the government lawyers subsequently decided not to use the tapes because they were so full of church members' testimony of their moral convictions and descriptions of atrocities from which refugees sought sanctuary that the prosecution feared the tapes would actually help the defense. Subsequently, the Presbyterian Church (U.S.A.), the American Lutheran Church, and four local congregations in Arizona filed suit over damages wrought by government infiltration of churches; the suit is supported by several other religious bodies, including the National Council of Churches.

conflicts over what is properly "religion" (and thus rightfully handled by religious groups) are also conflicts over what is properly "the state" (and thus rightfully under the aegis of the government).

**Religion and Politics.** The push of the state to expand the boundaries of its influence has brought a political response from numerous religious groups of varying theological and political persuasions. Public issues such as sex education in the schools, "gay rights," or "the right to die" have become arenas for political battles in which various religious groups assert their diverse moral judgments and argue that the state is not the proper agency for deciding the matter. Religion-oriented political activism (both conservative and liberal) is one response to the state's expansion of the scope of its control (see Bourg, 1980). Direct political power is not the only impact of modern religion, however. Religion often has significant power through its cultural influence, by which it exercises symbolic roles, shaping the terms of political discourse, for example (Demerath and Williams, 1990). Nevertheless, in a pluralistic society such as the United States, even pervasive cultural influence cannot be monolithic as in premodern societies.

More important than the specific legislative and regulatory influences of the modern government apparatus, however, is the sense that the modern state (in all its forms and agencies) is increasingly deciding the very purpose and direction of the society (Robertson, 1981:194). Citizens perceive the state as having greater power and influence over their lives and everyday well-being. Furthermore, many citizens sense that the extent and method of potential state control and influence are unknown and unlimited. Concern over the growing power of the government appears to be linked with the political mobilization of churchgoers in the United States. For example, in the 1980 elections, active churchgoers were likely to become politically involved if their attitudes toward certain issues (e.g., defense, abortion, welfare services, the Equal Rights Amendment) were combined with the feeling that the government was generally becoming too powerful (Wuthnow, 1988:318).

A related aspect of the interface of U.S. religion and politics is the variety of civil religions and nationalisms put forth for the American people. Internal attention to civil religion and nationalism reflect concerns about national identity, especially in a world context: Who are we, as a nation?

Answers to this question have profound religio-moral implications. Are we, as a nation, the kind of people who protect the "little guy," support ruthless dictators, retaliate against terrorists, invade countries that disagree with us, stand up for our values—and, if so, what are the values that our nation collectively holds dear enough to fight for? And internally, should ours be the kind of nation whose laws permit drinking, birth control, divorce, gambling, abortion, homosexuality, pornography, or possession of handguns? What values should we apply in deciding priorities for spending our national monies? Are we the kind of people who care most about health care, education, nuclear arms, space exploration, scientific research, decent housing, highways, or the environment? Public issues thus reflect the working out of national value judgments, but rarely are the specific moral and civil religious implications of such judgments made explicit.

Numerous international events, such as the holding of American hostages in the Mideast, the flexing of foreign investment "muscle" in America, and the so-called oil crisis, have made Americans more acutely aware of problems in their national identity. Many Americans had taken for granted their nation's role as an economically comfortable superpower, but events of the 1970s and 1980s threatened that self-image and, perhaps for the first time since World War II, made Americans aware of their global interdependence and consequent vulnerability. This awareness is heightened due to the immediacy of people's experience of remote events through mass media, especially television. As Americans must grapple, increasingly, with their situation in a global setting, religious matters are highlighted: in particular church–state conflicts, in religious groups' counterassertions against the expanding power and influence of the central state, and in civil religious and nationalistic formulations.

### Power and Order in the Global Context

The twin themes of power and order have been virtually ignored in much of the study of the sociology of religion. These aspects may, however, be the most important for understanding the linkage of religion to other spheres of life at all levels: individual, societal, or global (McGuire, 1983). "Power" has a range of meanings, all of which can be applied to religion in the modern world; in understanding religion, the most important advantage of an emphasis on the concept of power is that it highlights the *intentional* production of foreseen events (Beckford, 1983). Such a focus avoids overly reductionistic interpretations and recognizes the extent to which religious action often involves the deliberate assertion of power.[4]

In the global context, the concepts of power and order highlight the underlying problem of *legitimation* (comparable to societal problems of legitimation discussed on pp. 253–61). Legitimation involves the justification of power and its uses in terms other than the simple fact of holding power. Legitimation must therefore point to values, beliefs, and ideologies; the exercise of legitimate power cannot be merely a matter of technical procedures or rationalized operations of due process, but it always functions in the context of symbols held meaningful by those who accept the exercise of power as legitimate (Kokosalakis, 1985). One good example of the interface of religion and political life at the global level is the widespread attention to issues of human rights, justice, and abuses of power. Thus, problems of moral order and the wish for transnational or transcultural norms are evident. Religious groups are uniquely poised to question the legitimacy of regimes whose continued power rests on gross violations of norms of human rights.

Another issue is relative poverty and wealth among the nations. Robertson

---

[4]This use of the concept of power thus differs from the more reductionistic, materialistic interpretation of the global situation given by "world systems" theorists. "World systems" theorizing has been applied to religious phenomena by Wuthnow (1980); for a critique of this approach, see Robertson (1985).

(1987) suggests that the world economy suffers *de*legitimation precisely as the strong national state gains legitimation because it becomes more difficult to justify the existence of enormous economic inequalities among the nations. Because it refers to values and sources of authority that transcend the economic situation, religion is one of the foremost vehicles for both expressing dissatisfaction with the global economic situation and proferring alternatives.

To understand the role of religion in the modern world, it is therefore useful to look beyond the scope of societal changes to global changes. The foremost problematic issue appears to be legitimation of the global economy—the very system that promotes the globalization of societal and cultural issues. Related religion–politics tensions are also promoted by global interdependence; thus, this global perspective provides another window for understanding intrasocietal religious developments, such as church–state legal conflicts, nationalisms, the New Right, and liberation theologies.

## Summary

The social location and role of religion (including nonchurch religion) in contemporary society suggest that important social changes have occurred. These changes have apparently resulted in qualitatively different patterns of relationships and meanings. Religion, as well as other institutions, has been profoundly affected by larger changes in society. Although religion appears adaptable and vital in its numerous social forms, these larger changes raise some important questions about the continued ability of religion—and other institutions of the private sphere—to influence other important aspects of society. Particularly apparent is that major changes in the relationship of the individual to society have occurred, and any examination of the nature of religious change must be linked with an understanding of other major changes in the structure of society.

These processes are particularly important because of their implications for the relationship between the individual and society. Several interpretations of this relationship are represented in the key themes of religious change theories: institutional differentiation, competing sources of legitimacy, rationalization, privatization, and individuation. Assessing these projections about the individual-to-society relationship is one of the most important research tasks of modern sociology of religion.

Understanding the role of religion in modern societies requires looking beyond the level of individual or society. The interface of religion with power and order, in the subsocietal, societal, and international arenas, constitutes one very important item on the sociological research agenda. Due to increasing transnational economic interdependence, a global context is a helpful perspective for understanding religion and modernity. From this viewpoint, certain international and intranational developments can be understood. Recent history suggests that religion is one of the foremost forces speaking to issues of legitimation of power and moral order at the global level.

# Recommended Readings

### Articles

James Beckford. "The Restoration of 'Power' to the Sociology of Religion." *Sociological Analysis* 44 (1), 1983:11–32.

Robert N. Bellah. "Religious Evolution." *American Sociological Review* 29 (3), 1964:358–374.

N. J. Demerath, III, and Rhys H. Williams. "The Mythical Past and Uncertain Future." Pp. 77–90 in T. Robbins and R. Robertson (eds.), *Church–State Relations: Tensions and Transitions*. New Brunswick, N.J.: Transaction, 1987.

Thomas Luckmann. "Shrinking Transcendence, Expanding Religion?" *Sociological Analysis* 50 (2), 1990:127–138.

Roland Robertson. "Church–State Relations and the World System." Pp. 39–51 in T. Robbins and R. Robertson (eds.), *Church–State Relations: Tensions and Transitions*. New Brunswick, N.J.: Transaction, 1987.

### Books

James Beckford. *Religion and Advanced Industrial Societies*. London: Unwin Hyman, 1989. A critical synthesis of sociological theories about religion, modernization, and the nature of advanced industrial societies.

Peter Berger. *The Sacred Canopy: Elements of a Sociological Theory of Religion*. Garden City, N.Y.: Doubleday, 1967. Berger analyzes the impact of rationalization, privatism, and especially pluralism on traditional religious meaning systems.

Phillip E. Hammond, ed. *The Sacred in a Secular Age*. Berkeley: University of California Press, 1985. This collection of essays examines the usefulness of the secularization thesis for understanding numerous subdisciplinary areas, such as conversion, gender and family issues, the clergy, religious organizations, politics, and new religious movements.

Thomas Luckmann. *The Invisible Religion: The Problem of Religion in Modern Society*. New York: Macmillan, 1967. Luckmann's essay explores the changing social bases of religion, suggesting that modern society's chief mode of religiosity is a privatized "invisible religion."

# Epilogue

Weber closed his work on the Protestant ethic with an almost prophetic statement, wondering if society were heading toward the ultimate nullity of a totally rationalized, disenchanted world. In the preceding chapter, we explored this and related theses about the future of religion. Following Weber and Durkheim, we have used our examination of religion to provide a picture of the modern world and the individual's place in it.

Events of recent years make the sociology of religion a particularly helpful approach to understanding this relationship between individual and society. In the second half of the twentieth century, just as some observers were predicting that religion was declining and would virtually disappear, new religious movements and new patterns of spirituality arose among the very sectors of society that were supposed to be most secularized. New forms of religiosity highlight a number of questions about the future of religion: Do they presage new modes of individual-to-society relationship? Do they highlight individual autonomy, or are they new forms of authoritarianism? Do they represent religion's potential for bringing about social change, or are they so privatized that they exemplify the disempowerment of individuals in modern society? These questions remain to be addressed by further sociological research.

Similarly, the issue of religion's role in modern societal coherence is critical. Studies of civil religions, nationalism, political messianism, and church–state relations in modern and modernizing societies will be useful to our understanding of this issue. We need to know more about fundamental processes such as legitimacy and authority in the context of modern societies and global economic interdependence. Contemporary sociology of religion has the potential to contribute to our understanding of these important topics.

Although these themes suggest that our focus must be broader than the narrow, institutionally specific sociology of religion, much is yet to be learned from the study of church-oriented religion. Religious organizations continue to be important in the lives of many individuals; their membership, organization, and basis of social support make religious institutions perhaps the most important kind of voluntary association in the United States, and their direct and indirect influences in society are still worth studying. Besides official religion and religiosity, however, sociologists also need to study nonofficial religion and religiosity. We have been too quick to accept the official model of religion in our research (just as sociology of medicine uncritically accepted the medical profession's definition of health and illness as the basis of its approach). More important, the sociology of religious institutions needs to ask larger questions. Rather than focus on religious organizations (e.g., parishes or denominations) as ends in themselves, we must ask: Of what larger phenomenon is this an example? The model of religious collectivities presented in Chapter 5 illustrates such a broader perspective. Thus, the concepts of church and sect can be seen as broad developmental models of *any* group in particular kinds of tension with its social environment.

Sociology has strived for an objective approach to social reality. But sociologists of religion, while guided by this ideal, sometimes find themselves uncomfortable with its implications. Specifically, we cannot assume the superiority or necessity of religious worldviews; at the same time, however, we wonder about the direction society would take if it were utterly nonreligious. Without higher sources of authority for appeal, is legitimacy in modern society to be based on raw power and domination? What becomes of human freedom in the face of powerful public sphere institutions, especially those of modern scope such as huge multinational corporations, which are effectively free of societal normative constraints? There is no necessary religious "solution" to these issues, but the sociology of religion is uniquely poised to raise them.

# Doing a Literature Search in the Sociology of Religion

1. The primary current sociology of religion journals with articles in English are:
   *Archives de Sciences Sociales des Religions.* France, 1956 to date. Many articles in English.
   *Journal for the Scientific Study of Religion.* United States, 1961 to date.
   *Review of Religious Research.* United States, 1959 to date.
   *Social Compass.* Belgium, 1953 to 1989; United States, 1990 to date. Articles in English and French.
   *Sociological Analysis.* United States, 1939 to date.
   Also, *Current Sociology* (International Sociological Association) has occasional entire issues with annotated bibliographies on religion.
   Potentially useful are two new journals (beginning 1991): *The International Journal for the Psychology of Religion* and *Religion and American Culture.*
2. These journals are indexed or abstracted in:
   *Religion Index* (previously *Index to Religious Periodical Literature*).
   *Social Sciences Index* (previously *Social Science and Humanities Index*).
   *Sociological Abstracts.*
3. Newspapers with frequent good coverage of religion include:
   *Christian Science Monitor.*
   *Los Angeles Times.*
   *New York Times*, indexed in *New York Times Index.*
   *National Catholic Reporter*, indexed in *Catholic Periodical and Literature Index.*
4. Dictionaries and encyclopedias with specialized coverage of religion:
   *Encyclopedia of American Religions* (2 volumes), 1978, J. G. Melton (ed.). Wilmington, Del.: Consortium.
   *Encyclopedia of the American Religious Experience: Studies of Traditions and Movements* (3 volumes), 1988, C. H. Lippy and P. W. Williams (eds.). New York: Scribner.
   *Encyclopedia of Religion* (16 volumes), 1987, M. Eliade (ed.). New York: Macmillan.
   *Encyclopedic Handbook of Cults in America*, 1986, J. G. Melton (ed.). New York: Garland.
   *New Age Encyclopedia*, 1990, J. G. Melton, J. Clark, and A. A. Kelly. Detroit: Gale Research.
5. Bibliographies:
   Baer, Hans A. "Bibliography of social science literature on Afro-American religion in the United States." *Review of Religious Research* 29 (4), 1988:413–430.
   Blasi, Anthony J. and Michael W. Cuneo, eds. *Issues in the Sociology of Religion: A Bibliography.* New York: Garland, 1986. This bibliography, while containing little annotation, is an excellent *catalogue raisonné*, well organized by sociological themes. It includes good coverage of international sources as well as the main English-language journals, but some references (such as unpublished dissertations) will be difficult to obtain.

Blasi, Anthony J. and Michael W. Cuneo, eds. *The Sociology of Religion: An Organizational Bibliography*. New York: Garland, 1990. This second volume is categorized according to religious traditions (e.g., Asian, Christian, etc.) and organizations.

Brunkow, Robert de V., ed. *Religion and Society in North America: An Annotated Bibliography*. Santa Barbara, Calif.: Clio Press, 1983. This bibliography contains useful sources mainly for historical studies of religion.

Choquette, Diane, ed. *New Religious Movements in the United States and Canada: A Critical Assessment and Annotated Bibliography*. Westport, Conn.: Greenwood, 1985. Despite a very narrow definition of "new religious movements," this bibliography is not well focused or selective. The material was already sadly dated by the time of publication. Proportionately little emphasis on scientific studies of new religious movements; lots of pop/journalistic treatment.

Fraker, Anne T., ed. *Religion and American Life: Resources*. Urbana: University of Illinois Press, 1989. This bibliography focuses on the social history of American religion, especially Protestantism; useful sources on church–state issues. Because the annotations are long (and useful), the volume covers relatively few sources.

Homan, Roger, ed. *The Sociology of Religion: A Bibliographical Survey*. Westport, Conn.: Greenwood, 1986. A poorly selected list, with dated material (most from the 1970s and earlier) and spotty coverage of key issues, but the annotations are useful. The introductory essay contains seriously inaccurate generalizations about journals, sociologists, and areas of sociological research.

Lippy, Charles H. *Bibliography of Religion in the South*. Macon, Ga.: Mercer, 1985. Large number of references, but lacks depth and focus.

*Social Compass* (see journal listing above) publishes (usually in number 3 of each volume) an annual international bibliography of journal sources in sociology of religion; about two-thirds of the articles indexed are not in English. Arranged by broad subject areas, these references are useful for comparative, cross-cultural research and for current theory in the field.

Wilson, John F. *Church and State in America—A Bibliographic Guide* (2 volumes). Westport, Conn.: Greenwood, 1986 and 1987. Very useful compendium of references, topically organized, on the history of church–state relations in the United States.

Wolcott, Roger T. and Dorita Bolger. *Church and Social Action: A Critical Assessment and Bibliographical Survey*. New York: Greenwood, 1990. An excellent bibliography of comparative and historical studies on religious groups in social action; emphasis on Christian groups, but includes others (such as Southeast Asian religious activism).

# Teaching Resources

*Battle for the Bible.* PBS ("God and Politics" series), 1987, 60 minutes, color. Illustrates a contemporary conflict between Christians who want to enforce a conservative orthodoxy in their denomination versus others who want to maintain their denomination's tradition of freedom of conscience for individual believers.

*The Bible Belt: A CBS White Paper—The Politics of the Second Coming.* Canadian Broadcasting Company, 1972, 90 minutes, color. Examines the rise of fundamentalist Protestant sects in Western Canada during the 1920s and 1930s and their impact on Canadian politics then and now.

*Black Delta Religion.* Center for Southern Folklore (Memphis), 1974, 15 minutes, black and white. A brief but lively depiction of a wide range of Southern black religiosity, including ecstatic dance, singing, preaching styles, social-community context.

*Born Again: Life in a Fundamentalist Baptist Church.* James Ault Films, 1987, 87 and 58 minute versions, color. A sensitive documentary portraying the worldview and daily life of a fundamentalist community in Massachusetts.

*Bronx Baptism.* Icarus Films, 1980, 27 minutes, color. Film document, without narration, of a Pentecostal worship service in a Puerto Rican community in the South Bronx. Fine presentation of Pentecostal worship styles.

*Flight of the Dove.* University of California Media, 1989, 29 minutes, color. Religion and ethnic identity among Portuguese-Americans in rural southern California.

*Holy Ghost People.* McGraw-Hill, 1968, 53 minutes, black and white. Describes beliefs and practices of a snake-handling Pentecostal church and shows candid shots of congregation during service, including snake handling and glossolalia. Dramatic ending when the leader is bitten by a rattlesnake.

*Holy Terror.* CGuild, 1986, 58 minutes, color. Portrays the religious legitimation of New Right (especially antiabortion) activism of the 1980s.

*The Hutterites.* National Film Board of Canada, 1964, 28 minutes, black and white. Documentary (without shooting restrictions) of life in a Hutterite colony in western Canada (see also a color documentary, *Hutterites,* Canadian Broadcasting Company, Buller, 1984, 59 minutes).

*The Journey: From Faith to Action in Brazil.* Icarus, 1984, 29 minutes, color. Shows the actions of Basic Ecclesial Communities in a poor neighborhood of Rio de Janeiro and includes interviews with the bishops of the area, as well as pastoral agents and residents.

*Keeping the Faith.* PBS (Frontline series), 1987, 58 minutes, color. Depicts black churches (with particular focus on one middle-class congregation) in a midwestern city as sources of vitality, activism, community, and identity.

*The Long Search.* BBC/Time/Life, series of 13 videotapes, 1977, 52 minutes each, color. A documentary on world religions and new religions, narrated by Donald Eyre (who is irritatingly obtrusive in several instances). Especially useful are *Orthodox Christianity—The Rumanian Solution,* which is very useful for illustrating rich symbolism and ritual, the pervasiveness of religion in people's lives, and the place of religion in one Communist country; and *African Religion: Zulu Zion*, which focuses on new religions in South Africa, emphasizing the importance of dreams, ancestors, and place. Other useful films in the series include *Protestant Spirit: U.S.A.; Catholicism;* and *Judaism.* The film on *Alternate Lifestyles in California* is disappointingly shallow.

*Lubavitch.* WGBH TV ("Religion in America" series), 1974, 28 minutes, color. Brief documentary about Brooklyn's Hassidic Jews.

*Martin Luther King: A Man of Peace.* Journal Films, 1968, 30 minutes, color. Shows portions of sermons, speeches, and interviews with Martin Luther King, Jr., linking his philosophy of nonviolent resistance with biblical themes.

*My Town—Mio Paese.* University of California Media, 1989, 26 minutes, color. Religion and ethnic identity among Massachusetts immigrants from Palermiti, Italy; focuses on the celebration of the patron saint's feast day.

*The New Klan.* Corinth Films, 1978, 58 minutes, color. Vivid images of the contemporary Ku Klux Klan. Useful for discussion of symbols and rituals in a quasi-religious secret society.

*Onward Christian Soldiers.* Icarus, 1989, 52 minutes, color. Portrays inroads made into traditionally Catholic Latin American communities by evangelical Protestant preaching through the mass media.

*The Performed Word.* Center for Southern Folklore (Memphis), 1982, 58 minutes, color. Black religious styles in their cultural context—especially sermon music and preaching styles.

*The Radio Priest.* PBS, 1989, 58 minutes, color and black and white. An interesting documentary about Charles Coughlin, the controversial right-wing Catholic radio preacher of the 1930s.

*Romero.* Cinecom Distributors, 1989, 105 minutes, color. Feature film based on the story of Salvadoran Archbishop Oscar Romero, 1980 victim of a political assassination.

*Sanctuary.* PBS ("Frontline" series), 1983, 60 minutes, color. Examines the 1980s Sanctuary Movement, which engaged in civil disobedience to Immigration Service regulations by giving assistance to refugees from political oppression in Central America.

*The Shrine.* University of California Media, 1989, 46 minutes, color. Depicts the Holy Week pilgrimage to El Santuario de Chimayo in New Mexico; very useful for appreciation of folk religion, the time–space structure of pilgrimages, and the importance of sacred place both in America's Hispanic and Indian cultural heritages.

*Shrine Under Siege.* Icarus, 1987, 50 minutes, color. The Temple Mount in Jerusalem is the site of increasingly violent clashes between Moslems (whose centuries-old mosque there is one of the holiest sites in Islam) and Jews (who wish to rebuild the Jewish Temple on the same site); into the volatile mixture also go the money and energies of many American fundamentalists (who believe that Christ's second coming will follow the rebuilding of the Temple).

*Solidaridad: Faith, Hope and Haven.* Icarus, 1989, 57 minutes, color. A documentary about the work of the Vicaria de la Solidaridad, a Catholic organization in Chile that has provided support for victims of human rights abuses.

*Thank God and the Revolution.* Icarus, 1981, 50 minutes, color. Outstanding illustration of many theories about the role of religion in social change in Nicaragua. It includes Liberation Theology, basic Christian communities, brief historical background, and interviews with people from all strata of society—rural villagers to cabinet officials. The folk hymns in the soundtrack are moving.

*Triumph of the Will.* McGraw-Hill, 1936, 120 minutes, black and white. A Nazi propaganda piece documenting the Sixth Party Congress in 1934. Very useful for discussing the nature of ritual, political messianism, and whether movements like Nazism under Hitler are "religion."

*Two Black Churches.* Center for Southern Folklore (Memphis), 21 minutes, color. Images of black religious expression in two very different churches: one in rural Mississippi, the other in urban New Haven, Connecticut. Includes inspired preaching, faith healing, glossolalia, ecstatic dance.

# References

Abercrombie, Nicholas, John Baker, Sebastian Brett and Jane Foster
    1970    "Superstition and religion: The god of gaps." Pp. 93–129 in D. Martin and M. Hill (eds.), Sociological Yearbook of Religion in Britain. Volume 3. London: SCM.

Aberle, David
    1962    "A note on relative deprivation theory as applied to millenarian and other cult movements." Pp. 209–214 in S. Thrupp (ed.), Millennial Dreams in Action. The Hague: Mouton.
    1966    The Peyote Religion Among the Navaho. Chicago: Aldine.

Abramson, Harold
    1973    Ethnic Diversity in Catholic America. New York: Wiley.

Adriance, Madeleine
    1984    "Organic intellectuals and basic communities." Unpublished paper presented to the Society for the Scientific Study of Religion.
    1985a  "Opting for the poor: A social-historical analysis of the changing Brazilian Catholic church." Sociological Analysis 46, 2:131–146.
    1985b  "Religion and revolutionary ideology: The theoretical contribution of Antonio Gramsci." Unpublished paper presented to the Society for the Scientific Study of Religion.
    1986    Opting for the Poor: Brazilian Catholicism in Transition. Kansas City, Mo.: Sheed and Ward.
    1990    "Agents of change: The roles of clergy, women religious, and lay workers in the grassroots Catholic church in Brazil." Unpublished paper presented to the Society for the Scientific Study of Religion.

Ahlstrom, Sidney
    1972    A Religious History of the American People. New Haven, Conn.: Yale University Press.
    1978    "From Sinai to the Golden Gate: The liberation of religion in the Occident." Pp. 3–22 in J. Needleman and G. Baker (eds.), Understanding the New Religions. New York: Seabury Press.

Aho, James A.
    1990    The Politics of Righteousness: Idaho Christian Patriotism. Seattle: University of Washington Press.

Albanese, Catherine
    1990    Nature Religion in America: From the Algonkian Indians to the New Age. Chicago: University of Chicago Press.
    1991    America: Religions and Religion. Belmont, Calif.: Wadsworth.

Alfred, Randall
    1976    "The Church of Satan." Pp. 180–202 in C. Glock and R. Bellah (eds.), The New Religious Consciousness. Berkeley: University of California Press.

Alley, Robert S. (ed.)
    1988    The Supreme Court on Church and State. New York: Oxford University Press.

Allon, Natalie
    1973    "Group dieting rituals." Society 10, 2:36–42.

Alston, John P.
    1972    "Review of the polls." Journal for the Scientific Study of Religion 11, 2:180–186.

Aman, Kenneth
1988   "Popular culture and religion in Chile." Christian Century 105:766–769.
Aman, Kenneth and C. Parker (eds.)
1991   Popular Culture in Chile: Resistance and Survival. Boulder, Colo.: Westview.
AmaraSingham, Lorna Rhodes
1978   "The misery of the embodied." Pp. 101–126 in J. Hoch-Smith and A. Spring (eds.), Women in Ritual and Symbolic Roles. New York: Plenum.
Ammerman, Nancy Tatom
1987   Bible Believers: Fundamentalists in the Modern World. New Brunswick, N.J.: Rutgers University Press.
1990   Baptist Battles: Social Change and Religious Conflict in the Southern Baptist Convention. New Brunswick, N.J.: Rutgers University Press.
Anderson, Alan and Raymond Gordon
1978   "Witchcraft and the status of women—the case of England." British Journal of Sociology 29:171–184.
Annals of the American Academy of Political and Social Science
1979   Volume 446: The Uneasy Boundary: Church and State.
Anthony, Dick
1990   "Religious movements and brainwashing litigation: Evaluating key testimony." Pp. 295–344 in T. Robbins and D. Anthony (eds.), In Gods We Trust: New Patterns of Religious Pluralism in America. New Brunswick, N.J.: Transaction.
Anthony, Dick and Thomas Robbins
1975   "From symbolic realism to structuralism." Journal for the Scientific Study of Religion 14:403–414.
1978   "The effect of detente on the rise of new religions: The Unification Church of Reverend Sun Myung Moon." Pp. 80–100 in J. Needleman and G. Baker (eds.), Understanding the New Religions. New York: Seabury Press.
1990   "Civil religion and recent American religious ferment." Pp. 475–502 in T. Robbins and D. Anthony (eds.), In Gods We Trust: New Patterns of Religious Pluralism in America. New Brunswick, N.J.: Transaction.
Arjomand, Saïd Amir
1984   The Shadow of God and the Hidden Imam: Religion, Political Order and Societal Change in Shi'ite Iran from the Beginning to 1890. Chicago: University of Chicago Press.
1988   The Turban for the Crown: The Islamic Revolution in Iran. New York: Oxford University Press.
1989   "The emergence of Islamic political ideologies." Pp. 109–123 in J. A. Beckford and T. Luckmann (eds.), The Changing Face of Religion. London: ISA/Sage.
Atchley, Robert
1980   The Social Forces in Later Life. Belmont, Calif.: Wadsworth.
Atwood, Margaret
1985   Handmaid's Tale. New York: Fawcett Crest.
Ayubi, Nazih N. M.
1983   "The politics of militant Islamic movements in the Middle East." Journal of International Affairs 36:271–283.
Bacon, Margaret Hope
1986   Mothers of Feminism: The Story of Quaker Women in America. San Francisco: Harper & Row.
Baer, Hans A.
1984   The Black Spiritual Movement: A Religious Response to Racism. Knoxville: University of Tennessee Press.
1988   Recreating Utopia in the Desert: A Sectarian Challenge to Modern Mormonism. Albany: State University of New York Press.
Bahr, Howard
1970   "Aging and religious disaffiliation." Social Forces 49:59–71.

Bahr, Howard M. and Stan L. Albrecht
   1989   "Strangers once more: Patterns of disaffiliation from Mormonism." Journal for the Scientif-
           ic Study of Religion 28, 2:180–200.
Bainbridge, William S.
   1978   Satan's Power: A Deviant Psychotherapy Cult. Berkeley: University of California Press.
Bainbridge, William S. and Rodney Stark
   1984   "Formal explanation of religion: A progress report." Sociological Analysis 45:145–158.
Balch, Robert W. and David Taylor
   1976   "Salvation in a U.F.O." Psychology Today 10:58–66.
   1977   "Seekers and saucers: The role of the cultic milieu in joining a U.F.O. cult." American
           Behavioral Scientist 20, 6:839–860.
Barkat, Anwar
   1983   "Churches combatting racism in South Africa." Journal of International Affairs 36:297–306.
Barker, Eileen
   1982   (ed.), New Religious Movements: A Perspective for Understanding Society. New York:
           Edwin Mellen.
   1983   "New religious movements in Britain: The context and membership." Social Compass 30,
           1:33–48.
   1984   The Making of a Moonie: Choice or Brainwashing. Oxford, England: Basil Blackwell.
   1988a  "Defection from the Unification Church: Some statistics and distinctions." Pp. 166–184 in
           D. Bromley (ed.), Falling from the Faith: Causes and Consequences of Religious Apostasy.
           Newbury Park, Calif.: Sage.
   1988b  "Kingdoms of heaven on earth: New religious movements and political orders." Pp. 17–39
           in A. Shupe and J. Hadden (eds.), The Politics of Religion and Social Change. New York:
           Paragon.
Barkun, Michael
   1974   Disaster and the Millennium. New Haven, Conn.: Yale University Press.
   1985   "The awakening-cycle controversy." Sociological Analysis 46, 4:425–443.
   1986   Crucible of the Millennium: The Burned-Over District of New York in the 1840s. Syracuse,
           N.Y.: Syracuse University Press.
Barnes, Nancy Schuster
   1987   "Buddhism." Pp. 105–135 in A. Sharma (ed.), Women in World Religions. Albany: State
           University of New York Press.
Barnett, Homer G.
   1957   Indian Shakers: A Messianic Cult of the Pacific Northwest. Carbondale: Southern Illinois
           University Press.
Barrett, Stanley R.
   1987   Is God a Racist? The Right Wing in Canada. Toronto: University of Toronto Press.
Beach, Stephen
   1977   "Religion and political change in Northern Ireland." Sociological Analysis 38, 1:37–48.
Beck, Rosalie and David W. Hendon
   1990   "Notes on church-state affairs." Journal of Church and State 32, 3:684–685.
Becker, Howard
   1932   Systematic Sociology on the Basis of the Beziehungslehre and Gebildelehre of Leopold Van
           Wiese. New York: Wiley.
   1963   Outsiders: Studies in the Sociology of Deviance. New York: Free Press.
Beckett, J. C.
   1966   A Short History of Ireland. London: Hutchinson.
Beckford, James A.
   1975a  Religious organization: Current Sociology 21, 2. The Hague: Mouton.
   1975b  The Trumpet of Prophecy: A Sociological Study of the Jehovah's Witnesses. New York:
           Halsted-Wiley.
   1978a  "Accounting for conversion." British Journal of Sociology 29, 2:249–262.
   1978b  "Cults and cures." Unpublished paper presented to Ninth World Congress for Sociology,
           Research Committee for the Sociology of Religion.
   1983   "The restoration of 'power' to the sociology of religion." Sociological Analysis 44, 1:11–32.

1984 "Holistic imagery and ethics in new religious and healing movements." Social Compass 31, 2–3: 259–272.
1985 Cult Controversies: The Societal Response to the New Religious Movements. London: Tavistock.
1989 Religion and Advanced Industrial Societies. London: Unwin Hyman.
1990 "The sociology of religion, 1945–1989." Social Compass 37, 1:45–64.

Beckford, James A. and James T. Richardson
1983 "A bibliography of social scientific studies of new religious movements." Social Compass 30, 1:111–135.

Bednarowski, Mary Farrell
1983 "Women in occult America." Pp. 177–195 in H. Kerr and C. L. Crow (eds.), The Occult in America: New Historical Perspectives. Urbana: University of Illinois Press.

Bellah, Robert N.
1964 "Religious evolution." American Sociological Review 29, 3:358–374.
1967 "Civil religion in America." Daedalus 96:1–21.
1970 "Christianity and symbolic realism." Journal for the Scientific Study of Religion 9, 2:89–96.
1973 Emile Durkheim on Morality and Society. Chicago: University of Chicago Press.
1976 "New religious consciousness and the crisis of modernity." Pp. 333–352 in C. Glock and R. Bellah (eds.), The New Religious Consciousness. Berkeley: University of California Press.
1978 "Religion and legitimation in the American Republic." Society 15:16–23.
1987 "Legitimation processes in politics and religion." Current Sociology 35, 2:89–99.

Bendix, Reinhard
1960 Max Weber: An Intellectual Portrait. Garden City, N.Y.: Doubleday.

Bennett, David
1969 Demagogues in the Depression. New Brunswick, N.J.: Rutgers University Press.

Benson, Peter L., Michael J. Donahue, and Joseph A. Erickson
1989 "Adolescence and religion: A review of the literature from 1970 to 1986." Research in the Social Scientific Study of Religion 1:153–181.

Ben-Yehuda, Nachman
1985 Deviance and Moral Boundaries: Witchcraft, the Occult, Science Fiction, Deviant Sciences and Scientists. Chicago: University of Chicago Press.

Berger, Bennett M.
1969 "The new stage of American man—almost endless adolescence." New York Times Magazine, November 2, 1969:32, 33.

Berger, Joseph
1985 "Catholic dissent on church rules found." The New York Times, November 25, 1985: A7.

Berger, Peter
1958 "Sectarianism and religious sociation." American Journal of Sociology 64:41–44.
1961 The Noise of Solemn Assemblies: Christian Commitment and the Religious Establishment in America. Garden City, N.Y.: Doubleday.
1964 The Human Shape of Work. New York: Macmillan.
1967 The Sacred Canopy: Elements of a Sociological Theory of Religion. Garden City, N.Y.: Doubleday.

Berger, Peter, Brigitte Berger, and Hansfried Kellner
1973 The Homeless Mind: Modernization and Consciousness. New York: Random House.

Berger, Peter and Thomas Luckmann
1963 "Sociology of religion and sociology of knowledge." Sociology and Social Research 47:417–427.
1966 The Social Construction of Reality: A Treatise in the Sociology of Knowledge. Garden City, N.Y.: Doubleday.

Berger, Stephen
1971 "The sects and the breakthrough into the modern world: On the centrality of the sects in Weber's Protestant ethic thesis." Sociological Quarterly 12:486–499.

Bettis, Joseph and S. K. Johannesen (eds.)
1984 The Return of the Millennium. Barrytown, N.Y.: International Religious Foundation.

Bew, Paul and Henry Patterson
  1983   "Protestant and Catholic conflict in Ulster." Journal of International Affairs 36:223–234.
Bibby, Reginald W.
  1983   "Searching for the invisible thread: Meaning systems in contemporary Canada." Journal for the Scientific Study of Religion 22, 2:101–119.
  1987   Fragmented Gods: The Poverty and Potential of Religion in Canada. Toronto: Irwin.
Bibby, Reginald and Harold R. Weaver
  1985   "Cult consumption in Canada: A further critique of Stark and Bainbridge." Sociological Analysis 46, 4:445–460.
Bird, Frederick
  1978   "Charisma and ritual in new religious movements." Pp. 173–189 in J. Needleman and G. Baker (eds.), Understanding the New Religions. New York: Seabury Press.
  1979   "The pursuit of innocence: New religious movements and moral accountability." Sociological Analysis 40, 4:335–346.
Bird, Frederick and Bill Reimer
  1982   "Participation rates in new religious and para-religious movements." Journal for the Scientific Study of Religion 21, 1:1–14.
Birnbaum, Norman
  1973   "Beyond Marx in the sociology of religion?" Pp. 5–70 in C. Glock and P. Hammond (eds.), Beyond the Classics? Essays in the Scientific Study of Religion. New York: Harper & Row.
Bittner, Egon
  1968   "The structure of psychiatric influence." Mental Hygiene 52:423–430.
Blasi, Anthony J.
  1989   "Sociological implications of the great Western schism." Social Compass 36, 3:311–325.
Borker, Ruth
  1978   "To honor her head: Hats as a symbol of women's position in three evangelical churches in Edinburgh, Scotland." Pp. 55–73 in J. Hoch-Smith and A. Spring (eds.), Women in Ritual and Symbolic Roles. New York: Plenum.
Bouma, Gary D.
  1973   "Beyond Lenski: A critical review of recent 'Protestant ethic' research." Journal for the Scientific Study of Religion 12, 2:141–155.
Bourdieu, Pierre
  1977   Outline of a Theory of Practice. New York: Cambridge University Press.
Bourg, Carroll J.
  1980   "Politics and religion." Sociological Analysis 41, 4:297–316.
Braude, Ann D.
  1985   "Spirits defend the rights of women: Spiritualism and changing sex roles in nineteenth-century America." Pp. 419–431 in Y. Y. Haddad and E. B. Findly (eds.), Women, Religion and Social Change. Albany: State University of New York Press.
  1989   Spiritualism and Women's Rights in Nineteenth Century America. Boston: Beacon Press.
Brittan, Arthur
  1977   The Privatised World. London: Routledge & Kegan Paul.
Bromley, David and James T. Richardson (eds.)
  1983   The Brainwashing/Deprogramming Controversy: Sociological, Psychological, Legal and Historical Perspectives. New York: Edwin Mellen.
Bromley, David and Anson D. Shupe
  1979a  "Moonies" in America: Cult, Church and Crusade. Beverly Hills, Calif.: Sage.
  1979b  "The Tnevnoc cult." Sociological Analysis 40, 4:361–366.
Brooke, James
  1989   "Two archbishops, old and new, symbolize conflict in the Brazilian church." The New York Times, November 12.
Brown, Peter
  1981   The Cult of the Saints: Its Rise and Function in Latin Christianity. Chicago: University of Chicago Press.
Brozan, Nadine
  1990   "Telling the story of the Seder in a woman's voice." The New York Times, April 9.

Bruce, Steve
  1986   "Protestantism and politics in Scotland and Ulster." Pp. 410–429 in J. K. Hadden and A. Shupe (eds.), Prophetic Religions and Politics: Religion and the Political Order, Vol. I. New York: Paragon.
  1987   "The Moral Majority: The politics of fundamentalism in secular society." Pp. 177–194 in L. Caplan (ed.), Fundamentalism. London: Macmillan.
Brundage, James A.
  1982   "Concubinage and marriage in medieval Canon Law." Pp. 118–128 in V. L. Bullough and J. Brundage (eds.), Sexual Practices and the Medieval Church. Buffalo: Prometheus.
Bruneau, Thomas C.
  1979   "Basic Christian Communities in Latin America: Their nature and significance especially in Brazil." Pp. 225–237 in D. Levine (ed.), Churches and Politics in Latin America. Beverly Hills, Calif.: Sage.
  1982   The Church in Brazil: The Politics of Religion. Austin: University of Texas Press.
  1985   "Church and politics in Brazil: The genesis of change." Journal of Latin American Studies 17:271–293.
  1986   "Brazil: The Catholic Church and Basic Christian Communities." Pp. 106–123 in D. H. Levine (ed.), Religion and Political Conflict in Latin America. Chapel Hill: University of North Carolina Press.
  1988   "The role and response of the Catholic Church in the redemocratization of Brazil." Pp. 87–109 in A. Shupe and J. Hadden (eds.), The Politics of Religion and Social Change. New York: Paragon.
Bubis, Gerald B., Harry Wasserman with Alan Lert
  1983   Synagogue Havurot: A Comparative Study. Washington, D.C.: University of America Press.
Bullough, Vern L.
  1982   "Introduction: The Christian inheritance." Pp. 1–12 in V. L. Bullough and J. Brundage (eds.), Sexual Practices and the Medieval Church. Buffalo: Prometheus.
Bumpass, Larry and James Sweet
  1970   "Differentials in marital instability, 1970." American Sociological Review 37:754–766.
Burke, Kenneth
  1935   Permanence and Change. New York: New Republic.
  1953   A Rhetoric of Motives. Englewood Cliffs, N.J.: Prentice-Hall.
Burkholder, John Richard
  1974   "The law knows no heresy: Marginal religious movements and the courts." Pp. 27–52 in I. Zaretsky and M. Leone (eds.), Religious Movements in Contemporary America. Princeton, N.J.: Princeton University Press.
Bynum, Caroline Walker
  1986   "' . . . And woman his humanity': Female imagery in the religious writing of the later Middle Ages." Pp. 257–288 in C. W. Bynum, S. Harrell, and P. Richman (eds.), Gender and Religion: On the Complexity of Symbols. Boston: Beacon Press.
Callahan, Daniel
  1987   Setting Limits: Medical Goals in an Aging Society. New York: Simon & Schuster.
  1989   What Kind of Life: The Limits of Medical Progress. New York: Simon & Schuster.
Calvo, Robert
  1979   "The Church and the Doctrine of National Security." Pp. 135–154 in D. Levine (ed.), Churches and Politics in Latin America. Beverly Hills, Calif.: Sage.
Camara, Evandro M.
  1988   "Afro-American religious syncretism in Brazil and the United States." Sociological Analysis 48, 4:299–318.
Campbell, Colin
  1971   Toward a Sociology of Irreligion. London: Macmillan.
  1977   "Clarifying the cult." British Journal of Sociology 28, 3:375–388.
  1978   "The secret religion of the educated classes." Sociological Analysis 39, 2:146–156.

Campbell, Colin and Shirley McIver
  1987    "Cultural sources of support for contemporary occultism." Social Compass 34, 1:41–60.
Campbell, Ernest Q. and Thomas F. Pettigrew
  1959    Christians in Racial Crisis: A Study of Little Rock's Ministry. Washington, D.C.: Public Affairs Press.
Carnes, Mark C.
  1989    Secret Ritual and Manhood in Victorian America. New Haven, Conn.: Yale University Press.
Carroll, Jackson W., Barbara Hargrove, and Adair Lummis
  1983    Women of the Cloth: A New Opportunity for the Churches. New York: Harper & Row.
Carroll, Jackson W., Douglas W. Johnson, and Martin Marty
  1979    Religion in America: 1950 to the Present. San Francisco: Harper & Row.
Castelli, Jim and Joseph Gremillion
  1987    The Emerging Parish: The Notre Dame Study of Catholic Life Since Vatican II. San Francisco: Harper & Row.
Cherry, Conrad
  1969    "Two American sacred ceremonies." American Quarterly 21:739–754.
Chidester, David
  1988    Salvation and Suicide: An Interpretation of Jim Jones, the Peoples Temple, and Jonestown. Bloomington: Indiana University Press.
Chodak, Szymon
  1987    "People and the church versus the state: The case of the Roman Catholic Church in Poland." Pp. 280–307 in R. L. Rubenstein (ed.), Spirit Matters: The Worldwide Impact of Religion on Contemporary Politics. New York: Paragon.
Christ, Carol P.
  1987    The Laughter of Aphrodite. San Francisco: Harper & Row.
Chrypinski, Vincent C.
  1989    "Church and nationality in postwar Poland." Pp. 241–263 in P. Ramet (ed.), Religion and Nationalism in Soviet and East European Politics. Durham, N.C.: Duke University Press.
Clarke, Aidan
  1967    "The colonisation of Ulster and the rebellion of 1641." Pp. 189–203 in T. W. Moody and F. X. Martin (eds.), The Course of Irish History. Cork, Ireland: Mercier.
Clayton, Richard and James Gladden
  1974    "The five dimensions of religiosity: Toward demythologizing a sacred artifact." Journal for the Scientific Study of Religion 13, 2:135–143.
Cohen, Erik
  1990    "The missionary as stranger: A phenomenological analysis of Christian missionaries' encounter with the folk religions of Thailand." Review of Religious Research 31, 4:337–350.
Cohn, Norman
  1970    The Pursuit of the Millennium: Revolutionary Millenarians and Mystical Anarchists of the Middle Ages. New York: Oxford University Press.
Coleman, James S.
  1956    "Social cleavage and religious conflict." The Journal of Social Issues 12:44–56.
Coleman, John A.
  1970    "Civil religion." Sociological Analysis 31:67–77.
Comaroff, Jean
  1984    "Medicine, time and the perception of death." Listening 19, 2:155–169.
  1985    Body of Power, Spirit of Resistance: The Culture and History of a South African People. Chicago: University of Chicago Press.
Conrad, Peter and Joseph W. Schneider
  1980    Deviance and Medicalization: From Badness to Sickness. St. Louis: Mosby.
Conway, Flo and Jim Siegelman
  1979    Snapping: America's Epidemic of Sudden Personality Change. New York: Lippincott.
  1982    "Information disease: Have cults created a new mental illness?" Science Digest 90, 1:86–92.

Cornwall, Marie
  1987  "The social bases of religion: A study of factors influencing religious belief and commitment." Review of Religious Research 29, 1:44–68.
Cornwall, Marie and Perry H. Cunningham
  1989  "Surveying Latter-Day Saints: A review of methodological issues." Review of Religious Research 31, 2:162–174.
Coser, Lewis
  1973  "The militant collective: Jesuits and Leninists." Social Research 40, 1:110–128.
  1974  Greedy Institutions: Patterns of Undivided Commitment. New York: Free Press.
Coward, Harold and Leslie Kawamura (eds.)
  1978  Religion and Ethnicity. Waterloo, Ontario: Wilfred Laurier University Press.
Cowell, Alan
  1985  "Churches on cutting edge of Apartheid battle." New York Times, March 15, 1985:A2.
Crapanzano, Vincent and Vivian Garrison (eds.)
  1977  Case Studies in Spirit Possession. New York: Wiley.
Crittendon, Ann
  1988  Sanctuary: A Story of American Conscience and the Law in Collision. New York: Weidenfeld and Nicolson.
Cross, Robert D.
  1958  The Emergence of Liberal Catholicism in America. Cambridge: Cambridge University Press.
Culpepper, Emily
  1978  "The spiritual movement in radical feminist consciousness." Pp. 220–234 in J. Needleman and G. Baker (eds.), Understanding the New Religions. New York: Seabury Press.
Currie, Raymond and Wsevolod W. Isajiw
  1985  "The unravelling fabric: Religion and ethnic identity in Canada." Unpublished paper presented to the Society for the Scientific Study of Religion.
Daly, Mary
  1973  Beyond God the Father. Boston: Beacon Press.
  1978  Gyn/Ecology: The Meta-Ethics of Radical Feminism. Boston: Beacon Press.
D'Antonio, William V. and Joan Aldous (eds.)
  1983  Families and Religions: Conflict and Changes in Modern Society. Beverly Hills, Calif.: Sage.
D'Antonio, William V., James D. Davidson, Dean R. Hoge, and Ruth A. Wallace
  1989  American Catholic Laity in a Changing Church. New York: Sheed and Ward.
Danzger, M. Herbert
  1989  Returning to Tradition: The Contemporary Revival of Orthodox Judaism. New Haven, Conn.: Yale University Press.
Davidman, Lynn
  1990a  "Accommodation and resistance to modernity: A comparison of two contemporary Orthodox Jewish groups." Sociological Analysis 51, 1:35–51.
  1990b  "Women's search for family and roots: A Jewish religious solution to a modern dilemma." Pp. 385–407 in T. Robbins and D. Anthony (eds.), In Gods We Trust: New Patterns of Religious Pluralism in America. New Brunswick, N.J.: Transaction.
Demerath, N.J., III
  1965  Social Class in American Protestantism. Chicago: Rand McNally.
  1989  "Religious capital and capital religions: Cross-cultural and non-legal factors in the separation of church and state." Unpublished paper presented to the Societie Internationale de Sociologie des Religions.
Demerath, N. J., III, and Rhys H. Williams
  1987  "The mythical past and uncertain future." Pp. 77–90 in T. Robbins and R. Robertson (eds.), Church–State Relations: Tensions and Transitions. New Brunswick, N.J.: Transaction.
  1990  "Religion and power in the American experience." Pp. 427–448 in T. Robbins and D. Anthony (eds.), In Gods We Trust: New Patterns of Religious Pluralism in America. New Brunswick, N.J.: Transaction.

Derks, Frans
1983    "Uittreding uit nieuwe religieuze bewegingen." Unpublished research report in Psychology of Religion, Katholieke Universiteit Nijmegen (Netherlands).
Desroche, Henri
1969    "Socialism and the sociology of Christianity." Pp. 215–225 in N. Birnbaum and G. Lenzer (eds.), Sociology and Religion. Englewood Cliffs, N.J.: Prentice-Hall.
Diaz, Mariano
1980    María Lionza: Religiosidad Mágica de Venezuela. Caracas, Venezuela: Grupo Univensa.
Dobbelaere, Karel
1989    "The secularization of society? Some methodological suggestions." Pp. 27–44 in J. K. Hadden and A. Shupe (eds.), Religion and the Political Order: Secularization and Fundamentalism Reconsidered. New York: Paragon.
Dodson, Michael
1990    "The sword and the cross: Church–state conflict in Latin America." Pp. 161–182 in E. Sahliyeh (ed.), Religious Resurgence and Politics in the Contemporary World. Albany: State University of New York Press.
Doimo, Ana Maria
1989    "Social movements and the Catholic Church in Vitória, Brazil." Pp. 193–223 in S. Mainwaring and A. Wilde (eds.), The Progressive Church in Latin America. Notre Dame, Ind.: University of Notre Dame Press.
Dollard, John
[1937]  Caste and Class in a Southern Town. Garden City, N.Y.: Doubleday.
1949
Dougherty, Molly C.
1978    "Southern lay midwives as ritual specialists." Pp. 151–164 in J. Hoch-Smith and A. Spring (eds.), Women in Ritual and Symbolic Roles. New York: Plenum.
Douglas, Mary
1966    Purity and Danger: An Analysis of Concepts of Pollution and Taboo. London: Routledge & Kegan Paul.
1970    Natural Symbols: Explorations in Cosmology. New York: Random House.
Downton, James V.
1979    Sacred Journeys: The Conversion of Young Americans to Divine Light Mission. New York: Columbia University Press.
Dreitzel, Hans P.
1981    "The socialization of nature: Western attitudes towards body and emotions." Pp. 205–223 in P. Heelas and A. Lock (eds.), Indigenous Psychologies: The Anthropology of the Self. New York: Academic Press.
Ducey, Michael H.
1977    Sunday Morning: Aspects of Urban Ritual. New York: Free Press.
Dudley, Donald M. and Margaret G. Dudley
1986    "Transmission of religious values from parents to adolescents." Review of Religious Research 28, 1:3–15.
Durkheim, Emile
1938    The Rules of the Sociological Method. Chicago: University of Chicago Press.
[1897]  Suicide: A Study in Sociology. Tr. J. A. Spaulding and G. Simpson. Glencoe, Ill.: Free Press.
1951
[1915]  Elementary Forms of the Religious Life. Tr. J. W. Swain. New York: Free Press.
1965
[1898]  "Individualism and the intellectuals." Tr. S. Lukes. Political Studies 17:14–30.
1969
Earle, John, Dean Knudsen, and Donald Shriver
1976    Spindles and Spires: A Re-Study of Religion and Social Change in Gastonia. Atlanta: John Knox.
Ebaugh, Helen Rose
1988    "Leaving Catholic convents: Toward a theory of disengagement." Pp. 100–121 in D. Bromley (ed.), Falling from the Faith: Causes and Consequences of Religious Apostasy. Newbury Park, Calif.: Sage.

Eckhardt, Kenneth W. and Gerry E. Hendershot
  1984  "Religious preference, religious participation and sterilization decisions: Findings from the National Survey of Family Growth, Cycle II." Review of Religious Research 25, 3:232–246.
Eco, Umberto
  1980  The Name of the Rose. New York: Harcourt Brace Jovanovich.
Ehrenreich, Barbara and Dierdre English
  1973  Witches, Midwives and Nurses: A History of Women Healers. New York: Feminist Press.
Eisenstadt, S. N.
  1968  (ed.), The Protestant Ethic and Modernization: A Comparative View. New York: Basic Books.
  1969  "The Protestant ethic thesis." Pp. 297–317 in R. Robertson (ed.), Sociology of Religion. Baltimore: Penguin Books.
Eisenstadt, S. N. and Stein Rokkan (eds.)
  1973  Building States and Nations. Beverly Hills, Calif.: Sage.
Eisinga, Rob, Albert Felling, and Jan Peters
  1990  "Religious belief, church involvement, and ethnocentrism in the Netherlands." Journal for the Scientific Study of Religion 29, 1:54–75.
Eliade, Mircea
  1958  Rites and Symbols of Initiation: The Mysteries of Birth and Rebirth. New York: Harper & Row.
  1964  Shamanism: Archaic Techniques of Ecstasy. Tr. W. R. Trask. New York: Pantheon Books.
Elkind, David
  1964  "Age changes in the meaning of religious identity." Review of Religious Research 6, 1:36–40.
Ellwood, Robert S., Jr.
  1978  "Emergent religion in America: An historical perspective." Pp. 267–284 in J. Needleman and G. Baker (eds.), Understanding the New Religions. New York: Seabury Press.
  1986  "The several meanings of cult." Thought 61, 241:212–224.
Elzey, Wayne
  1975  "Liminality and symbiosis in popular American Protestantism." Journal of the American Academy of Religion 43:741–756.
  1976  "The most unforgettable magazine I've ever read: Religion and social hygiene in The Reader's Digest." Journal of Popular Culture 10:181–190.
  1988  "Popular culture." Pp. 1727–1741 in C. H. Lippy and P. W. Williams (eds.), The Encyclopedia of American Religious Experience, Vol. III. New York: Scribner.
Eversley, David
  1989  Religion and Employment in Northern Ireland. London: Sage.
Fabian, Johannes
  1974  "Genres in an emerging tradition: An anthropological approach to religious communication." Pp. 249–272 in A. Eister (ed.), Changing Perspectives in the Scientific Study of Religion. New York: Wiley.
Faulkner, J. E. and Gordon DeJong
  1966  "Religiosity in 5-D: An empirical analysis." Social Forces 45:246–254.
Feltey, Kathryn M. and Margaret M. Poloma
  1990  "From sex differences to gender role beliefs: Exploring effects on six dimensions of religiosity." Unpublished paper presented to the Association for the Sociology of Religion.
Fenn, Richard
  1972  "Toward a new sociology of religion." Journal for the Scientific Study of Religion 11, 1:16–32.
  1974  "Religion and the legitimation of social systems." Pp. 143–161 in A. Eister (ed.), Changing Perspectives in the Scientific Study of Religion. New York: Wiley.
  1978  Toward a Theory of Secularization. Monograph Series 1. Storrs, Conn.: Society for the Scientific Study of Religion.
  1982  Liturgies and Trials: The Secularization of Religious Language. Oxford, England: Basil Blackwell.
Fernandes, Luiza
  1984  "Basic Christian Communities in the Brazilian context." Unpublished paper presented to the Society for the Scientific Study of Religion.

Festinger, Leon
1957   A Theory of Cognitive Dissonance. Palo Alto, Calif.: Stanford University Press.
Festinger, Leon, Henry W. Riecken, and Stanley Schachter
1956   When Prophecy Fails. New York: Harper & Row.
Fichter, Joseph
1951   Dynamics of a City Church. Chicago: University of Chicago Press.
1954   Social Relations in the Urban Parish. Chicago: University of Chicago Press.
1975   The Catholic Cult of the Paraclete. New York: Sheed and Ward.
Fields, Karen E.
1982   "Charismatic religion as popular protest." Theory and Society 11, 3:321–360.
1985   Revival and Rebellion in Colonial Central Africa. Princeton, N.J.: Princeton University Press.
Fischler, Claude
1974   "Astrology and French society: The dialectic of archaism and modernity." Pp. 281–293 in E. Tiryakian (ed.), On the Margin of the Visible. New York: Wiley.
Frankenberg, Ronald
1986   "Sickness as cultural performance: Drama, trajectory and pilgrimage, root metaphors and the making social of disease." International Journal of Health Services 16, 4:603–626.
Frazier, E. Franklin
1974   The Negro Church in America. New York: Schocken.
Freeman, James M.
1980   "The ladies of Lord Krishna: Rituals of middle-aged women in Eastern India." Pp. 110–126 in N. A. Falk and R. M. Gross (eds.), Unspoken Worlds: Women's Religious Lives in Non-Western Cultures. San Francisco: Harper & Row.
Freidson, Eliot
1970   Profession of Medicine. New York: Dodd, Mead.
Freund, Peter E. S. and Meredith B. McGuire
1991   Health, Illness and the Social Body: A Critical Sociology. Englewood Cliffs, N.J.: Prentice-Hall.
Friedl, Ernestine
1975   Women and Men: An Anthropologist's View. New York: Holt, Rinehart & Winston.
Friedman, Thomas L.
1986   "No illusions: Israel reassesses its chances for peace." New York Times Magazine, Jan. 26, 1986:13–18ff.
1987   "Fight builds over the shape of religious future in Israel." New York Times, June 29.
Fukuyama, Yoshio
1961   "The major dimensions of church membership." Review of Religious Research 2:154–161.
Fulton, John
1987a   "State, religion and law in Ireland." Pp. 253–268 in T. Robbins and R. Robertson (eds.), Church–State Relations: Tensions and Transitions. New Brunswick, N.J.: Transaction.
1987b   "Religion and politics in Gramsci: An introduction." Sociological Analysis 48, 3:197–216.
Gallup, George, Jr., and Jim Castelli
1989   The People's Religion: American Faith in the 90s. New York: Macmillan.
Gallup Poll
1976   "Americans taking up religious, spiritual experimentation." Princeton, N.J.: Gallup International. See also New York Times, November 18, 1976.
1984   Religion in America: 1984. Report 222. Princeton, N.J.: The Princeton Religion Research Center.
Gallup Poll and Princeton Religion Research Center
1978   The Unchurched Americans, Princeton, N.J.: Gallup International.
Gallup Poll/Social Surveys, Limited
1984   Human Values and Beliefs. London: European Value Systems Study Group; Gallup International.
Gannon, Thomas M.
1981   "The New Christian Right in America." Archives de Sciences des Religions 52, 1:69–83.
Gardner, Hugh
1978   The Children of Prosperity. New York: St. Martin's Press.

Garrett, William R.
1975   "Maligned mysticism: The maledicted career of Troeltsch's third type." Sociological Analysis 36, 3:205–223.
Geertz, Clifford
1957   "Ritual and social change: A Javanese example." American Anthropologist 59:23–54.
1963   "The integrative revolution: Primordial sentiments and civil politics in the new states." Pp. 105–157 in C. Geertz (ed.), Old Societies and New States. New York: Free Press.
1964   "Ideology as a cultural system." Pp. 47–76 in D. Apter (ed.), Ideology and Discontent. New York: Free Press.
1966   "Religion as a cultural system." Pp. 1–46 in M. Banton (ed.), Anthropological Approaches to the Study of Religion. London: Tavistock.
1973   The Interpretation of Cultures. New York: Basic Books.
Gellner, Ernest
1974   Legitimation of Belief. Cambridge, England: Cambridge University Press.
Gerlach, Luther
1974   "Pentecostalism: Revolution or counter-revolution." Pp. 669–699 in I. Zaretsky and M. Leone (eds.), Religious Movements in Contemporary America. Princeton, N.J.: Princeton University Press.
Gerlach, Luther and Virginia Hine
1968   "Five factors crucial to the growth and spread of a modern religious movement." Journal for the Scientific Study of Religion 7, 1:23–40.
1970   People, Power, and Change: Movements of Social Transformation. Indianapolis: Bobbs-Merrill.
Gismondi, Michael
1988   "Conceptualizing religion from below: The Central American experience." Social Compass 35, 2–3:343–370.
Glanz, David and Michael Harrison
1977   "Varieties of identity transformation: The case of newly Orthodox Jews." Unpublished paper presented to Israeli Sociological Association.
Glazer, Barney G. and Anselm L. Strauss
1971   Status Passage. London: Routledge & Kegan Paul.
Glazer-Malbin, Nona
1976   "Housework." Signs 1:905–922.
Glenn, Norval D.
1982   "Interreligious marriage in the United States: Patterns and recent trends." Journal of Marriage and the Family 44, 3:555–566.
Glock, Charles Y.
[1962]   "On the study of religious commitment." Research Supplement, Religious Education 57, 4.
1965   Reprinted in C. Glock and R. Stark, Religion and Society in Tension. Chicago: Rand McNally.
1973   (ed.), Religion in Sociological Perspective. Belmont, Calif.: Wadsworth.
Glock, Charles and Robert Bellah (eds.)
1976   The New Religious Consciousness. Berkeley: University of California Press.
Glock, Charles and Rodney Stark
1966   Christian Beliefs and Anti-Semitism. New York: Harper & Row.
Glock, Charles and Robert Wuthnow
1979   "Departures from conventional religion: The nominally religious, the nonreligious, and the alternatively religious." Pp. 47–68 in R. Wuthnow (ed.), The Religious Dimension: New Directions in Quantitative Research. New York: Academic Press.
Goffman, Erving
1961   Asylums. Garden City, N.Y.: Doubleday.
Gold, Penny Schine
1985   The Lady and the Virgin: Image, Attitude and Experience in Twelfth-Century France. Chicago: University of Chicago Press.

Goldenberg, Judith
1974 "Epilogue: The coming of Lilith." Pp. 341–343 in R. Ruether (ed.), Religion and Sexism. New York: Simon & Schuster.
Goldsmith, Peter
1989 When I Rise Cryin' Holy: African-American Denominationalism on the Georgia Coast. New York: AMS Press.
Goode, Erich
1967 "Some critical observations on the church–sect dimension." Journal for the Scientific Study of Religion 6:69–77. Followed by commentaries, N.J. Demerath and Allan Eister, pp. 77–90.
Gorsuch, Richard and D. Aleshire
1974 "Christian faith and ethnic prejudice." Journal for the Scientific Study of Religion 13:281–307.
Gravely, Will B.
1989 "The rise of African churches in America (1786–1822): Re-examining the contexts." Pp. 301–317 in G. S. Wilmore (ed.), African American Religious Studies. Durham, N.C.: Duke University Press.
Gray, Robert and David O. Moberg
1977 The Church and the Older Person. Grand Rapids, Mich.: Eerdmans.
Greeley, Andrew
1963 Religion and Career: A Study of College Graduates. New York: Sheed and Ward.
1964 "The Protestant ethic: Time for a moratorium." Sociological Analysis 25:20–33.
1971 Why Can't They Be Like Us?: America's White Ethnic Groups. New York: Dutton.
1972a The Denominational Society: A Sociological Approach to Religion in America. Glenview, Ill.: Scott, Foresman.
1972b Unsecular Man: The Persistence of Religion. New York: Schocken.
1975 The Sociology of the Paranormal: A Reconnaissance. Beverly Hills, Calif.: Sage.
1977 The American Catholic: A Social Portrait. New York: Basic Books.
1979a "Towards a secular theory of religious behavior." Unpublished paper presented to American Sociological Association.
1979b "Ethnic variations in religious commitment." Pp. 113–134 in R. Wuthnow (ed.), The Religious Dimension: New Directions in Quantitative Research. New York: Academic Press.
1987 "Mysticism goes mainstream." American Health, Jan./Feb.:47–49.
Green, Robert W. (ed.)
1959 Protestantism and Capitalism: The Weber Thesis and Its Critics. Boston: Heath.
Greenhouse, Linda
1990 "Use of illegal drugs as part of religion can be prosecuted, High Court says." New York Times, April 18.
Greil, Arthur
1977 "Previous disposition and conversion to perspectives of social and religious movements." Sociological Analysis 38, 2:115–125.
Greil, Arthur and David Rudy
1984 "What have we learned from process models of conversion? An examination of ten case studies." Sociological Focus 17, 4:305–323.
Guizzardi, Gustavo
1989 "Religion in the television era." Social Compass 36, 3:337–353.
Gutiérrez, Gustavo
1973 A Theology of Liberation: History, Politics and Salvation. Maryknoll, N.Y.: Orbis.
Hadaway, C. Kirk
1989 "Will the real Southern Baptist please stand up: Methodological problems in surveying Southern Baptist congregations and members." Review of Religious Research 31, 2:149.
Hadden, Jeffrey K.
1969 The Gathering Storm in the Churches: The Widening Gap Between Clergy and Laymen. Garden City, N.Y.: Doubleday.

Hadden, Jeffrey K. and Anson Shupe
  1986   (eds.), Prophetic Religions and Politics: Religion and the Political Order. New York: Paragon.
  1987   "Televangelism in America." Social Compass 34, 1:61–75.
Hadden, Jeffrey K. and Charles E. Swann
  1981   Prime-Time Preachers: The Rising Power of Televangelism. Reading, Mass.: Addison-Wesley.
Hall, C. Margaret
  1985   "Religion and Aging." Journal of Religion and Health 24, 1:70–78.
Hamilton, Charles
  1972   The Black Preacher in America. New York: Morrow.
Hammond, John L.
  1979   The Politics of Benevolence: Revival Religion and American Voting Behavior. Norwood, N.J.: Ablex.
  1983   "The reality of revivals." Sociological Analysis 44:111–116.
Hammond, Phillip E.
  1974   "Religious pluralism and Durkheim's integration thesis." Pp. 115–142 in A. Eister (ed.), Changing Perspectives in the Scientific Study of Religion. New York: Wiley.
  1976   "The sociology of American civil religion: A bibliographical essay." Sociological Analysis 37, 2:169–182.
  1985   (ed.), The Sacred in a Secular Age. Berkeley: University of California Press.
  1987   "The courts and secular humanism." Pp. 91–101 in T. Robbins and R. Robertson (eds.), Church–State Relations: Tensions and Transitions. New Brunswick, N.J.: Transaction.
  1988   "Religion and the persistence of identity." Journal for the Scientific Study of Religion 27, 1:1–11.
Happold, F. C.
  1970   Mysticism. Baltimore: Penguin Books.
Hargrove, Barbara
  1978   "Integrative and transformative religions." Pp. 257–266 in J. Needleman and G. Baker (eds.), Understanding the New Religions. New York: Seabury Press.
Hargrove, Barbara, Jean Miller Schmidt, and Sheila Greeve Davaney
  1985   "Religion and the changing role of women." Annals of the American Academy of Political and Social Science 480:117–131.
Harper, Charles L. and Kevin Leicht
  1984   "Explaining the New Religious Right: Status politics and beyond." Pp. 101–110 in D. Bromley and A. Shupe (eds.), New Christian Politics. Macon, Ga.: Mercer University Press.
Harrell, David Edwin
  1984   "Dispensational premillennialism and the religious right." Pp. 9–34 in J. Bettis and S. K. Johannesen (eds.), The Return of the Millennium. Barrytown, N.Y.: New ERA.
Harris, James H.
  1987   Black Ministers and Laity in the Urban Church: An Analysis of Political Social Expectations. New York: University Press of America.
Harris, Louis and Associates
  1975   The Myths and Reality of Aging in America. Washington, D.C.: National Council on the Aging.
Harrison, Michael
  1974   "Sources of recruitment to Catholic pentecostalism." Journal for the Scientific Study of Religion 13:49–64.
  1975   "The maintenance of enthusiasm: Involvement in a new religious movement." Sociological Analysis 36, 2:150–160.
Hay, David and Ann Morisy
  1985   "Secular society, religious meanings: A contemporary paradox." Review of Religious Research 26, 3:213–227.
Haywood, Carol Lois
  1983   "The authority and empowerment of women among spiritualist groups." Journal for the Scientific Study of Religion 22, 2:157–166.

Heilman, Samuel C.
1976   Synagogue Life: A Study in Symbolic Interaction. Chicago: University of Chicago Press.
1990   "The Jews: Schism or division?" Pp. 185–198 in T. Robbins and D. Anthony (eds.), In Gods We Trust: New Patterns of Religious Pluralism in America. New Brunswick, N.J.: Transaction.
Heilman, Samuel C. and Steven M. Cohen
1989   Cosmopolitans and Parochials: Modern Orthodox Jews in America. Chicago: University of Chicago Press.
Heirich, Max
1977   "Change of heart: A test of some widely held theories about religious conversion." American Journal of Sociology 83, 3:653–680.
Heller, David
1986   The Children's God. Chicago: University of Chicago Press.
Herberg, Will
1960   Protestant-Catholic-Jew: An Essay in American Religious Sociology. Garden City, N.Y.: Doubleday.
Herbert, Frank
1965   Dune. New York: Berkley Books.
1969   Dune Messiah. New York: Berkley Books.
1976   Children of Dune. New York: Berkley Books.
1981   God Emperor of Dune. New York: Putnam.
1984   Heretics of Dune. New York: Putnam.
1985   Chapterhouse Dune. New York: Putnam.
Hermassi, Elbaki
1978   "Politics and culture in the Middle East." Social Compass 25, 3–4:445–464.
Hernández, Edwin I.
1988   "Eschatological hope and the Hispanic evangelical experience." Unpublished paper presented to the Society for the Scientific Study of Religion.
Hertel, Bradley R.
1980   "Inconsistency of beliefs in the existence of heaven and afterlife." Review of Religious Research 21, 2:171–183.
Hervieu-Léger, Danièle
1990   "Religion and modernity in the French context: For a new approach to secularization." Sociological Analysis 51, S:S15–25.
Hewitt, W. E.
1987   "Basic Christian Communities of the middle-classes in the Archdiocese of São Paulo." Sociological Analysis 48, 2:158–166.
1990   "Religion and the consolidation of democracy in Brazil: The role of the Communidades Eclesiais de Base (CEBs)." Sociological Analysis 50, 2:139–152.
Hill, Herbert
1973   "Anti-Oriental agitation and the rise of working-class racism." Society 10, 2:43–54.
Hill, Michael
1973a   The Religious Order: A Study of Virtuoso Religion and Its Legitimation in the Nineteenth Century Church of England. London: Heinemann.
1973b   A Sociology of Religion. New York: Basic Books.
Himmelstein, Jerome L.
1983   "The New Right." Pp. 13–30 in R. C. Liebman and R. Wuthnow (eds.), The New Christian Right. New York: Aldine.
1986   "The social basis of antifeminism: Religious networks and culture." Journal for the Scientific Study of Religion 25, 1:1–15.
Hobsbawm, Eric J.
1959   Primitive Rebels. Manchester, England: Manchester University Press.
Hochschild, Arlie Russell
1983   The Managed Heart: Commercialization of Human Feeling. Berkeley: University of California Press.

Hoch-Smith, Judith and Anita Spring (eds.)
1978   Women in Ritual and Symbolic Roles. New York: Plenum.
Hoge, Dean R.
1981   Converts, Dropouts, Returnees: A Study of Religious Change Among Catholics. New York: Pilgrim.
1987   The Future of Catholic Leadership: Responses to the Priest Shortage. Kansas City, Mo.: Sheed and Ward.
Holt, John
1940   "Holiness religion: Cultural shock and social reorganization." American Sociological Review 5:740–747.
Holt, Michael
1973   "The Antimasonic and Know-Nothing Parties." Pp. 575–620 in A. M. Schlesinger, Jr. (ed.), History of U.S. Political Parties, Vol. I. New York: Chelsea House.
Hood, Ralph W., Jr.
1985   "Mysticism." Pp. 285–297 in P. Hammond (ed.), The Sacred in a Secular Age. Berkeley: University of California Press.
Hood, Ralph W., Jr., Ronald J. Morris, and Paul J. Watson
1989   "Prayer experience and religious orientation." Review of Religious Research 31, 1:39–45.
Houtart, François
1977   "Theravada Buddhism and political power—construction and destruction of its ideological function." Social Compass 24, 2–3:207–246.
1980   Religion and Ideology in Sri Lanka. Maryknoll, N.Y.: Orbis.
1988   "Towards a sociology of Marxist atheism." Social Compass 35, 2–3:161–173.
Houtart, François and Geneviève Lemercinier
1984   Hai Van: Life of a Vietnamese Commune. London: Zed Press.
Hunter, James Davison
1983   American Evangelicalism: Conservative Religion and the Quandary of Modernity. New Brunswick, N.J.: Rutgers University Press.
1985   "Conservative Protestantism." Pp. 150–166 in P. Hammond (ed.), The Sacred in a Secular Age. Berkeley: University of California Press.
1987a  "American Protestantism: Sorting out the present, looking toward the future." This World 17:53–76.
1987b  Evangelicalism: The Coming Generation. Chicago: University of Chicago Press.
Hurh, Won Moo and Kwang Chung Kim
1990   "Religious participation of Korean immigrants in the United States." Journal for the Scientific Study of Religion 29, 1:19–34.
Hyman, Paula
1973   "The other half: Women in the Jewish tradition." Response: A Contemporary Jewish Review 8, 18:67–76.
Iadarola, Antoinette
1985   "The American Catholic bishops and woman: From the Nineteenth Amendment to ERA." Pp. 457–476 in Y. Y. Haddad and E. B. Findly (eds.), Women, Religion and Social Change. Albany: State University of New York Press.
Ireland, Rowan
1989   "Catholic Base Communities, spiritist groups and the deepening of democracy in Brazil." Pp. 224–250 in S. Mainwaring and A. Wilde (eds.), The Progressive Church in Latin America. Notre Dame, Ind.: University of Notre Dame Press.
Irwin, Susan and Brigitte Jordan
1987   "Knowledge, practice and power: Court-ordered cesarean section." Medical Anthropology Quarterly 1, 3:319–334.
Isichei, Elizabeth
1967   "From sect to denomination among English Quakers." Pp. 161–181 in B. Wilson (ed.), Patterns of Sectarianism. London: Heinemann. Also, "Organization and power in the Society of Friends, 1852–59," pp. 182–212.

Jacobs, Janet L.
  1989a  Divine Disenchantment: Deconverting from New Religions. Bloomington: Indiana University Press.
  1989b  "The effects of ritual healing on female victims of abuse: A study of empowerment and transformation." Sociological Analysis 50, 3:265–279.
  1990   "Women-centered healing rites: A study of alienation and reintegration." Pp. 373–383 in T. Robbins and D. Anthony (eds.), In Gods We Trust: New Patterns of Religious Pluralism in America. New Brunswick, N.J.: Transaction.
Jacobson, Cardell K., Tim B. Heaton, and Rutledge M. Dennis
  1990   "Black-white differences in religiosity: Item analysis and a formal structural test." Sociological Analysis 51, 3:257–270.
Jahoda, Gustav
  1969   The Psychology of Superstition. London: Allen Lane.
James, Janet Wilson
  1980   Women in American Religion. Philadelphia: University of Pennsylvania Press.
James, William
  [1902] The Varieties of Religious Experience. New York: New American Library.
  1958
Jamison, A. Leland
  1961   "Religions on the Christian perimeter." Pp. 162–231 in J. W. Smith and A. L. Jamison (eds.), The Shaping of American Religion. Princeton, N.J.: Princeton University Press.
Johnson, Benton
  1961   "Do holiness sects socialize in dominant values?" Social Forces 39:309–316.
  1963   "On church and sect." American Sociological Review 28:539–549.
  1971   "Church–sect revisited." Journal for the Scientific Study of Religion 10:124–137.
  1977   "Sociological theory and religious truth." Sociological Analysis 38, 4:368–388.
Johnson, Doyle Paul
  1979   "Dilemmas of charismatic leadership: The case of the People's Temple." Sociological Analysis 40, 4:315–323.
Johnson, Stephen D.
  1987   "Factors related to intolerance of AIDS victims." Journal for the Scientific Study of Religion 26, 1:105–110.
Johnson, Stephen D. and Joseph B. Tamney
  1984   "Support for the Moral Majority: A test of a model." Journal for the Scientific Study of Religion 23, 2:183–196.
Johnson, Weldon T.
  1971   "The religious crusade: Revival or ritual?" American Journal of Sociology 76, 5:873–880.
Johnston, Hank and Josef Figa
  1988   "The church and political opposition: Comparative perspectives on mobilization against authoritarian regimes." Journal for the Scientific Study of Religion 27, 1:32–47.
Jones, R. Kenneth
  1975   "Some sectarian characteristics of therapeutic groups." Pp. 190–210 in R. Wallis (ed.), Sectarianism: Analyses of Religious and Non-Religious Sects. London: Peter Owen.
  1978   "Paradigm shifts and identity theory: Alternation as a form of identity management." Pp. 59–82 in H. Mol (ed.), Identity and Religion. London: Sage.
Jubber, Ken
  1985   "The prodigal church: South Africa's Dutch Reformed Church and the apartheid policy." Social Compass 32, 2–3:273–285.
Kanter, Rosabeth Moss
  1972   Commitment and Community: Communes and Utopias in Sociological Perspective. Cambridge, Mass.: Harvard University Press.
Kearl, Michael
  1980   "Time, identity, and the spiritual needs of the elderly." Sociological Analysis 41, 2:172–180.
Kelley, Dean M.
  1987   "The Supreme Court redefines tax exemption." Pp. 115–124 in T. Robbins and R. Robertson (eds.), Church–State Relations: Tensions and Transitions. New Brunswick, N.J.: Transaction.

Kephart, William M. and William W. Zellner
  1991  Extraordinary Groups: An Examination of Unconventional Life-Styles. New York: St. Martin's Press.
Kerr, Howard and Charles L. Crow (eds.)
  1983  The Occult in America: New Historical Perspectives. Urbana-Champagne: University of Illinois Press.
Kersevan, Marko
  1975  "Religion and the Marxist concept of social formation." Social Compass 22, 3–4:323–352.
Kertzer, David
  1988  Ritual, Politics and Power. New Haven, Conn.: Yale University Press.
Khalsa, Kirpal Singh
  1986  "New religious movements turn to worldly success." Journal for the Scientific Study of Religion 25, 2:233–247.
Kilbourne, Brock and James T. Richardson
  1989  "Paradigm conflict, types of conversion, and conversion theories." Sociological Analysis 50, 1:1–21.
Kimmel, Michael
  1989  " 'New prophets' and 'old ideals': Charisma and tradition in the Iranian revolution." Social Compass 36, 4:493–510.
Klapp, Orrin
  1969  Collective Search for Identity. New York: Holt, Rinehart & Winston.
Kleiman, Dena
  1985  "Changing way of death: Some agonizing choices." New York Times, January 24.
Kleinman, Arthur
  1973  "Some issues for a comparative study of medical healing." International Journal of Social Psychiatry 19, 3–4:159–165.
  1978  "The failure of Western medicine." Human Nature, Nov.:63–68.
  1988  The Illness Narratives: Suffering, Healing and the Human Condition. New York: Basic Books.
Knudsen, Dean, John Earle and D. W. Shriver
  1978  "The conception of sectarian religion: An effort at clarification." Review of Religious Research 20, 1:44–60.
Koenig, Barbara
  1988  "The technological imperative in medical practice: The social creation of 'routine' treatment." Pp. 465–496 in M. Lock and D. R. Gordon (eds.), Biomedicine Examined. Dordrecht, Netherlands: Kluwer.
Kokosalakis, Nikos
  1985  "Legitimation power and religion in modern society." Sociological Analysis 46, 4:367–376.
  1986  "The political significance of popular religion in Greece." Unpublished paper presented to the XIth World Congress of Sociology, International Sociological Association.
Kraybill, Donald B.
  1989  The Riddle of Amish Culture. Baltimore: Johns Hopkins University Press.
Kroll-Smith, Stephen
  1980  "Testimony as performance." Journal for the Scientific Study of Religion 19, 1:16–25.
Kuhn, Thomas S.
  1970  The Structure of Scientific Revolutions. Chicago: University of Chicago Press.
Kurtz, Lester
  1986  The Politics of Heresy: The Modernist Crisis in Roman Catholicism. Berkeley: University of California Press.
Labaki, Boutros
  1988  "Confessional communities, social stratification and wars in Lebanon." Social Compass 35, 4:533–561.
Laeyendecker, Leo
  1972  "The Netherlands." Pp. 325–363 in H. Mol (ed.), Western Religion: A Country by Country Sociological Inquiry. The Hague: Mouton.
Lakoff, Robin
  1975  Language and Women's Place. New York: Harper & Row.

L'Alive D'Epinay, Christian
   1969   Haven of the Masses: A Study of the Pentecostal Movement in Chile. London: Lutter-
          worth.
Lancaster, Roger N.
   1988   Thanks to God and the Revolution: Popular Religion and Class Consciousness in the New
          Nicaragua. New York: Columbia University Press.
Lanternari, Vittorio
   1963   The Religions of the Oppressed: A Study of Modern Messianic Cults. Tr. L. Sergio. New
          York: Knopf.
Larson, Lyle E. and J. Walter Goltz
   1989   "Religious participation and marital commitment." Review of Religious Research 30, 4:387–
          400.
Larson, Lyle E. and Brenda Munro
   1985   "Religious intermarriage in Canada, 1974–1982." International Journal of the Sociology of
          the Family 15:31–49.
Laumann, Edward O.
   1969   "The social structure of religious and ethno-religious groups in a metropolitan communi-
          ty." American Sociological Review 34:182–197.
Lavender, Abraham (ed.)
   1977   A Coat of Many Colors: Jewish Sub-Communities in the United States. Westport, Conn.:
          Greenwood.
Lazerwitz, Bernard and Michael Harrison
   1979   "American Jewish denominations: A social and religious profile." American Sociological
          Review 44:656–666.
Lazerwitz, Bernard, J., Alan Winter, and Arnold Dashefsky
   1988   "Localism, religiosity, orthodoxy and liberalism: The case of Jews in the United States."
          Social Forces 67, 1:229–242.
Lebra, Sugiyama
   1972   "Millenarian movements and resocialization." American Behavioral Scientist 16, 2:195–217.
Lechner, Frank J.
   1990   "Fundamentalism revisited." Pp. 77–97 in T. Robbins and D. Anthony (eds.), In Gods We
          Trust: New Patterns of Religious Pluralism in America. New Brunswick, N.J.: Transaction.
Lehman, Edward C., Jr.
   1985   Women Clergy: Breaking Through Gender Barriers. New Brunswick, N.J.: Transaction.
   1987   Women Clergy in England: Sexism, Modern Consciousness, and Church Viability. Lewis-
          ton, N.Y.: Edwin Mellen.
Lemercinier, Geneviève
   1981   "Relationship between means of production, caste, and religion." Social Compass 28,
          2–3:163–199.
Lemert, Charles
   1975   "Defining non-church religion." Review of Religious Research 16, 3:186–197.
Lenski, Gerhard
   1963   The Religious Factor: A Sociological Study of Religion's Impact on Politics, Economics and
          Family Life. Garden City, N.Y.: Doubleday.
Lerner, Max
   1937   "The Constitution and the Court as symbols." Yale Law Journal 46:1290–1319.
Levine, Daniel H.
   1986   (ed.), Religion and Political Conflict in Latin America. Chapel Hill: University of North
          Carolina Press.
   1990   "From church and state to religion and politics and back again." Social Compass 37,
          3:331–351.
Levine, Daniel H. and Scott Mainwaring
   1989   "Religion and popular protest in Latin America: Contrasting experiences." Pp. 203–240 in
          S. Eckstein (ed.), Power and Popular Protest: Latin American Social Movements. Berkeley:
          University of California Press.

Lewis, James
  1986    "Reconstructing the 'cult' experience: Post-involvement attitudes as a function of mode of exit and post-involvement socialization." Sociological Analysis 47, 2:151–159.
  1989    "Apostates and the legitimation of repression: Some historical and empirical perspectives on the cult controversy." Sociological Analysis 49, 4:386–396.
Lewis, James and David G. Bromley
  1987    "The cult withdrawal syndrome: A case of misattribution of cause?" Journal for the Scientific Study of Religion 26, 4:508–522.
Lewis, Robert E., Mark W. Fraser, and Peter J. Pecora
  1988    "Religiosity among Indochinese refugees in Utah." Journal for the Scientific Study of Religion 27, 2:272–283.
Lewy, Guenther
  1974    Religion and Revolution. New York: Oxford University Press.
Liebman, Charles S.
  1988    Deceptive Images: Toward a Redefinition of American Judaism. New Brunswick, N.J.: Transaction.
Liebman, Charles S. and Eliezar Don-Yehiya
  1983    Civil Religion in Israel: Traditional Judaism and Political Culture in the Jewish State. Berkeley: University of California Press.
Lifton, Robert J.
  1963    Thought Reform and the Psychology of Totalism. New York: Norton.
Lincoln, C. Eric
  1974    "The power in the black church." Cross Currents 14, 1:3–21.
  1989    "The Muslim Mission in the context of American social history." Pp. 340–356 in G. S. Wilmore (ed.), African American Religious Studies. Durham, N.C.: Duke University Press.
Lincoln, C. Eric and Lawrence H. Mamiya
  1990    The Black Church in the African American Experience. Durham, N.C.: Duke University Press.
Lipset, Seymour M.
  1963    "Three decades of the radical right: Coughlinites, McCarthyites, and Birchers." Pp. 373–446 in D. Bell (ed.), The Radical Right. Garden City, N.Y.: Doubleday.
Lipset, Seymour M. and Earl Raab
  1981    "The election and the Evangelicals." Commentary 71 (March):25–32.
Lofland, John
  1966    Doomsday Cult: A Study of Conversion, Proselytization and Maintenance of Faith. Englewood Cliffs, N.J.: Prentice-Hall.
  1977    "Becoming a world-saver revisited." American Behavioral Scientist 20, 6:805–818.
Lofland, John and James T. Richardson
  1984    "Religious movement organizations: Elemental forms and dynamics." Research in Social Movements, Conflict and Change 7:29–51.
London Sunday Times Insight Team
  1972    Northern Ireland: A Report on the Conflict. New York: Vintage.
Long, Theodore E. and Jeffrey K. Hadden
  1983    "Religious conversion and the concept of socialization: Integrating the brainwashing and drift models." Journal for the Scientific Study of Religion 22, 1:1–14.
Longhurst, John E.
  1962    The Age of Torquemada. Lawrence, Kans.: Coronado.
Lorentzen, Louise J.
  1980    "Evangelical life-style concerns expressed in political action." Sociological Analysis 41, 2:144–154.
Luckmann, Thomas
  1967    The Invisible Religion: The Problem of Religion in Modern Society. New York: Macmillan.
  1973    "Comments on the Laeyendecker et al. research proposal." Pp. 55–68 in The Contemporary Metamorphosis of Religion: Acts of the 12th Conference Internationale de Sociologie Religieuse. Lille, France: CISR.

1977    "Theories of religion and social change." Annual Review of the Social Sciences of Religion 1:1–28.

1987    "Comments on legitimation." Current Sociology 35, 2:109–117.

1990    "Shrinking transcendence, expanding religion?" Sociological Analysis 50, 2:127–138.

Luker, Kristen

1984    Abortion and the Politics of Motherhood. Berkeley: University of California Press.

Lustick, Ian S.

1988    The Land and the Lord: Jewish Fundamentalism in Israel. New York: Council on Foreign Relations.

McAllister, Ian

1982    "The devil, miracles, and the afterlife: The political sociology of religion in Northern Ireland." British Journal of Sociology 33, 3:330–347.

McAvoy, Thomas

1957    The Great Crisis in American Catholic History, 1895–1900. Chicago: Henry Regnery.

McCarthy, James

1979    "Religious commitment, affiliation, and marriage dissolution." Pp. 179–197 in R. Wuthnow (ed.), The Religious Dimension: New Directions in Quantitative Research. New York: Academic Press.

McClelland, David C.

1961    The Achieving Society. Princeton, N.J.: Princeton University Press.

McCoy, Charles S.

1964    "The churches and protest movements for racial justice." Pp. 37–54 in R. Lee and M. Marty (eds.), Religion and Social Conflict. New York: Oxford University Press.

McCracken, J. L.

1967    "Northern Ireland, 1921–1966." Pp. 313–323 in T. W. Moody and F. X. Martin (eds.), The Course of Irish History. Cork, Ireland: Mercier.

McCready, William and Andrew Greeley

1976    The Ultimate Values of the American Population. Beverly Hills, Calif.: Sage.

McCutcheon, Alan L.

1988    "Denominations and religious intermarriage: Trends among white Americans in the twentieth century." Review of Religious Research 29, 3:213–227.

McDonnell, Colleen

1986    The Christian Home in Victorian America, 1840–1900. Bloomington: Indiana University Press.

MacEoin, Gary

1974a   "Irish Catholicism: What Protestant Christians fear." Cross Currents 23:397–417.

1974b   Northern Ireland. New York: Holt, Rinehart & Winston.

1987    "Catholics, Catholicism, and the Northern Ireland crisis." Pp. 124–140 in A. J. Ward (ed.), Northern Ireland: Living with the Crisis. New York: Praeger.

1991    "Struggle for Latin America's soul quickens: How Vatican applies bureaucratic brake to lively local churches." National Catholic Reporter, February 22.

McFarland, Sam G.

1989    "Religious orientations and the targets of discrimination." Journal for the Scientific Study of Religion 28, 3:324–336.

McGaw, Douglas B.

1980    "Meaning and belonging in a Charismatic congregation: An investigation into sources of neo-Pentecostal success." Review of Religious Research 21:284–301.

McGuire, Meredith B.

1972    "Toward a sociological interpretation of the Underground Church movement." Review of Religious Research 14, 1:41–47.

1974    "An interpretive comparison of elements of the pentecostal and Underground Church movements in American Catholicism." Sociological Analysis 35, 1:57–65.

1977    "Testimony as a commitment mechanism in Catholic pentecostal prayer groups." Journal for the Scientific Study of Religion 16, 2:165–168.

1982 Pentecostal Catholics: Power, Charisma and Order in a Religious Movement. Philadelphia: Temple University Press.
1983 "Discovering religious power." Sociological Analysis 44, 1:1–10.
1985 "Religion and healing." Pp. 268–284 in P. Hammond (ed.), The Sacred in a Secular Age. Berkeley: University of California Press.
1988 Ritual Healing in Suburban America. New Brunswick, N.J.: Rutgers University Press.

MacIntyre, Alisdair
1967 Secularisation and Moral Change. London: Oxford University Press.

MacIver, Martha Abele
1989 "A clash of symbols in Northern Ireland: Division between extremist and modern Protestant elites." Review of Religious Research 30, 4:360–374.

McKinney, William and Wade Clark Roof
1990 "Liberal Protestantism's struggle to recapture the heartland." Pp. 167–176 in T. Robbins and D. Anthony (eds.), In Gods We Trust: New Patterns of Religious Pluralism in America. New Brunswick, N.J.: Transaction.

McKown, Delos B.
1975 The Classical Marxist Critiques of Religion: Marx, Engels, Lenin, Kautsky. The Hague: Martinus Nijhoff.

McLaughlin, Eleanor C.
1974 "Equality of souls, inequality of sexes: Women in medieval theology." Pp. 213–266 in R. Ruether (ed.), Religion and Sexism: Images of Women in the Jewish and Christian Traditions. New York: Simon & Schuster.

McLoughlin, William G.
1983 "Timepieces and butterflies: A note on the Great-Awakening-Construct and its critics." Sociological Analysis 44:103–110.

McNamara, Jo Ann
1982 "Chaste marriage and clerical celibacy." Pp. 22–33 in V. L. Bullough and J. Brundage (eds.), Sexual Practices and the Medieval Church. Buffalo: Prometheus.

McNamara, Patrick
1985 "Conservative Christian families and their moral world view." Sociological Analysis 46, 2:93–99.

Maduro, Otto
1975 "Marxist analysis and the sociology of religion." Social Compass 22, 3–4:305–322.
1977 "New Marxist approaches to the relative autonomy of religion." Sociological Analysis 38, 4:359–367.
1982 Religion and Social Conflicts. Tr. Robt. R. Barr. Maryknoll, N.Y.: Orbis.
1989 "Latin America: Apocalypse now?" Unpublished paper presented to the Conference on Religious Traditions and States in the Modern World, American Academy of Arts and Sciences.

Mainwaring, Scott
1986a "Brazil: The Catholic Church and the Popular Movement in Nova Iguaçu, 1974–1985." Pp. 124–155 in D. H. Levine (ed.), Religion and Political Conflict in Latin America. Chapel Hill: University of North Carolina Press.
1986b The Catholic Church and Politics in Brazil, 1916–1985. Stanford, Calif.: Stanford University Press.
1989 "Grass-roots Catholic groups and politics in Brazil." Pp. 151–192 in S. Mainwaring and A. Wilde (eds.), The Progressive Church in Latin America. Notre Dame, Ind.: University of Notre Dame Press.

Malinowski, Bronislaw
[1925] "Magic, science and religion." Pp. 17–92 in Magic, Science and Religion and Other Essays.
1948 New York: Free Press.

Mamiya, Lawrence H.
1982 "From Black Muslim to Bilalian: The evolution of a movement." Journal for the Scientific Study of Religion 21, 2:138–152.

Mamiya, Lawrence and C. Eric Lincoln
  1988    "Black militant and separatist movements." Pp. 755–771 in C. H. Lippy and P. W. Williams
          (eds.), Encyclopedia of the American Religious Experience, Vol. II. New York: Scribner.
Mansueto, Anthony
  1988    "Religion, solidarity and class struggle: Marx, Durkheim, and Gramsci on the religion
          question." Social Compass 35, 2–3:261–277.
Manwaring, David
  1962    Render unto Caesar: The Flag Salute Controversy. Chicago: University of Chicago Press.
Marable, Manning
  1984    "The rainbow coalition: Jesse Jackson and the politics of ethnicity." Cross Currents Quar-
          terly 34, 1:21–42.
Marglin, Frédérique Apffel
  1985    "Female sexuality in the Hindu world." Pp. 39–60 in C. W. Atkinson, C. H. Buchanan, and
          M. R. Miles (eds.), Immaculate and Powerful: The Female in Sacred Image and Social
          Reality. Boston: Beacon Press.
Markoff, John and Daniel Regan
  1981    "The rise and fall of civil religion: Comparative perspectives." Sociological Analysis 42,
          4:333–352.
Martin, Bernard (ed.)
  1978    Movement and Issues in American Judaism: An Analysis and Sourcebook of De-
          velopments since 1945. Westport, Conn.: Greenwood.
Martin, David A.
  1962    "The denomination." British Journal of Sociology 13, 2:1–14.
  1965    Pacificism: An Historical and Sociological Study. New York: Schocken.
  1978    A General Theory of Secularization. New York: Harper & Row.
Marty, Martin
  1970    "The occult establishment." Social Research 37:212–230.
  1974    "Two kinds of two kinds of civil religion." Pp. 139–160 in R. Richey and D. Jones (eds.),
          American Civil Religion. New York: Harper & Row.
  1980    (ed.), Where the Spirit Leads: American Denominations Today. Atlanta: John Knox.
Marx, Gary T.
  1967    "Religion: Opiate or inspiration of civil rights militancy among Negroes?" American
          Sociological Review 32, 1:64–72.
Marx, Karl
  [1844]  "Contribution to the critique of Hegel's philosophy of right." Pp. 43–59 in T. B. Bottomore
  1963    (ed.), Early Writings. New York: McGraw-Hill.
Maryknoll Fathers (eds.)
  1957    Daily Missal of the Mystical Body. New York: Kennedy.
Maslow, Abraham
  1964    Religions, Values and Peak-Experiences. New York: Viking.
Mayrl, William
  1976    "Marx' theory of social movements and the church–sect typology." Sociological Analysis
          37, 1:19–31.
Mead, George Herbert
  1918    "The psychology of punitive justice." American Journal of Sociology 23:577–602.
Melton, J. Gordon, Jerome Clark, and Aidan A. Kelly
  1990    New Age Encyclopedia. Detroit: Gale Research.
Menendez, A. J.
  1973    The Bitter Harvest: Church and State in Northern Ireland. Washington, D.C.: Luce.
Merton, Robert K.
  1957    Social Theory and Social Structure. Glencoe, Ill.: Free Press.
Mess, Zulkarnaina M. and W. Barnett Pearce
  1986    "Dakwah Islamiah: Islamic revivalism in the politics of race and religion in Malaysia." Pp.
          196–220 in J. K. Hadden and A. Shupe (eds.), Prophetic Religions and Politics: Religion and
          the Political Order. New York: Paragon.

Meyer, Donald
  1965   The Positive Thinkers. Garden City, N.Y.: Doubleday.
Michaelson, Robert S.
  1987   "Civil rights, Indian rights." Pp. 125–133 in T. Robbins and R. Robertson (eds.), Church–State Relations: Tensions and Transitions. New Brunswick, N.J.: Transaction.
Miller, Walter M., Jr.
  1959   A Canticle for Liebowitz. New York: Harold Matson.
Mindel, Charles H. and C. Edwin Vaughan
  1978   "A multidimensional approach to religiosity and disengagement." Journal of Gerontology 33:103–108.
Mintz, Jerome
  1977   "Brooklyn's Hasidim." Natural History 86:46–59.
Mojzes, Paul
  1987   "The impact of the Eastern European church upon their own societies." Pp. 258–279 in R. L. Rubenstein (ed.), Spirit Matters: The Worldwide Impact of Religion on Contemporary Politics. New York: Paragon.
Mol, Hans J. (ed.)
  1976   Identity and the Sacred. New York: Free Press.
Moodie, T. Dunbar
  1978   "The Afrikaner civil religion." Pp. 203–228 in H. Mol (ed.), Identity and Religion. London: Sage.
Moody, Edward
  1971   "Urban witches." Pp. 280–290 in J. Spradley and D. McCurdy (eds.), Conformity and Conflict: Readings in Cultural Anthropology. Boston: Little, Brown.
  1974   "Magical therapy: An anthropological investigation of contemporary Satanism." Pp. 355–382 in I. Zaretsky and M. Leone (eds.), Religious Movements in Contemporary America. Princeton, N.J.: Princeton University Press.
Mooney, James
  1965   The Ghost-Dance Religion and the Sioux Outbreak of 1890. Chicago: University of Chicago Press.
Morioka, Kiyomi and William H. Newell
  1968   The sociology of Japanese Religion. International Studies in Sociology and Social Anthropology 6. Leiden, Netherlands: Brill.
Mörth, Ingo
  1987   "Elements of religious meaning in science-fiction literature." Social Compass 34, 1:87–108.
Moses, Wilson Jeremiah
  1982   Black Messiahs and Uncle Toms: Social and Literary Manipulations of a Religious Myth. University Park: Pennsylvania State University Press.
Mosqueda, Lawrence
  1986   Chicanos, Catholicism and Political Ideology. New York: University Press of America.
Moxon-Browne, Edward
  1983   Nation, Class and Creed in Northern Ireland. Aldershot, England: Gower.
Muller, Jessica H. and Barbara A. Koenig
  1988   "On the boundary of life and death: The definitions of dying by medical residents." Pp. 351–374 in M. Lock and D. R. Gordon (eds.), Biomedicine Examined. Dordrecht, Netherlands: Kluwer.
Murvar, Vatro
  1971   "Messianism in Russia: Religious and revolutionary." Journal for the Scientific Study of Religion 10, 4:277–338.
Myerhoff, Barbara
  1974   The Peyote Hunt: The Sacred Journey of the Huichol Indians. Ithaca, N.Y.: Cornell University Press.
  1978   "Bobbes and Zeydes: Old and new roles for elderly Jews." Pp. 207–244 in J. Hoch-Smith and A. Spring (eds.), Women in Ritual and Symbolic Roles. New York: Plenum.

Nason-Clark, Nancy
  1987a  "Are women changing the image of ministry? A comparison of British and American realities." Review of Religious Research 28, 4:330–340.
  1987b  "Ordaining women as priests: Religious vs. sexist explanations for clerical attitudes." Sociological Analysis 48, 3:259–273.
Neitz, Mary Jo
  1986  Charisma and Community: A Study of Religious Commitment Within the Charismatic Renewal. New Brunswick, N.J.: Transaction.
  1990  "In goddess we trust." Pp. 353–372 in T. Robbins and D. Anthony (eds.), In Gods We Trust: New Patterns of Religious Pluralism in America. New Brunswick, N.J.: Transaction.
Neitz, Mary Jo and James V. Spickard
  1990  "Steps toward a sociology of religious experience: The theories of Mihaly Csikszentmihalyi and Alfred Schutz." Sociological Analysis 51, 1:15–33.
Nelsen, Hart M., Neil H. Cheek, Jr., and Paul Au
  1985  "Gender differences in images of God." Journal for the Scientific Study of Religion 24, 4:396–402.
Nelsen, Hart M. and Anne Kusner Nelsen
  1975  The Black Church in the Sixties. Lexington: University of Kentucky Press.
Nelson, Benjamin
  1949  The Idea of Usury: From Tribal Brotherhood to Universal Otherhood. Princeton, N.J.: Princeton University Press.
Nelson, Geoffrey
  1969  Spiritualism and Society. New York: Schocken.
Niebuhr, H. Richard
  1929  The Social Sources of Denominationalism. New York: Meridian.
Noonan, John T., Jr.
  1965  Contraception: A History of Its Treatment by Catholic Theologians and Canonists. Cambridge, Mass.: Harvard University Press.
O'Brien, Conor Cruise
  1974  States of Ireland. London: Granada.
  1987  The Siege: The Saga of Israel and Zionism. New York: Simon & Schuster.
O'Connor, Mary
  1989  "The Virgin of Guadalupe and the economics of symbolic behavior." Journal for the Scientific Study of Religion 28, 2:105–119.
O'Dea, Thomas
  1961  "Five dilemmas in the institutionalization of religion." Journal for the Scientific Study of Religion 1:30–39.
Oomen, T. K.
  1989  "State and religion in multi-religious nation-states: The case of South Asia." Unpublished paper presented to the Conference Internationale de Sociologie des Religions.
"Orthodox Eastern Church"
  1943  Encyclopaedia Brittanica 16:938–942.
O'Toole, Roger
  1976  "Underground traditions in the study of sectarianism: Non-religious uses of the concept 'sect'." Journal for the Scientific Study of Religion 15, 2:145–156.
  1984  Religion: Classic Sociological Approaches. Toronto: McGraw-Hill-Ryerson.
Ozorak, Elizabeth Weiss
  1989  "Social and cognitive influences on the development of religious beliefs and commitment in adolescence." Journal for the Scientific Study of Religion 28, 4:448–463.
Pace, Enzo
  1989  "Pilgrimage as spiritual journey: An analysis of pilgrimage using the theory of V. Turner and the resource mobilization approach." Social Compass 36, 2:229–244.
Page, Ann L. and Donald A. Clelland
  1978  "The Kanawha County textbook controversy: A study of the politics of lifestyle concern." Social Forces 57:265–281.

Paris, Arthur E.
   1982   Black Pentecostalism: Southern Religion in an Urban World. Amherst: University of Massachusetts Press.
Parsons, Talcott
   1944   "The theoretical development of the sociology of religion." Journal of the History of Ideas 5:176–190.
   1969   "Family and church as 'boundary' structures." Pp. 423–429 in N. Birnbaum and G. Lenzer (eds.), Sociology and Religion. Englewood Cliffs, N.J.: Prentice-Hall.
   1971   "Belief, unbelief and disbelief." Pp. 207–245 in R. Caporale and A. Grumelli (eds.), The Culture of Unbelief. Berkeley: University of California Press.
   1972   "Definitions of health and illness in the light of American values and social structure." Pp. 107–127 in E. G. Jaco (ed.), Patients, Physicians and Illness. New York: Macmillan.
Parsons, Talcott, Edward Shils, Kaspar D. Naegele, and Jesse R. Pitts
   1961   Theories of Society: Foundations of Modern Sociological Theory. New York: Free Press.
Pepper, George B.
   1988   The Boston Heresy Case in View of the Secularization of Religion: A Case Study in the Sociology of Religion. Lewiston, N.Y.: Edwin Mellen.
Peshkin, Alan
   1986   God's Choice: The Total World of a Fundamentalist Christian School. Chicago: University of Chicago Press.
Peter, Karl A.
   1987   The Dynamics of Hutterite Society. Edmonton: University of Alberta Press.
Peters, Victor
   1971   All Things Common: The Hutterite Way of Life. New York: Harper & Row.
Pfeffer, Leo
   1974   "The legitimation of marginal religions in the United States." Pp. 9–26 in I. Zaretsky and M. Leone (eds.), Religious Movements in Contemporary America. Princeton, N.J.: Princeton University Press.
   1987   "Religious exemption." Pp. 103–114 in T. Robbins and R. Robertson (eds.), Church–State Relations: Tensions and Transitions. New Brunswick, N.J.: Transaction.
Pfieffer, Jeffrey
   1989   "The psychological framing of cults: Schematic presentations and cult evaluations." Unpublished paper presented to the Society for the Scientific Study of Religion.
Pickering, W.S.F.
   1984   Durkheim's Sociology of Religion: Themes and Theories. London: Routledge & Kegan Paul.
Pilarzyk, Thomas
   1978   "The origin, development, and decline of a youth culture religion: An application of sectarianization theory." Review of Religious Research 20, 1:23–43.
Pinsky, Mark I.
   1986   "Havurah: A new spirit in Judaism." Los Angeles Times, October 11.
Poll, Solomon
   1969   The Hasidic Community in Williamsburg: A Study in Sociology of Religion. New York: Schocken.
Poloma, Margaret
   1989   The Assemblies of God at the Crossroads: Charisma and Institutional Dilemmas. Knoxville: University of Tennessee Press.
Poloma, Margaret and Brian F. Pendleton
   1989   "Religious experiences, evangelism, and institutional growth within the Assemblies of God." Journal for the Scientific Study of Religion 28, 4:415–431.
Pope, Barbara Corrado
   1985   "Immaculate and powerful: The Marian revival in the nineteenth century." Pp. 173–200 in C. W. Atkinson, C. H. Buchanan, and M. R. Miles (eds.), Immaculate and Powerful: The Female in Sacred Image and Social Reality. Boston: Beacon Press.
Pope, Liston
   1942   Millhands and Preachers. New Haven, Conn.: Yale University Press.

Potvin, Raymond H. and C. F. Lee
  1982    "Adolescent religion: A developmental approach." Sociological Analysis 43, 2:131–144.
Priesand, Sally
  1975    Judaism and the New Woman. New York: Behrman.
Prokesch, Steven
  1990    "With the world made over, can even Belfast change?" New York Times, April 2.
Raab, Earl (ed.)
  1964    Religious Conflict in America. New York: Doubleday.
Ramet, Pedro
  1989    "The interplay of religious policy and nationalistic policy in the Soviet Union and Eastern
           Europe." Pp. 3–41 in P. Ramet (ed.), Religion and Nationalism in Soviet and East European
           Politics. Durham, N.C.: Duke University Press.
Regan, Daniel
  1989    "Islam as a new religious movement in Malaysia." Pp. 124–146 in J. A. Beckford and T.
           Luckmann (eds.), The Changing Face of Religion. London: ISA/Sage.
Ribeiro de Oliveira, Pedro A.
  1979    "The 'Romanization' of Catholicism and agrarian capitalism in Brazil." Social Compass 26,
           2–3:309–329.
Ribuffo, Leo P.
  1983    The Old Christian Right: The Protestant Far Right from the Great Depression to the Cold
           War. Philadelphia: Temple University Press.
  1988    "Religious prejudice and nativism." Pp. 1525–1546 in C. H. Lippy and P. W. Williams
           (eds.), Encyclopedia of the American Religious Experience, Vol. III. New York: Scribner.
Richardson, James
  1979    "From cult to sect: Creative eclecticism in new religious movements." Pacific Sociological
           Review 22, 2:139–166.
  1981    (ed.), The Deprogramming Controversy: Sociological, Psychological, Legal and Historical
           Perspectives. New Brunswick, N.J.: Transaction.
  1985a   "The active vs. passive convert: Paradigm conflict in conversion/recruitment research."
           Journal for the Scientific Study of Religion 24, 2:119–236.
  1985b   "Studies of conversion: Secularization or re-enchantment." Pp. 104–121 in P. Hammond
           (ed.), The Sacred in a Secular Age. Berkeley: University of California Press.
  1990a   "Coercion in court: Brainwashing allegations and religious proselytizing." Unpublished
           paper presented to the Society for the Scientific Study of Religion.
  1990b   "Definitions of cult: From sociological-technical to popular-negative." Unpublished paper
           presented to the American Psychological Association.
  1991    "Cult/brainwashing cases and freedom of religion." Journal of Church and State 33, 1:55–74.
Richardson, James, Mary Harder, and Robert B. Simmonds
  1979    Organized Miracles: A Sociological Study of a Jesus Movement Organization. New Bruns-
           wick, N.J.: Transaction.
Richardson, James T., Jan van der Lans, and Frans Derks
  1986    "Leaving and labeling: Voluntary and coerced disaffiliation from religious social move-
           ments." Research in Social Movements, Conflicts and Change 9:97–126.
Riding, Alan
  1988    "In Brazil's northeast, misery molded by man and nature." New York Times, May 3.
Riis, Ole
  1988    "The role of religion in legitimating the modern structuration of society." Paper presented
           to the Nordic Conference on Sociology of Religion.
Robbins, Thomas
  1979/   "Religious movements, the State, and the Law: Reconceptualizing 'the cult problem'."
  1980    New York University Review of Law and Social Change IX, 1:33–49.
  1981    "Church, state, and cult." Sociological Analysis 42, 3:209–226.
  1984    "Constructing cultist 'mind control'." Sociological Analysis 45, 3:241–256.
  1985    "Government regulatory powers and church autonomy: Deviant groups as test cases."
           Journal for the Scientific Study of Religion 24, 3:237–252.

1987   "Church–state tensions and marginal movements." Pp. 135–149 in T. Robbins and R. Robertson (eds.), Church–State Relations: Tensions and Transitions. New Brunswick, N.J.: Transaction.

Robbins, Thomas and Dick Anthony
1972   "Getting straight with Meher Baba." Journal for the Scientific Study of Religion 11, 2:122–140.
1979   "Cults, brainwashing, and counter-subversion." Annals of the American Academy of Political and Social Science 446 (Nov.): 78–90.
1990   (eds.), In Gods We Trust: New Patterns of Religious Pluralism in America. New Brunswick, N.J.: Transaction.

Robbins, Thomas, Dick Anthony and Thomas Curtis
1975   "Youth culture religious movements: Evaluating the integrative hypothesis." Sociological Quarterly 16, 1:48–64.

Robbins, Thomas, Dick Anthony, Madeleine Doucas, and Thomas Curtis
1976   "The last civil religion: The Unification Church of the Reverend Sun Myung Moon." Sociological Analysis 37, 2:111–125.

Robbins, Thomas, Dick Anthony, and James Richardson
1978   "Theory and research on today's 'new religions'." Sociological Analysis 39, 2:95–122.

Robbins, Thomas and David Bromley
1991   "Social experimentation and the significance of American new religions: A focused review essay." Forthcoming in Research in the Social Scientific Study of Religion, Vol. 4.

Robbins, Thomas and Roland Robertson (eds.)
1987   Church–State Relations: Tensions and Transitions. New Brunswick, N.J.: Transaction.

Robbins, Thomas, William Shepherd, and J. McBride (eds.)
1986   Cults, Culture and the Law. Chico, Calif.: Scholars Press.

Robertson, Roland
1967   "The Salvation Army: The persistence of sectarianism." Pp. 49–105 in B. Wilson (ed.), Patterns of Sectarianism. London: Heinemann.
1970   The Sociological Interpretation of Religion. New York: Schocken.
1977   "Individualism, societalism, worldliness, universalism: Thematizing theoretical sociology of religion." Sociological Analysis 38, 4:281–308.
1978   Meaning and Change: Explorations in the Cultural Sociology of Modern Societies. New York: New York University Press.
1979   "Religious movements and modern societies: Toward a progressive problemshift." Sociological Analysis 40, 4:297–314.
1981   "Consideration from within the American context on the significance of church–state tension." Sociological Analysis 42, 3:193–208.
1985   "The sacred and the world system." Pp. 347–358 in P. Hammond (ed.), The Sacred in a Secular Age. Berkeley: University of California Press.
1987   "Church–state relations and the world system." Pp. 39–51 in T. Robbins and R. Robertson (eds.), Church–State Relations: Tensions and Transitions. New Brunswick, N.J.: Transaction.
1988   "Christian Zionism and Jewish Zionism: Points of contact." Pp. 239–258 in A. Shupe and J. Hadden (eds.), The Politics of Religion and Social Change. New York: Paragon.
1989   "Globalization, politics and religion." Pp. 10–23 in J. Beckford and T. Luckmann (eds.), The Changing Face of Religion. London: ISA/Sage.

Robinson, John A.
1963   Honest to God. Philadelphia: Westminster.

Rochford, E. Burke, Jr.
1985   Hare Krishna in America. New Brunswick, N.J.: Rutgers University Press.
1989   "Factionalism, group defection, and schism in the Hare Krishna movement." Journal for the Scientific Study of Religion 28, 2:162–179.

Roebuck, Julian and Robert Quan
1976   "Health care practices in the American deep South." Pp. 141–161 in R. Wallis and P. Morley (eds.), Marginal Medicine. New York: Free Press.

Rokeach, Milton
  1960    The Open and Closed Mind. New York: Basic Books.
Roof, Wade Clark
  1978    Commitment and Community: Religious Plausibility in a Liberal Protestant Church. New
          York: Elsevier.
  1979    "Concepts and indicators of religious commitment: A critical review." Pp. 17–45 in R.
          Wuthnow (ed.), The Religious Dimension: New Directions in Quantitative Research. New
          York: Academic Press.
  1982    "America's voluntary establishment: Mainline religion in transition." Pp. 130–149 in M.
          Douglas and S. Tipton (eds.), Religion and America: Spirituality in a Secular Age. Boston:
          Beacon Press.
  1989    "Multiple religious switching: A research note." Journal for the Scientific Study of Religion
          28, 4:530–535.
Roof, Wade Clark and William McKinney
  1987    American Mainline Religion: Its Changing Shape and Future. New Brunswick, N.J.: Rut-
          gers University Press.
Roof, Wade Clark and Jennifer L. Roof
  1984    "Review of the polls: Images of God among Americans." Journal for the Scientific Study of
          Religion 23, 2:201–205.
Rose, Richard
  1971    Governing Without Consensus. London: Faber and Faber.
Rose, Susan D.
  1988    Keeping Them Out of the Hands of Satan: Evangelical Schooling in America. New York:
          Routledge, Chapman and Hall.
Rosenberg, Ellen M.
  1989    The Southern Baptists: A Subculture in Transition. Knoxville: University of Tennessee
          Press.
Rowthorn, Bob and Naomi Wayne
  1988    Northern Ireland: The Political Economy of Conflict. Boulder, Colo.: Westview.
Rudy, David R. and Arthur L. Greil
  1987    "Taking the pledge: The commitment process in Alcoholics Anonymous." Sociological
          Focus 20, 1:45–60.
  1989    "Is Alcoholics Anonymous a religious organization? Meditations on marginality." Sociolog-
          ical Analysis 50, 1:41–51.
Ruether, Rosemary R.
  1974    (ed.), Religion and Sexism: Images of Women in Jewish and Christian Traditions. New
          York: Simon & Schuster.
  1975    New Woman/New Earth. New York: Seabury Press.
  1989    "Sexism and God-language." Pp. 151–162 in J. Plaskow and C. P. Christ (eds.), Weaving
          the Visions: New Patterns in Feminist Spirituality. San Francisco: Harper & Row.
Ruether, Rosemary R. and Rosemary Skinner Keller (eds.)
  1981    Women and Religion in America: The Nineteenth Century. New York: Harper & Row.
Ruether, Rosemary R. and Eleanor McLaughlin
  1979    Women of Spirit: Female Leadership in the Jewish and Christian Traditions. New York:
          Simon & Schuster.
Ruether, Rosemary Radford and Herman Ruether
  1989    The Wrath of Jonah: The Crisis of Religious Nationalism in the Israeli-Palestinian Conflict.
          San Francisco: Harper & Row.
Ryan, Joseph
  1978    "Ethnoscience and problems of method in the social scientific study of religion." Sociologi-
          cal Analysis 39, 3:241–249.
Ryan, Mary P.
  1983    Womanhood in America: From Colonial Times to the Present. New York: Franklin
          Watts.

Sacks, Karen
  1974  "Engels revisited: Women, the organization of production and private property." Pp. 207–222 in M. S. Rosaldo and L. Lamphere (eds.), Woman, Culture and Society. Stanford, Calif.: Stanford University Press.

Sacred Congregation for the Doctrine of the Faith
  1977  "Declaration on the question of the admission of women to the ministerial priesthood." Origins 6, Feb. 3:524.

Samuelsson, Kurt
  1964  Religion and Economic Action. New York: Harper & Row.

Sargant, William
  1957  Battle for the Mind. Garden City, N.Y.: Doubleday.

Sarna, Jonathan
  1986  The American Jewish Experience. New York: Holmes and Meier.

Scanzoni, John and Cynthia Arnett
  1987  "Enlarging the understanding of marital commitment via religious devoutness, gender role preferences, and locus of control." Journal of Family Issues 8, 1:135–156.

Scherer, Ross P.
  1980  American Denominational Organization: A Sociological View. Pasadena, Calif.: Wm. Carey Library.

Schmitt, David
  1973  The Irony of Irish Democracy. Lexington, Mass.: Heath.

Schneider, Louis and Sanford M. Dornbusch
  1957  "Inspirational religious literature: From latent to manifest functions of religion." American Journal of Sociology 62:476–481.

Schoenfeld, Stuart
  1978  "The Jewish religion in North America: Canadian and American comparisons." Canadian Journal of Sociology 3:209–231.

Schumann, Howard
  1971  "The religious factor in Detroit: Review, replication and reanalysis." American Sociological Review 36, 1:30–48.

Schutz, Alfred
  [1951]  "Making music together: A study in social relationship." Pp. 135–158 in A. Broderson (ed.)
  1964  Collected Papers II: Studies in Social Theory. The Hague: Martinus Nijhoff.

Schweitzer, Albert
  1936  Indian Thought and Its Development. Boston: Beacon Press.

Seidler, John and Katherine Meyer
  1989  Conflict and Change in the Catholic Church. New Brunswick, N.J.: Rutgers University Press.

Shaffir, William
  1974  Life in a Religious Community: Lubavitcher Chassidim in Montreal. Toronto: Holt, Rinehart & Winston.
  1978  "Witnessing as identity consolidation." Pp. 39–57 in H. Mol (ed.), Identity and Religion. London: Sage.

Shannon, William V.
  1963  The American Irish. New York: Macmillan.

Sharot, Stephen
  1976  Judaism: A Sociology. London: David & Charles.

Sheehy, Michael
  1969  Is Ireland Dying? Culture and the Church in Modern Ireland. New York: Taplinger.

Shinn, Larry D.
  1989  "The search for meaning in conversions to ISKCON." Pp. 117–134 in D. Bromley and L. Shinn (eds.), Krishna Consciousness in the West. Lewisburg, Pa.: Bucknell University Press.

Shupe, Anson D., Jr., and David G. Bromley
    1980   The New Vigilantes: Deprogrammers, Anti-Cultists and the New Religions. Beverly Hills, Calif.: Sage.

Shupe, Anson and William Stacey
    1982   Born Again Politics and the Moral Majority: What Social Surveys Really Show. New York: Edwin Mellen.

Simmel, Georg
    1906   "The sociology of secrecy and of secret societies." American Journal of Sociology 11:441–498.
    1955   Conflict: The Web of Group Affiliations. Glencoe, Ill.: Free Press.
    [1906] Sociology of Religion. New York: Philosophical Library.
    1959
    [1908] On Individuality and Social Forms. Chicago: University of Chicago Press.
    1971

Simms, J. G.
    1967   "The restoration and the Jacobite war." Pp. 204–216 in T. W. Moody and F. X. Martin (eds.), The Course of Irish History. Cork, Ireland: Mercier.

Simpson, John H.
    1983   "Moral issues and status politics." Pp. 187–205 in R. C. Liebman and R. Wuthnow (eds.), The New Christian Right. New York: Aldine.

Sizer, Sandra
    1979   Gospel Hymns and Social Religion: The Rhetoric of Nineteenth Century Revivalism. Philadelphia: Temple University Press.

Sklare, Marshall
    1972   Conservative Judaism: An American Religious Movement. New York: Schocken.

Slater, Philip
    1966   Microcosm: Structural, Psychological and Religious Evolution in Groups. New York: Wiley.

Smith, Brian H.
    1979   "Churches and human rights in Latin America: Recent trends on the subcontinent." Pp. 155–193 in D. Levine (ed.), Churches and Politics in Latin America. Beverly Hills, Calif.: Sage.
    1982   The Church and Politics in Chile: Challenges to Modern Catholicism. Princeton, N.J.: Princeton University Press.

Smith, Jonathan Z.
    1987   To Take Place: Toward a Theory in Ritual. Chicago: University of Chicago Press.

Smith, Timothy L.
    1983   "My rejection of a cyclical view of 'Great Awakenings'." Sociological Analysis 44:97–102.

Snow, David
    1987   "Organization, ideology and mobilization: The case of Nichiren Shoshu of America." Pp. 153–172 in D. Bromley and P. Hammond (eds.), The Future of New Religious Movements. Macon, Ga.: Mercer University Press.

Snow, David A. and Richard Machalek
    1983   "The convert as a social type." Pp. 259–289 in R. Collins (ed.), Sociological Theory: 1983. San Francisco: Jossey-Bass.

Snow, David A., Louis Zurcher, Jr., and Sheldon Ekland-Olson
    1980   "Social networks and social movements: A micro-structural approach to differential recruitment." American Sociological Review 45:787–801.

Solomon, Trudy
    1981   "Integrating the 'Moonie' experience: A survey of ex-members of the Unification Church." Pp. 275–294 in T. Robbins and D. Anthony (eds.), In Gods We Trust (first edition). New Brunswick, N.J.: Transaction.

Spickard, James V.
    1988   "Experiencing religious rituals: A Schutzian analysis of Navajo ceremonies." Unpublished paper presented to the Society for the Scientific Study of Religion.

Spilka, Bernard, Richard L. Gorsuch and Ralph W. Hood, Jr.
 1985   The Psychology of Religion: An Empirical Approach. Englewood Cliffs, N.J.: Prentice-Hall.
Spillers, Hortense J.
 1971   "Martin Luther King and the style of the black sermon." Black Scholar 3, 1:14–27.
Spiro, Melford
 1966   "Religion: Problems of definition and explanation." Pp. 85–126 in M. Banton (ed.), Anthropological Approaches to the Study of Religion. London: Tavistock.
 1970   Kibbutz. New York: Schocken.
Sprenger, Reverend J. and H. Kramer
 [1486] Malleus Mallificarum. Tr. M. Summers. London: Pushkin Press.
 1948
Stacey, William and Anson Shupe
 1982   "Correlates of support for the electronic church." Journal for the Scientific Study of Religion 21, 4:291–303.
Staples, Clifford L. and Armand L. Mauss
 1987   "Conversion or commitment? A reassessment of the Snow and Machalek approach to the study of conversion." Journal for the Scientific Study of Religion 26, 2:133–147.
Stark, Rodney and William Sims Bainbridge
 1979   "Of churches, sects, and cults: Preliminary concepts for a theory of religious movements." Journal for the Scientific Study of Religion 18, 2:117–133.
 1980   "Toward a theory of religion: Religious commitment." Journal for the Scientific Study of Religion 19, 2:114–128.
Stark, Rodney and Charles Glock
 1968   American Piety: The Nature of Religious Commitment. Berkeley: University of California Press.
Stauffer, Robert E.
 1974   "Radical symbols and conservative functions." Unpublished paper presented to Society for the Scientific Study of Religion.
Steeman, Theodore M.
 1975   "Church, sect, mysticism, denomination: Periodological aspects of Troeltsch's types." Sociological Analysis 36, 3:181–204.
Steinberg, Stephen
 1965   "Reform Judaism: The origin and evolution of a 'church movement'." Journal for the Scientific Study of Religion 5:117–129.
Steinfels, Peter
 1990a  "Mormons drop rites opposed by women." New York Times, May 3.
 1990b  "Split, Catholic bishops shun proposals on women's issues." New York Times, October 14.
Steinitz, Lucy Y.
 1980   "Religiosity, well-being, and Weltanschauung among the elderly." Journal for the Scientific Study of Religion 19, 1:60–67.
Straus, Roger
 1979   "Religious conversion as a personal and collective accomplishment." Sociological Analysis 40, 2:158–165.
Swatos, William, Jr.
 1975   "Monopolism, pluralism, acceptance, and rejection: An integrated model for church–sect theory." Review of Religious Research 16, 3:174–185.
 1979   Into Denominationalism: The Anglican Metamorphosis. Monograph Series 2. Storrs, Conn.: Society for the Scientific Study of Religion.
Szasz, Thomas S.
 1970   The Manufacture of Madness. New York: Dell.
Takayama, K. Peter
 1988   "Revitalization movement of modern Japanese civil religion." Sociological Analysis 48, 4:328–341.

Talmon, Yonina
1966 "Millenarian movements." Archives Europeenes de Sociologie 7, 2:159–200.
Tamney, Joseph and Stephen Johnson
1983 "The Moral Majority in Middletown." Journal for the Scientific Study of Religion 22, 2:145–157.
Taves, Ann
1986 The Household of Faith: Roman Catholic Devotions in Mid-Nineteenth Century America. Notre Dame, Ind.: University of Notre Dame Press.
Tawney, R. H.
1926 Religion and the Rise of Capitalism. New York: Harcourt Brace.
Terkel, Studs
1975 Working. New York: Random House.
Thompson, E. P.
1968 The Making of the English Working Class. Harmondsworth, England: Penguin Books.
Thompson, Kenneth
1986 Beliefs and Ideology. Chichester, England: Ellis Horwood.
Thompson, Leonard
1990 A History of South Africa. New Haven, Conn.: Yale University Press.
Thompson, William Irwin
1967 The Imagination of an Insurrection: Dublin, Easter, 1916. New York: Harper & Row.
Tippett, Alan R.
1973 "The phenomenology of worship, conversion and brotherhood." Pp. 92–109 in W. Clark (ed.), Religious Experience: Its Nature and Function in the Human Psyche. Springfield, Ill.: Thomas.
Tipton, Steven M.
1982a Getting Saved from the Sixties: Moral Meaning in Conversion and Cultural Change. Berkeley: University of California Press.
1982b "The moral logic of alternative religions." Pp. 79–107 in M. Douglas and S. Tipton (eds.), Religion and America: Spirituality in a Secular Age. Boston: Beacon Press.
Tiryakian, Edward A.
1974 "Toward the sociology of esoteric culture." Pp. 257–280 in E. Tiryakian (ed.), On the Margins of the Visible. New York: Wiley.
Tobey, Alan
1976 "Summer solstice of the Happy-Healthy-Holy Organization." Pp. 5–30 in C. Glock and R. Bellah (eds.), The New Religious Consciousness. Berkeley: University of California Press.
Towler, Robert and Audrey Chamberlain
1973 "Common religion." Pp. 1–27 in M. Hill (ed.), Sociological Yearbook of Religion in Britain, Vol. 6. London: SCM.
Troeltsch, Ernst
[1931] The Social Teachings of the Christian Churches. Vols. 1 and 2. Tr. O. Wyon. New York:
1960 Harper & Row.
Truzzi, Marcello
1972 "The occult revival as popular culture: Some random observations in the old and nouveau witch." Sociological Quarterly 13:16–36.
1974a "Definition and dimension of the occult: Towards a sociological perspective." Pp. 243–255 in E. Tiryakian (ed.), On the Margins of the Visible. New York: Wiley.
1974b "Witchcraft and Satanism." Pp. 215–222 in E. Tiryakian (ed.), On the Margins of the Visible. New York: Wiley.
Turner, Bryan
1977 "Confession and social structure." Annual Review of the Social Sciences of Religion 1:29–58.
1980 "The body and religion: Towards an alliance of medical sociology and sociology of religion." Annual Review of the Social Sciences of Religion 4:247–286.
1983 Religion and Social Theory: A Materialistic Perspective. Atlantic Highlands, N.J.: Humanities Press.

Turner, Victor
  1969   The Ritual Process. Chicago: Aldine.
  1974a  Dramas, Fields, and Metaphors. Ithaca, N.Y.: Cornell University Press.
  1974b  "Metaphors of anti-structure in religious culture." Pp. 63–84 in A. Eister (ed.), Changing Perspectives in the Scientific Study of Religion. New York: Wiley.
  1979   "Betwixt and between: The liminal period in 'rites de passage'." Pp. 234–243 in E. Vogt (ed.), Reader in Comparative Religion: An Anthropological Approach. New York: Harper & Row.
Turner, Victor and Edith Turner
  1978   Image and Pilgrimage in Christian Culture: Anthropological Perspectives. New York: Columbia University Press.
Tyson, Ruel W.
  1988   "The testimony of Sister Annie Mae." Pp. 105–125 in R. W. Tyson, J. L. Peacock, and D. W. Patterson (eds.), Diversities of Gifts: Field Studies in Southern Religion. Urbana: University of Illinois Press.
Umansky, Ellen M.
  1985   "Feminism and the reevaluation of women's roles within American Jewish life." Pp. 477–494 in Y. Y. Haddad and E. B. Findly (eds.), Women, Religion and Social Change. Albany: State University of New York Press.
Underhill, Evelyn
  1961   Mysticism. New York: Dutton.
Utrecht, Ernst
  1978   "Religion and social protest in Indonesia." Social Compass 25, 3–4:395–418.
Van Gennep, A.
  1960   The Rites of Passage. Tr. M. Vikedom and G. Coffee. Chicago: University of Chicago Press.
Vaughn, William Preston
  1983   The Antimason Party in the United States, 1826–1843. Lexington: University of Kentucky Press.
Verdesi, Elizabeth Howell
  1976   In But Still Out: Women in the Church. Philadelphia: Westminster.
Voll, John Obert
  1982   Islam: Continuity and Change in the Modern World. Boulder, Colo.: Westview.
Vonnegut, Kurt, Jr.
  1963   Cat's Cradle. Baltimore: Penguin Books.
Wach, Joachim
  1944   Sociology of Religion. Chicago: University of Chicago Press.
Wagner, Melinda Bollar
  1990   God's Schools: Choice and Compromise in American Society. New Brunswick, N.J.: Rutgers University Press.
Wald, Kenneth D., Dennis E. Owen, and Samuel S. Hill, Jr.
  1989   "Evangelical politics and status issues." Journal for the Scientific Study of Religion 28, 1:1–16.
Walker, Andrew and James S. Atherton
  1971   "An Easter pentecostal convention: The successful management of a 'time of blessing'." Sociological Review 19, 3:367–387.
Wallace, Anthony
  1956   "Revitalization movements." American Anthropologist 58:264–281.
  1957   "Mazeway disintegration: The individual's perception of socio-cultural disorganization." Human Organization 16:23–27.
Wallace, Ruth A.
  1975   "Bringing women in: Marginality in the churches." Sociological Analysis 36, 4:291–303.
  1988   "Catholic women and the creation of a new social reality." Gender and Society 2, 1:24–38.
  1990   "Institutional constraints and role transition: The case of Catholic women pastors." Paper presented to the Association for the Sociology of Religion.
  1991   They Call Her Pastor. Albany: State University of New York Press.

Wallis, Roy
1973    "A comparative analysis of problems and processes of change in two manipulationist movements: Christian Science and Scientology." Pp. 407–422 in Contemporary Metamorphosis of Religion?: Acts of the 12th Conference Internationale de Sociologie Religieuse. Lille, France: CISR.
1974    "Ideology, authority and the development of cultic movements." Social Research 41:299–327.
1977    The Road to Total Freedom: A Sociological Analysis of Scientology. New York: Columbia University Press.
1978    The Rebirth of the Gods?: Reflections on the New Religions in the West. Belfast: Queens University New Lecture Series 108.
1985    "The dynamics of change in the Human Potential Movement." Pp. 129–156 in R. Stark (ed.), Religious Movements: Genesis, Exodus, and Numbers. New York: Paragon.
Wallis, Roy and Steve Bruce
1984    "The Stark–Bainbridge theory of religion: A critical analysis and counter proposal." Sociological Analysis 45, 1:11–28.
1986    Sociological Theory, Religion and Collective Action. Belfast: Queens University.
Walton, John
1989    "Debt, protest and the state in Latin America." Pp. 299–328 in S. Eckstein (ed.), Power and Popular Protest: Latin American Social Movements. Berkeley: University of California Press.
Wardwell, Walter
1972    "Orthodoxy and heterodoxy in medical practice." Social Science and Medicine 6:759–763.
Warner, R. Stephen
1988    New Wine in Old Wineskins: Evangelicals and Liberals in a Small-Town Church. Berkeley: University of California Press.
Warner, W. Lloyd
1953    American Life: Dream and Reality. Chicago: University of Chicago Press.
Washington, James M.
1985    "Jesse Jackson and the symbolic politics of Black Christendom." Annals of the American Academy of Political and Social Science, 480:89–105.
Washington, Joseph
1972    Black Sects and Cults. Garden City, N.Y.: Doubleday.
Waxman, Chaim I.
1983    America's Jews in Transition. Philadelphia: Temple University Press.
Weber, Max
[1920–21] From Max Weber: Essays in Sociology. Tr. and ed. H. H. Gerth and C. W. Mills. New
1946    York: Oxford University Press.
[1925] The Theory of Social and Economic Organization. Tr. A. M. Henderson and T. Parsons.
1947    New York: Oxford University Press.
[1920–21] The Religion of China. Tr. H. H. Gerth. New York: Free Press.
1951
[1920–21] Ancient Judaism. Tr. H. H. Gerth and D. Martindale. New York: Free Press.
1952
[1904] The Protestant Ethic and the Spirit of Capitalism. Tr. T. Parsons. New York: Scribner.
1958a
[1920–21] The Religion of India. Tr. H. H. Gerth and D. Martindale. New York: Free Press.
1958b
[1922] The Sociology of Religion. Tr. E. Fischoff. Boston: Beacon Press.
1963
[1925] Economy and Society, Vol. 3. Ed. G. Roth and C. Wittich. New York: Bedminster.
1968
Weigert, Andrew and Darwin Thomas
1969    "Religiosity in 5-D: A critical note." Social Forces 48:260–263.
Weil, Andrew
1973    The Natural Mind. Boston: Houghton Mifflin.

Welter, Barbara
  1966  "The cult of true womanhood: 1820–1860." American Quarterly (Summer), Part I:151–174.
  1976  Dimity Convictions: The American Woman in the Nineteenth Century. Athens: Ohio University Press.
Westhues, Kenneth
  1973  "The established church as an agent of change." Sociological Analysis 34, 2:106–123.
  1976  "The church in opposition." Sociological Analysis 37, 4:299–314.
  1978  "Stars and stripes, the maple leaf, and the papal coat of arms." Canadian Journal of Sociology 3, 2:245–261.
Westin, Alan
  1964  "The John Birch Society." Pp. 239–268 in D. Bell (ed.), The Radical Right. Garden City, N.Y.: Doubleday.
Westley, Frances
  1978  "The cult of man: Durkheim's predictions and new religious movements." Sociological Analysis 39, 2:135–145.
  1983  The Complex Forms of the Religious Life: A Durkheimian View of New Religious Movements. Chico, Calif.: Scholars Press.
Westoff, Charles F.
  1979  "The blending of Catholic reproductive behavior." Pp. 231–240 in R. Wuthnow (ed.), The Religious Dimension: New Directions in Quantitative Research. New York: Academic Press.
Whitney, Craig R.
  1988  "Belfast: Strife-wracked but no Beirut." New York Times, October 30.
Whyte, J. H.
  1980  Church and State in Modern Ireland, 1923–1979. Dublin: Gill.
Wiesel, Elie
  1960  Night. New York: Pyramid.
Willen, Richard S.
  1983  "Religion and law: The secularization of testimonial procedures." Sociological Analysis 44, 1:53–64.
Williams, Melvin D.
  1974  Community in a Black Pentecostal Church: An Anthropological Study. Pittsburgh: University of Pittsburgh Press.
Williams, Rhys
  1988  "American political culture and religion: Conflict and convergence." Paper presented to the Society for the Scientific Study of Religion.
Willits, Fern K. and Donald M. Crider
  1989  "Church attendance and traditional religious beliefs in adolescence and young adulthood: A panel study." Review of Religious Research 31, 1:68–81.
Wilmore, Gayraud S., Jr.
  1983  Black Religion and Black Radicalism. Maryknoll, N.Y.: Orbis.
Wilson, Bryan R.
  1967  (ed.), Patterns of Sectarianism: Organization and Ideology in Social and Religious Movements. London: Heinemann.
  1970  Religious Sects. New York: McGraw-Hill.
  1976  Contemporary Transformations of Religion. London: Oxford University Press.
Wilson, John F.
  1979  Public Religion in American Culture. Philadelphia: Temple University Press.
Wimberley, Ronald, Thomas Hood, C. M. Lipsey, Donald Clelland, and M. Hay
  1975  "Conversion in a Billy Graham crusade: Spontaneous event or ritual performance." Sociological Quarterly 16:162–170.
Wogaman, J. Philip
  1988  "The use of government infiltration in religious groups: A new/old issue." Pp. 123–137 in J. E. Wood (ed.), Ecumenical Perspectives on Church and State. Waco, Tex.: Baylor University, J. M. Dawson Institute on Church–State Studies.

Wood, James E., Jr.
  1988   "Religious pluralism and American society." Pp. 1–18 in J. E. Wood (ed.), Ecumenical Perspectives on Church and State. Waco, Tex.: Baylor University, J. M. Dawson Institute on Church–State Studies.
Wood, James R.
  1970   "Authority and controversial policy: The churches and civil rights." American Sociological Review 35, 6:1057–1069.
  1981   Leadership in Voluntary Organizations: The Controversy over Social Action in Protestant Churches. New Brunswick, N.J.: Rutgers University Press.
Woodcock, George and Ivan Avakumonic
  1977   The Doukhobors. Toronto: McClelland and Stewart.
Woodrum, Eric
  1982   "Religious organizational change: An analysis based on the TM movement." Review of Religious Research 24:89–103.
Worsley, Peter
  1968   The Trumpet Shall Sound: A Study of "Cargo" Cults in Melanesia. New York: Schocken.
Wright, Stuart A.
  1987   Leaving Cults: The Dynamics of Defection. Washington, D.C.: Society for the Scientific Study of Religion, Monograph Series.
  1988   "Leaving new religious movements: Issues, theory and research." Pp. 143–165 in D. Bromley (ed.), Falling from the Faith: Causes and Consequences of Religious Apostasy. Newbury Park, Calif.: Sage.
Wuthnow, Robert
  1976a  "Astrology and marginality." Journal for the Scientific Study of Religion 15, 2:157–168.
  1976b  The Consciousness Reformation. Berkeley: University of California Press.
  1978a  Experimentation in American Religion. Berkeley: University of California Press.
  1978b  "Religious movements and the transition in world culture." Pp. 63–79 in J. Needleman and G. Baker (eds.), Understanding the New Religions. New York: Seabury Press.
  1980   "World order and religious movements." Pp. 57–75 in A. Bergesen (ed.), Studies of the Modern World System. New York: Academic Press.
  1988   The Restructuring of American Religion: Society and Faith Since World War II. Princeton, N.J.: Princeton University Press.
Yinger, J. Milton
  1946   Religion in the Struggle for Power. Durham, N.C.: Duke University Press.
  1963   "The 1962 H. Paul Douglass lectures, I. Religion and social change: Functions and dysfunctions of sects and cults among the disprivileged." Review of Religious Research 4, 2:65–84.
  1970   The Scientific Study of Religion. New York: Macmillan.
Young, Allan
  1976   "Some implications of medical beliefs and practices for social anthropology." American Anthropologist 78:5–24.
Zablocki, Benjamin
  1971   The Joyful Community. Baltimore: Penguin Books.
Zald, Mayer
  1970   Organization Change: The Political Economy of the Y.M.C.A. Chicago: University of Chicago Press.
Zaretsky, Eli
  1976   Capitalism, the Family and Personal Life. London: Pluto Press.
Zeldin, Mary-Barbara
  1969   "The religious nature of Russian Marxism." Journal for the Scientific Study of Religion 8:100–111.

# AUTHOR INDEX

# SUBJECT INDEX

Abolitionist movement, 214, 222
Abortion, 66, 100, 217–218
Adolescence, 61–63
African Methodist Episcopal Church, 160
Afterlife, 100, 122, 216–217. *See also* Death and dying
AIDS, 192
Albigenses, 228
Alienation, 215–216
American Baptist Churches, 140
American Muslim Mission. *See* Black Muslims
Americanist heresy, 193–194
Amish, 38, 59, 140, 156, 186
Anabaptists, 140, 224
Ananda community, 60
Anglican church. *See* Church of England
Anomie, 34–35, 44–45, 77
Anti-Catholic movements, 76, 186–188, 256
Anti-communism, 181, 187
Anti-"cult" movement, 73–74, 88–90, 137
Anti-feminist movements, 217–218
Anti-Mason movement, 76
Anti-Mormon movement, 76
Anti-Oriental agitation, 181
Anti-Semitism, 187
Apartheid, 236, 238
Apostasy. *See* Disengagement.
Arica, 163
Asceticism, 224–225
Asian-Americans, 39
Astrology, 109–110, 129, 140
Australia, 177–178, 255
Authority, 154–156, 159, 184, 192–194, 227–229, 253–261, 281
Awakening. *See* Great Awakening; Revivalism

Ba'ale teshuvah, 72
Baptists, 9, 38, 64, 142, 146, 230, 256. *See also* American Baptist Churches; Southern Baptists
Basic Ecclesial Community (Comunidade Eclesial de Base), 230, 243–245, 329
Becker, Howard, 135
Belgium, 255
Bellah, Robert, 180–181, 184, 254, 266
Berger, Peter, 25–29, 35, 37, 255
Birth control, 65–66, 100
Black Muslims, 43, 47, 158, 220, 228, 260
Black religion, 85, 236–237
"Born again" Christians, 190. *See also* Evangelicalism; Pentecostalism
Brainwashing, 75–76, 88–90. *See also* Conversion
Brazil, 34, 119, 239–245
Brethren of the Common Life, 147
Brook Farm commune, 164
Brotherhood of the Spirit, 60
Bruderhof (Society of Brothers), 60, 80, 82, 147, 156, 159
Buddha, Gautama, 116–117, 137, 226
Buddhism
  in America, 140, 159, 163, 165, 168
  in Asia, 116–117, 122, 127, 137, 139–140, 148, 232

Calvinism, 134, 224–225
Canada, 40, 59–60, 64, 81, 109, 145, 161, 185, 190, 221, 255
Caste, 29–30, 113–117, 235
Catalonia, 230
Catholic Action, 240
Catholic Charismatic Renewal (Catholic Pentecostalism), 42, 60, 72–73, 127, 146, 151–152, 170, 272

Catholics. *See* Roman Catholics
Charisma, 127, 133, 148, 227–229
Childhood, 54–61
Children of God, 140. *See also* Jesus movement
Christadelphians, 47, 140
Christianity, 70, 115, 127, 137, 148, 156, 191, 221
Christian Science, 126, 159, 160, 270
Church. *See also* Church-sect theories
  churchly religious orientation, 143, 145–146
  definition, 139
  some variables, 142
  transformations of, 161, 230
Churches. *See* Official religion
Church of England, 118, 125–126, 161. *See also* Episcopal Church.
Church of Ireland, 197, 200, 208. *See also* Episcopal Church
Church of Latter Day Saints. *See* Mormons
Church of Religious Science, 269
Church-sect theories, 134–135, 137–139
Church-state relationship, 184, 219, 234–235, 256–260, 277–281. *See also* Civil religion
Cistercians, 148. *See also* Religious orders
Civil disobedience, 189, 254–255
Civil religion
  Bellah's thesis, 180–181
  definition, 179
  Durkheim's thesis, 177–179
  forms of, 184–185
  in Indonesia, 258
  in Israel, 183

*335*